Nanomaterials in Biological Milieu: Biomedical Applications and Environmental Sustainability

Edited by

Manoranjan Arakha & Arun Kumar Pradhan

Centre for Biotechnology
Siksha 'O' Anusandhan, (Deemed to be University)
Bhubaneswar, Odisha 751030
India

Nanomaterials in Biological Milieu: Biomedical Applications and Environmental Sustainability

Editors: Manoranjan Arakha & Arun Kumar Pradhan

ISBN (Online): 978-981-5313-26-0

ISBN (Print): 978-981-5313-27-7

ISBN (Paperback): 978-981-5313-28-4

need for a court order if at any point you breach any terms of this License Agreement. In no event will any delay or failure by Bentham Science Publishers in enforcing your compliance with this License Agreement constitute a waiver of any of its rights.

3. You acknowledge that you have read this License Agreement, and agree to be bound by its terms and conditions. To the extent that any other terms and conditions presented on any website of Bentham Science Publishers conflict with, or are inconsistent with, the terms and conditions set out in this License Agreement, you acknowledge that the terms and conditions set out in this License Agreement shall prevail.

Bentham Science Publishers Pte. Ltd.
80 Robinson Road #02-00
Singapore 068898
Singapore
Email: subscriptions@benthamscience.net

BENTHAM SCIENCE

CONTENTS

PREFACE

Inside the biological milieu, nanomaterials come in myriad shapes and sizes and interact with various biomolecules, forming a bio-nano interface. The fate of both nanomaterial and biomolecule depends on the type of interaction at the bio-nano interface. Tremendous research on the bio-nano interface has led to the development of nanomaterials based on various products and applications in biomedical science. In view of the above discussion, this book intends to discuss the interactions of nanomaterials with different biomolecules and the consequences of this interaction in the field of biomedical science. Hence, the book will initially discuss the current state-of-the-art techniques for various nanomaterial formulations. The later chapters of the book will discuss the potential applications of nanomaterials in biomedical sciences and environmental sustainability. The book is a contribution of experts from microbiology, cancer biology, pharmaceutical science, nanotechnology, plant biotechnology, and environmental sciences. The book will discuss some novel applications of nanomaterials in the field of biomedical and environment.

The book, in total, comprises 11 chapters written by experts working in the respective aspects of nanotechnology. For instance, Chapters 1, 2 and 3 describe various methods for the synthesis of nanomaterials for different applications. Chapter 4 describes the roles of nanomaterials in various disease diagnosis. The applications of nanomaterials in various biomedical sectors are discussed in Chapters 5, 6, and 7. Chapters 8 discuss the application of nanomaterials in cancer therapy. Chapter 9 interprets artificial intelligence and nanotechnology-integrated recent applications in early lung cancer detection and therapy. Various environmental applications of nanomaterials are discussed in Chapters 10 and 11.

We are very much thankful to all the contributing authors for their timely cooperation and support, without which this book would not have taken its final shape. We are also thankful to the series editors for their critical reviews and positive remarks that brought positive developments to the book.

Manoranjan Arakha & Arun Kumar Pradhan
Centre for Biotechnology
Siksha 'O' Anusandhan, (Deemed to be University)
Bhubaneswar, Odisha 751030
India

List of Contributors

Ashirbad Sarangi	Centre for Biotechnology, Siksha 'O' Anusandhan (Deemed to be University), Bhubaneswar, Odisha 751030, India
Arun Kumar Pradhan	Center for Biotechnology, School of Pharmaceutical Science, Siksha 'O' Anusandhan (Deemed to be University), Bhubaneswar, Odisha 751030, India
Akhlash P. Singh	Department of Biochemistry, GGDSD College (Panjab University), Chandigarh, India
Ankit Srivastava	Department of Biotechnology, Motilal Nehru National Institute of Technology, Allahabad, India
Ananya Bhattacharjee	Bio-Medical Imaging Laboratory (BIOMIL), National Institute of TeTechnology, Chnology Silchar-788010, Assam, India
Abhishek Bhattacharjee	Department of Pharmaceutical Sciences, Assam University (A Central University), Silchar-788011, Assam, India
Arunima Pandey	KIIT School of Biotechnology, Kalinga Institute of Industrial Technology (KIIT-DU), Bhubaneswar, Odisha 751024, India
Arka Ghosh	KIIT School of Biotechnology, Kalinga Institute of Industrial Technology (KIIT-DU), Bhubaneswar, Odisha 751024, India
Bhabani Shankar Das	Centre for Biotechnology, Siksha 'O' Anusandhan (Deemed to be University), Bhubaneswar, Odisha 751030, India
Banishree Sahoo	Center for Biotechnology, Siksha 'O' Anusandhan (Deemed to be University), Bhubaneswar, Odisha 751030, India
Bhangyashree Nanda	Department of Psychiatry, Siksha 'O' Anusandhan (Deemed to be University), Bhubaneswar, Odisha 751030, India
Biswajita Pradhan	School of Biological Sciences, AIPH University, Bhubaneswar, India
Bikash Chandra Behera	School of Biological Sciences, National Institute of Science Education and Research, Bhubaneswar, Odisha 752050, India
Debapriya Bhattacharya	Centre for Biotechnology, Siksha 'O' Anusandhan (Deemed to be University), Bhubaneswar, Odisha 751030, India Department of Biological Sciences, Indian Institute of Science Education and Research (IISER), Bhopal 462030, Madhya Pradesh, India
Gargi Balabantaray	Department of Immunology and Rheumatology, Institute of Medical Science, Sum Hospital, Siksha 'O' Anusandhan (Deemed to be University), Bhubaneswar, Odisha 751030, India
Gautam Mohapatra	Centre for Biotechnology, Siksha 'O' Anusandhan (Deemed to be University), Bhubaneswar, Odisha 751030, India Department of Biotechnology, Indian Institute of Technology Madras (IITM), Chennai, Tamil Nadu 600036, India
Jangmang Chongloi	Department of Emergency Medicine, JPNATC AIIMS, New Delhi, Delhi 110029, India
Kali Prasad Pattanaik	National Rice Research Institute, Cuttack, Odisha, India

Lopamudra Subudhi	Centre for Biotechnology, Siksha 'O' Anusandhan (Deemed to be University), Bhubaneswar, Odisha 751030, India
Manoranjan Arakha	Center for Biotechnology, School of Pharmaceutical Science, Siksha 'O' Anusandhan (Deemed to be University), Bhubaneswar, Odisha 751030, India
Manish Jaiswal	Department of Biochemistry, AIIMS, New Delhi, Delhi 110029, India
Nayan Ranjan Ghosh Biswas	Department of Pharmaceutical Sciences, Dibrugarh University, Dibrugarh-786004, Assam, India
Pradeepta Sekhar Patro	Department of Immunology and Rheumatology, Institute of Medical Science, Sum Hospital, Siksha 'O' Anusandhan (Deemed to be University), Bhubaneswar, Odisha 751030, India
Pratyush Kumar Behera	Department of Zoology, Maharaja Sriram Chandra Bhanja Deo University, Takatpur, Baripada-757003, Mayurbhanj, Odisha, India
Prolita Pattanayak	Centre for Biotechnology, Siksha 'O' Anusandhan (Deemed to be University), Bhubaneswar, Odisha 751030, India
Ruchi Bhuyan	Department of Medical Research, IMS and SUM Hospital, Siksha 'O' Anusandhan (Deemed to be University), Bhubaneswar, Odisha 751030, India Institute of Dental Science, Siksha 'O' Anusandhan (Deemed to be University), Bhubaneswar, Odisha 751030, India
Rangnath Ravi	Shivaji College, University of Delhi, New Delhi, Delhi 110021, India
Ranjeet Kumar Nirala	Saheed Raj Guru College of Applied Sciences for Women, University of Delhi, New Delhi, Delhi 110096, India
R. Murugan	Bio-Medical Imaging Laboratory (BIOMIL), National Institute of TeTechnology, Chnology Silchar-788010, Assam, India
Ranjit Prasad Swain	Department of Pharmaceutics, GITAM School of Pharmacy, GITAM (Deemed to be University), Visakhapatnam-530045, Andhra Pradesh, India
Ram Kumar Sahu	Department of Pharmaceutical Sciences, Hemvati Nandan Bahuguna University (A Central University), Garhwal-249161, Uttarakhand, India
Rajeswari Rath	Centre for Biotechnology, Siksha 'O' Anusandhan (Deemed to be University), Bhubaneswar, Odisha 751030, India
Sanjana Parida	Centre for Biotechnology, Siksha 'O' Anusandhan (Deemed to be University), Bhubaneswar, Odisha 751030, India
Soumya Subhashree Satapathy	Center for Biotechnology, School of Pharmaceutical Science, Siksha 'O' Anusandhan (Deemed to be University), Bhubaneswar, Odisha 751030, India
Sanat Kumar Bhuyan	Institute of Dental Science, Siksha 'O' Anusandhan (Deemed to be University), Bhubaneswar, Odisha 751030, India
Satabdi Mishra	Center for Biotechnology, Siksha 'O' Anusandhan (Deemed to be University), Bhubaneswar, Odisha 751030, India
Shivani Devi	Department of Biochemistry, GGDSD College (Panjab University), Chandigarh, India
Seema Tripathy	School of Biological Sciences, National Institute of Science Education and Research, Bhubaneshwar-752050, India

Soumyashree Rout   Department of Neurology, Siksha 'O' Anusandhan (Deemed to be University), Bhubaneswar, Odisha 751030, India

Snehalata Pradhan   Samanta Chandra Sekhar Autonomous College, Puri, Odisha 752001, India

Subham Preetam   Institute of Advanced Materials, IAAM, Gammalkilsvägen 18, Ulrika 59053, Sweden
Department of Robotics Engineering, Daegu Gyeongbuk Institute of Science and Technology (DGIST), Daegu 42988, Republic of Korea

Twinkle Rout   Department of Surgical Oncology, Siksha 'O' Anusandhan (Deemed to be University), Bhubaneswar, Odisha 751030, India

CHAPTER 1

The Current State-of-the-Art Technique of Nanomaterial Fabrication

Bhabani Shankar Das[1], Lopamudra Subudhi[1], Sanjana Parida[1], Ashirbad Sarangi[1] and Debapriya Bhattacharya[1,2,*]

[1] *Centre for Biotechnology, Siksha 'O' Anusandhan (Deemed to be University), Bhubaneswar, Odisha 751030, India*

[2] *Department of Biological Sciences, Indian Institute of Science Education and Research (IISER), Bhopal 462030, Madhya Pradesh, India*

Abstract: Nanomaterials are the finest materials that are being used in frontline research for their unique properties and potential applications in different fields, including medicine, electronics, and energy. To take advantage of these properties, the fabrication of nanomaterials provides a key resource with greater biocompatibility in reliable forms. In recent years, numerous advances in nanomaterial fabrication have made it possible to use them in different fields for precise control of material properties. However, there are still many challenges that need to be addressed to unlock the potential of nanomaterials. This book chapter will provide a brief overview of the challenges associated with nanomaterial fabrication and the potential future applications for them. The chapter will begin by discussing the different nanomaterial fabrication techniques, including physical, chemical, and biological approaches, as well as their advantages and disadvantages. Hereinafter, the chapter will also explicitly explore the growth of research into nanomaterial fabrication, including recent advances in the field and their potential applications for the future development of nanomaterials.

Keywords: Biocompatibility, Biological advantages, Nanofabrication, Nanoparticles.

INTRODUCTION

Fabrication has played a key role in human history since the dawn of time. The invention of man-made tools has enabled a better understanding of fabrication and predictions of advancements in nanomaterials. Raw materials used in the past had unknowingly served various purposes thousands of years ago. However, around

---

* **Corresponding author Debapriya Bhattacharya:** Department of Biological Sciences, Indian Institute of Science Education and Research (IISER), Bhopal 462030, Madhya Pradesh, India;
E-mail: debapriyabhattacharyab@gmail.com

Manoranjan Arakha & Arun Kumar Pradhan (Eds.)

4500 years ago, these materials were employed to fortify ceramic mixtures and utilized in nanoform for different functions [1]. As time has passed, the evaluation of nanomaterials has grown quicker and has been implemented in various sectors, but their use has been limited to a certain extent. Our advanced technology has made it possible to use these nanomaterials in a new and directed way. To explore newer forms of these nanomaterials, fabrication has made it possible to create artifacts on a nanoscale, giving a new perspective to our modern civilization. Although the fabrication of nanostructures with minimal dimensions has provided us with a great degree of precision in terms of material composition, it has also enabled the development of highly functional devices that are no more than 100 nm in size [2, 3]. This has led to a rapid increase in knowledge and a greater understanding of the characteristics of many nanostructures and their associated applications. Examples of these devices include solar cells [4], LEDs [5], laser diodes [6], hard drives [7, 8], self-cleaning handwash [9, 10], antibacterial materials for skin protection [11, 12], food preservation [13], *etc.*

These features are employed in hybridizing engineering fabrication processes with chemistry and biology to create new fabrication methods. With the right approach, nanoscale materials with bulk- and molecular-level features can be designed, manufactured, and integrated into functional devices [14]. This potential advancement in modern nanofabrication techniques is thought to be of great benefit in many areas. To make this possible, engineering nanofabrication approaches must be refined to manipulate the structural, mechanical, optical, and magnetic or electronic properties of the materials being processed [15]. This can be done through two main categories of nanofabrication: top-down [16] and bottom-up [17]. Top-down fabrication relies on molecular processing to produce atomically precise materials in bulk and at low cost. With this approach, a broad range of materials from both wafer and thin-film sources can be accessed [18 - 21]. Top-down fabrication is also used to manufacture devices with the desired shape and characteristics, such as cell sorting platforms, bead assemblies, biochemical sensors, and drug testing platforms [22 - 25]. Bottom-up nano-fabrication, on the other hand, makes and uses complex nanoscale assemblies and guided self-assemblies into different shapes and materials, like nanowires and nanotubes with multiple organic and inorganic elements [15].

Overall, this book chapter provides an overview of nanofabrication methods for the production of highly functional and structurally complex nanostructures with an emphasis on large-scale nanomanufacturing (Fig. **1**). Moreover, this chapter provides a deep insight into the fundamentals of synthesis methods and the potential opportunities associated with the broad and exciting area of nanomaterials.

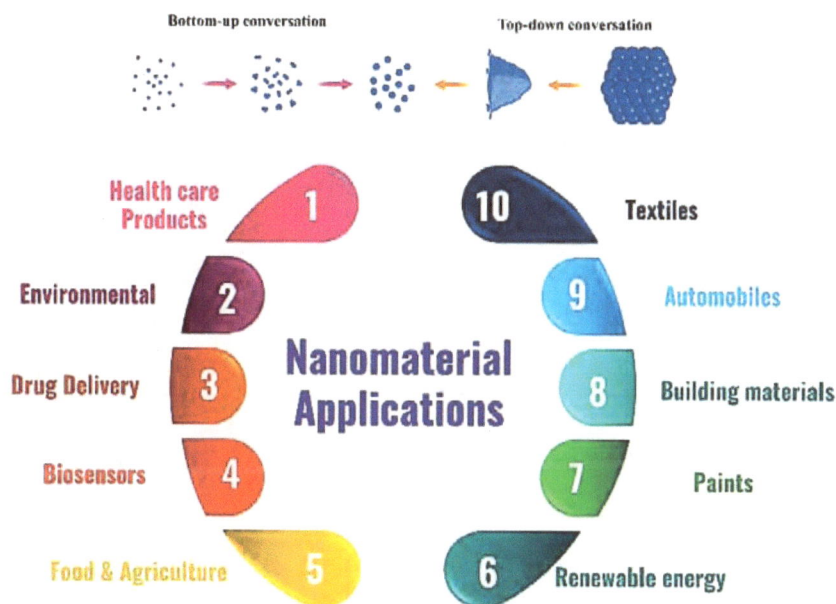

Fig. (1). Overview of nanomaterial fabrication with its applications.

DIFFERENT FORMS OF NANOMATERIALS

Different methods can be used to synthesize nanomaterials, such as carbon fullerenes using helium or neon atmosphere to generate an arc [26], carbon nanotubes through carbon-arc discharge, laser ablation of carbon, or chemical vapor deposition [27]. Fabrication of metals can be produced by different methods, including combustion synthesis, pyrolysis, mechanochemical processing, chemical precipitation, sol-gel processing, and laser ablation [28]. Other materials, such as biomolecules [29], polymer-based nanomaterials [30], and quantum dots [31], can be used in nanomaterial synthesis. In Table **1**, the synthesis of nanomaterials is described briefly.

Table 1. Different forms of nanomaterials with synthesis techniques.

Materials	Synthesis Technique	References
Carbon Fullerenes	Using two carbon electrodes in a helium or neon atmosphere, carbon-arc discharge, laser ablation, chemical vapor deposition	[26]
Carbon Nanotubes	Carbon-arc discharge, laser ablation of carbon, or chemical vapor deposition	[27]
Fabricated Metals	Combustion synthesis, pyrolysis, mechanochemical processing, chemical precipitation, sol-gel processing, and laser ablation	[28]
Biomolecules	Enzymatic synthesis, self-assembly, and bioconjugation	[29]
Polymer-Based Nanomaterials	Emulsion polymerization, micellar polymerization, and inverse miniemulsion polymerization	[30]
Quantum Dots	Colloidal synthesis, hot injection method, and solvothermal method	[31]

CLASSIFICATION OF FABRICATED NANOMATERIALS

Nanomaterial fabrication can be broadly classified into various categories based on the nature of the materials used. One such classification is based on the type of materials, namely organic nanoparticles, non-organic nanoparticles, nano-clays, and nanoemulsions. Organic nanoparticles are made up of organic compounds such as lipids, proteins, and carbohydrates and can be synthesized using various techniques such as self-assembly, emulsification, and solvent evaporation [32]. Non-organic nanoparticles, on the other hand, are made up of inorganic materials like metals, metal oxides, and semiconductors. They can be made using sol-gel, hydrothermal, and microwave-assisted methods [33]. Nano-clays are nanoparticle-sized layered silicate platelets that can be synthesized by exfoliating layered minerals using various methods, such as intercalation and sonication [34]. Nanoemulsions are a type of nanomaterial that consists of droplets of one liquid dispersed in another and can be synthesized using techniques such as high-pressure homogenization and microfluidization [35]. A comprehensive list of different fabrication techniques for each type of nanomaterial is presented in Table **2**.

Table 2. Nanomaterial types and their fabricated techniques.

Nanomaterial Types	Fabrication Techniques	References
Organic Nanoparticle	Self-assembly, emulsification, solvent evaporation	[32]
Non-organic Nanoparticle	Sol-gel, hydrothermal, microwave-assisted methods	[33]
Nanoclay	Intercalation, sonication	[34]
Nanoemulsion	High-pressure homogenization, micro fluidization	[35]

Classification of Nanomaterials Based on Dimensions

Dimensionality is a key property that can be used to separate distinct nanomaterial structures. Nanomaterials are classified based on their dimensionalities, which are zero-dimensional (0D), one-dimensional (1D), two-dimensional (2D), and three-dimensional (3D) structures (Table **3**). Zero-dimensional nanomaterials have a size of 1 to 100 nm and are spherical or near-spherical in shape. Examples of 0D nanomaterials include fullerenes, quantum dots, and nanospheres [36]. One-dimensional nanomaterials are distinguished by their elongated shape and nanoscale range of 1–100 nm. Carbon nanotubes and silicon nanowires are examples of 1D nanomaterials [37]. Two-dimensional nanomaterials are thin films that have a thickness of only a few nanometers but have significantly larger dimensions in the other two dimensions. Graphene, transition metal dichalcogenides, and metal-organic frameworks are examples of 2D nanomaterials [38]. Three-dimensional nanomaterials have nanoscale features throughout all three dimensions, and examples include mesoporous materials, aerogels, and nanocomposites [39].

Table 3. Classification of nanomaterials based on dimensionality.

Nanomaterial Dimensionality	Examples	Ranges	References
Zero-Dimensional	Quantum dots, metal nanoparticles	1-100 nm	[36]
One-Dimensional	Nanowires, nanorods, nanotubes	1-100 nm diameter, 100 nm to several micrometers in length	[37]
Two-Dimensional	Graphene, transition metal dichalcogenides, black phosphorus	Less than 100 nm in thickness, up to millimeters in lateral dimensions	[38]
Three-Dimensional	Mesoporous silica, aerogels	Characteristic dimension less than 100 nm in bulk materials	[39]

Synthesis of Nanomaterials Using Top-down and Bottom-up Approaches

Nanomaterial synthesis can be achieved through two main approaches: top-down and bottom-up. The top-down approach involves physically synthesizing nanomaterials by breaking down bulk materials into smaller sizes using techniques such as milling, lithography, and etching. This approach offers precise control over the size and shape of the resulting nanomaterials, making it suitable for applications such as electronics and optics. Examples of nanomaterials synthesized using the top-down approach include nanoparticles, nanowires, and thin films [40]. On the other hand, the bottom-up approach involves the chemical or biological synthesis of nanomaterials by building up materials atom-by-atom or molecule-by-molecule. This approach helps with the ability to synthesize complex structures with a high degree of control over their properties. Examples of nanomaterials synthesized using a bottom-up approach include quantum dots, carbon nanotubes, and DNA nanotechnology. The advantages of the bottom-up approach include its ability to produce highly uniform and well-defined nanomaterials with tailored properties, making it suitable for applications such as drug delivery and nanoelectronics [41]. Table **4**, summarizing the advantages, examples, and references for each approach, is presented below.

Table 4. Synthesis of nanomaterials by different approaches.

Synthesis Approach	Advantages	Examples	References
Top-Down (Physical Synthesis)	Precise control of particle size, ability to produce large quantities of nanoparticles	Silicon nanoparticles, Metal nanoparticles	[40]
Bottom-Up (Chemical and Biological Synthesis)	Synthesis of complex and functional nanoparticles with unique properties	Quantum dots, Carbon nanotubes, DNA nanotechnology	[41]

Physical Methods for Synthesis of Nanomaterials

Physical methods are commonly used to fabricate nanomaterials. Some of the widely used physical methods include mechanical milling, electrospinning, lithography, laser ablation, and fluorescent nanomaterials for bioimaging, which have been summarized in Table **5**.

Table 5. Synthesis of nanomaterials using physical techniques.

Physical Method	Advantages	Examples	References
Mechanical Milling	Controlled particle size and morphology, High purity	Metal nanoparticles, Oxide nanoparticles	[42]

(Table 5) cont.....

Physical Method	Advantages	Examples	References
Electrospinning	Controlled fiber diameter and surface functionalities, High surface area	Polymer nanofibers, Composite nanofibers	[43]
Lithography	High precision and reproducibility, Large-scale patterning	Electron beam lithography, Nanoimprint lithography	[44]
Laser Ablation	Controlled particle size and shape, High purity	Metal nanoparticles, Carbon nanoparticles	[45]
Fluorescent Nanomaterials	High brightness, Photostability, Biocompatibility	Quantum dots, Carbon dots, Upconversion nanoparticles	[46]

Mechanical Milling

Mechanical milling involves the use of high-energy milling processes to produce nanoparticles and nanostructured materials. During this procedure, balls collide with each other and the walls of the milling chamber, resulting in the fragmentation and refinement of the starting materials into smaller particles. This method has been used to fabricate a wide range of nanomaterials, including metals, alloys, oxides, and composites [47]. The mechanical milling technique offers several advantages, including scalability, simplicity, and cost-effectiveness, which have led to its widespread use in research and industrial applications. A number of studies have reported on the synthesis of nanomaterials using mechanical milling. Zhang and Wen (2020), for example, investigated the synthesis of Fe_3O_4 nanoparticles using mechanical milling and found that the size of the nanoparticles could be controlled by adjusting the milling time and speed [48]. For example, Bishoyi and Behera (2022) made Si nanoparticles using Al_2O_3 milling media and showed that large amounts of nanoparticle Si made from Al_2O_3 milling media can be used in industry [49]. These examples highlight the versatility and potential of mechanical milling for the synthesis of nanomaterials. The synthesis of fabrication nanomaterials using mechanical milling is a promising technique for the production of nanoparticles and nanostructured materials. The scalability, simplicity, and cost-effectiveness of the technique have made it a popular choice in both research and industrial settings. As more studies investigate the potential of mechanical milling for nanomaterial synthesis, it is likely that the technique will continue to gain popularity as a reliable method for easy nanomaterial production.

Electrospinning Method

Electrospinning is a popular technique for the synthesis and fabrication of nanomaterials due to its simplicity and ability to produce a wide range of materials with unique properties. Electrospinning is better than chemical methods for making nanomaterials because it lets you precisely control the size and shape of the particles that are made. It also lets you make materials with a lot of surface area and unique optical, mechanical, and electrical properties [50]. Furthermore, using this method has found useful applications in different fields, such as electronics, energy storage, and biomedicine [51]. These useful fields have proven to offer significant advantages over traditional chemical methods in terms of cost, efficiency, and environmental impact and hold the potential to revolutionize the field of nanomaterial synthesis. In conclusion, electrospinning is an efficient method for synthesizing nanomaterials with customized properties for various applications.

Lithography

The lithography technique uses electron beam light to transfer patterns onto a substrate, which can be used to create nanoscale features on the material's surface. This technique is particularly useful for producing complex, high-resolution patterns that are difficult to achieve using other methods [52]. Lithography techniques can be broadly classified into two categories: top-down and bottom-up approaches. In a top-down approach, bulk materials are etched or patterned using techniques such as electron beam lithography, photolithography, or focused ion beam milling to create nanoscale structures [53]. Another method that has gained popularity in recent years is nanoimprint lithography, which involves the use of a stamp or mold to create patterns on a substrate [53]. In summary, the synthesis of fabricated nanomaterials using lithography suggests precise control over nanomaterial size, shape, and structure, enabling the creation of novel materials with unique properties and functions.

Laser Ablation

The laser ablation technique is widely used as a physical method for synthesizing fabricated nanomaterials with high-quality nanoparticles of different sizes and shapes. Using this technique, a high-energy laser pulse is directed at a target material, causing it to vaporize and form plasma. The plasma cools rapidly, leading to the formation of nanoparticles that can be collected and characterized [54]. This method has been used to produce a wide range of nanomaterials, including metals, metal oxides, and metal carbides, with various applications in fields [55]. Recent research has shown that laser ablation can be used to make complex nanostructures like core-shell and heterostructured nanoparticles [56]. This can be done by changing the laser parameters and the materials that are being ablated. Furthermore, a study by Qian *et al.* (2006) reported the synthesis of carbon nanotubes using a pulsed laser ablation method, which yielded high-quality nanotubes with excellent electrical properties [57]. In combination with the advancement of these features, laser ablation could be a promising physical method for the synthesis of nanomaterials with continued development in different areas of nanotechnology.

Fluorescent Nanomaterial Synthesis

Fluorescent nanomaterials have unique optical properties that make them attractive for a range of applications, including imaging, sensing, and bioimaging [58]. This approach can be achieved through various physical methods, such as sol-gel, hydrothermal, and microwave-assisted methods [59]. For instance, a review report by Zhan *et al.* (2022) states the synthesis of fluorescent carbon dots using a microwave-assisted method. The synthesized carbon dots exhibited excellent photoluminescence properties and were found to be biocompatible, making them ideal for bioimaging applications [60]. The sol-gel method for synthesizing fluorescent silicon nanoparticles also makes them discoverable due to their high quantum yield and excellent stability, making them suitable for optical-electronic devices [61]. Overall, physical methods for synthesizing fluorescent nanomaterials promise a simple, scalable, and versatile approach for the fabrication of nanomaterials with unique optical properties, making them promising candidates for various applications.

Chemical Methods for Synthesis of Nanomaterials

Chemical synthesis was also found to be a key approach for the fabrication of nanomaterials. The different techniques used for synthesizing chemical nanomaterials are chemical vapor deposition (CVD), sol-gel synthesis, hydrothermal synthesis, and co-precipitation. Table **6** summarizes the chemical approach used to synthesize nanomaterials.

Table 6. Comparison of chemical synthesis methods for nanomaterial fabrication.

Method	Advantages	Examples	References
CVD	Thin, uniform films; control over composition and morphology	Graphene, carbon nanotubes	[62]
Sol-Gel	Control over composition, morphology, and porosity	Silica nanoparticles, metal oxides	[63]
Hydrothermal	Control over size, shape, and crystallinity	Zinc oxide, titanium dioxide, silver nanowires	[64]
Co-precipitation	Simple, low-cost, large-scale production	Iron oxide, cobalt ferrite	[65, 66]

Chemical Vapour Deposition (CVD)

CVD is a versatile and scalable method for the synthesis of various types of nanomaterials, including graphene, carbon nanotubes, metal oxides, and semiconductors. In CVD, precursor gases are introduced into a reaction chamber and heated, causing a chemical reaction that results in the formation of nanomaterials [67]. The properties of the resulting nanomaterials can be tuned by adjusting the temperature, pressure, and precursor gas composition. The use of CVD for the synthesis of nanomaterials has been widely reported in the literature, with numerous studies exploring the synthesis of various materials. For instance, Lei *et al.* (2016) successfully added nanomaterials that were made using the CVD method and made Bi_2T_3 nanosheets that were very uniform and crystalline using CVD [68]. However, Jain *et al.* (2019) also fabricated three-dimensional hierarchical NiFeP nanosheet arrays *via* CVD for efficient water splitting [69]. Moreover, CVD can be used to deposit materials onto a wide range of substrates, including metals, ceramics, and polymers, making it a popular choice for industrial applications [70]. With the growing demand for advanced materials, chemical methods such as CVD are expected to play an increasingly important role in the fabrication of next-generation nanomaterials.

SOL-GEL

The sol-gel technique is widely used as a chemical method for fabricating nanomaterials with exceptional properties. This process entails the formation of a sol, which is a stable colloidal suspension of nanoscale particles in a liquid, followed by gelation, during which the sol is transformed into a three-dimensional network of interconnected nanoparticles [71]. The sol-gel method provides many advantages, including precise control over the size, shape, and composition of the nanoparticles, as well as the ability to incorporate various functional groups and dopants [72]. One example of sol-gel synthesis is the fabrication of metal oxide nanomaterials such as titania, silica, and alumina. These materials have a wide range of applications in sensors, catalysis, and electronics [73]. In recent years, there has been a significant interest in the synthesis of hybrid nanomaterials using the sol-gel method. Hybrid materials are composed of two or more different components, resulting in novel properties not found in the individual components. To improve the efficiency and properties of sol-gel synthesis, researchers have developed approaches such as modified solvents, surfactants, and templates [74]. For instance, surfactants are often used to control the particle size and morphology of the nanomaterials by regulating the growth rate and direction of the particles. Additionally, templates can be used to create ordered structures and nanoarrays of the nanomaterials [75]. Overall, the sol-gel method is an efficient chemical technique for the fabrication of nanomaterials with unique properties. Its potential has been demonstrated in the fields of science and engineering, leading to more exciting advances in the synthesis of nanomaterials.

Hydrothermal Synthesis

Hydrothermal nanomaterial synthesis is a chemical method used to fabricate different types of nanomaterials with unique properties. This method involves the use of high-pressure and high-temperature water to initiate chemical reactions between precursor materials, leading to the formation of nanomaterials with well-controlled size, shape, and composition [76]. It is very flexible and can be used to make many different kinds of nanomaterials, such as nanoparticles, single crystals, hybrid inorganic and organic materials, zeolites, complex oxides, fluorides, and carbon-based nanomaterials [77]. The hydrothermal method has also been shown to make nanomaterials that are better in many ways than other methods, including having better crystallinity, more surface area, and better reactivity [78]. Some research has looked at how to make nanomaterials using hydrothermal methods. These include ZnO nanoparticles for photocatalysis [79] and TiO_2 nanotubes for electrochemical sensors [80]. Overall, the hydrothermal technique shows a promising approach for the synthesis of nanomaterials with significant properties, and ongoing research is expected to yield further advancements in this area.

Co-Precipitation

The co-precipitation technique involves the simultaneous precipitation of two or more components from a solution to form a solid nanoparticle. Co-precipitation offers a simple and cost-effective approach to producing nanomaterials with controlled size, shape, and composition [81]. Moreover, it allows for the synthesis of a wide range of materials, including metal oxides, magnetic nanoparticles, and alloys [82]. Some studies use co-precipitation nanomaterial synthesis, which, as described in a study published by Syu *et al.* (2019), used a co-precipitation method to synthesize iron oxide nanoparticles with a size range of 150 nm. They found that by adjusting the reaction parameters, such as pH and reaction time, they could control the size and shape of the nanoparticles. Additionally, they observed that the synthesized nanoparticles exhibited good magnetic properties, making them suitable for use in magnetic hyperthermia [83]. Additionally, Dey *et al.* (2020) created $CuFe_2O_4$ nanoparticles using the co-precipitation method and showed that they were very good at reducing 4-nitrophenol [84]. Likewise, Yalcin *et al.* (2021) also reported the synthesis of cobalt ferrite nanoparticles by the co-precipitation method and studied their magnetic and photocatalytic properties [85]. Thus, the co-precipitation approach is an effective and versatile method for nanomaterial synthesis. With control of reaction parameters, it is possible to control the size, shape, and properties of the resulting nanoparticles for specific applications.

BIOLOGICAL SYNTHESIS FABRICATION OF NANOMATERIALS

The biological synthesis of nanomaterials has emerged as a promising alternative to conventional methods, offering numerous advantages such as low cost, eco-friendliness, scalability, and biocompatibility. Different biological sources, such as plant extracts, algae, fungi, and microorganisms, have been used for the synthesis of nanomaterials (Table **7**). For example, plant extracts have been used as reducing agents and capping agents in the synthesis of metallic nanoparticles. The reducing agents present in the extracts reduce the metal ions to form nanoparticles, while the capping agents stabilize the nanoparticles and prevent agglomeration [86]. The use of plant extracts for nanomaterial synthesis has several advantages, including the abundance of plant sources, the low cost of extraction, and the wide variety of phytochemicals present in the extracts that can be used for the synthesis of different types of nanoparticles. Examples of metallic nanoparticles synthesized using plant extracts include gold, silver, copper, and iron oxide nanoparticles [87]. Algae has also been explored as a source of biogenic nanomaterials due to their ability to synthesize a wide variety of biomolecules. Algal extracts contain polysaccharides, proteins, and other biomolecules that can be used for the synthesis of metallic and metal oxide nanoparticles [88]. On the other hand, fungi are also shown to be excellent sources for the synthesis of metal and metal oxide nanoparticles such as silver, gold, and zinc oxide. Fungi are known to produce a variety of extracellular enzymes, such as laccase, which can act as reducing agents in nanoparticle synthesis [89]. Microorganisms such as bacteria are used to synthesize nanomaterials like silver and gold nanoparticles [90]. Bacteria such as *Escherichia coli* and *Bacillus subtilis* have been shown to produce extracellular proteins such as cytochrome c, which can act as reducing agents in nanoparticle synthesis [91, 92]. As a result, biological methods appear to be a promising alternative to traditional chemical methods for nanomaterial synthesis. Plant extracts, algae, fungi, and microorganisms offer several advantages over chemical methods, including low toxicity, biocompatibility, and sustainability. Furthermore, biogenic nanomaterials synthesized using these methods have unique properties that make them suitable for a wide range of applications. Further research is needed to optimize the synthesis methods and explore the potential applications of these biogenic nanomaterials.

Table 7. Biological sources and potential applications for nanomaterial synthesis.

Biological Source	Type of Nanomaterial	Potential Applications
Plant Extracts	Silver, gold, copper	Antimicrobial, Catalysis
Algae	Iron, silver, gold	Biomedical, Photocatalysis
Fungi	Gold, silver, selenium	Biomedical, Catalysis
Microorganisms	Silver, gold, titanium	Electronics, Catalysis

CONCLUSION

Nanomaterial fabrication has seen remarkable advancements in recent years, driven by the growing demand for new materials with improved properties and functions. The current state-of-the-art techniques for nanomaterial fabrication include top-down and bottom-up approaches, each having its own strengths and limitations. Top-down approaches are ideal for high-precision and high-resolution fabrication. While bottom-up approaches such as chemical and biological synthesis are better suited for creating complex nanostructures and assembling them into functional materials, significant challenges remain in achieving large-scale production and maintaining the quality and consistency of nanomaterials. Despite significant challenges in achieving large-scale production and maintaining consistency in quality, nanomaterial fabrication presents promising opportunities for creating novel materials with enhanced properties and functionalities. Therefore, advancements in nanomaterial creation through nanomaterial fabrication represent an ideal way to unlock new applications and opportunities in the field of nanotechnology, as well as provide continued research and development in nanomaterial fabrication, which will lead to new and exciting innovations in the field of nanotechnology.

ACKNOWLEDGEMENTS

We are thankful to the esteemed editors for the invitation to contribute a book chapter in the book titled: *"Nanomaterials in Biological Milieu: Biomedical Applications and Environmental Sustainability"*. We acknowledge the Centre for Biotechnology, Siksha O Anusandhan University, Bhubaneswar, Odisha, for the overwhelming support.

REFERENCES

[1]   Heiligtag FJ, Niederberger M. The fascinating world of nanoparticle research. Mater Today 2013; 16(7-8): 262-71.
       [http://dx.doi.org/10.1016/j.mattod.2013.07.004]

[2]   Schrittwieser S, Haslinger MJ, Mitteramskogler T, *et al.* Multifunctional nanostructures and nanopocket particles fabricated by nanoimprint lithography. Nanomaterials (Basel) 2019; 9(12): 1790.

[http://dx.doi.org/10.3390/nano9121790] [PMID: 31888231]

[3]   Nayfeh MH. Fundamentals and applications of nano silicon in plasmonics and fullerines: current and future trends. United States: Elsevier Publishing 2018; pp. 1-586.

[4]   Peter Amalathas A, Alkaisi MM. Efficient light trapping nanopyramid structures for solar cells patterned using UV nanoimprint lithography. Mater Sci Semicond Process 2017; 57: 54-8.
[http://dx.doi.org/10.1016/j.mssp.2016.09.032]

[5]   Lee YC, Tu SH. Improving the light-emitting efficiency of GaN LEDs using nanoimprint lithography. In: Cui B, Ed. Recent Advances in Nanofabrication Techniques and Applications. Canada: Intech Open 2011; pp. 173-96.
[http://dx.doi.org/10.5772/20712]

[6]   Yanagisawa M. Application of nanoimprint lithography to distributed feedback laser diodes. In: Cui B, Ed. Recent Advances in Nanofabrication Techniques and Applications. Canada: Intech Open 2011; pp. 211-24.
[http://dx.doi.org/10.5772/22678]

[7]   Sullivan DB, Boonstra T, Kief MT, Youtt L, Jayashankar S, Van Dorn C, *et al.* Hard disk drive thin film head manufactured using nanoimprint lithography. In: Douglas JR, Christopher B, Eds. Alternative Lithographic Technologies. SPIE publisher 2014; pp. 183-99.

[8]   Sullivan DB, Boonstra T, Kief MT, *et al.* Hard disk drive thin film head manufactured using nanoimprint lithography. J Micro Nanolithogr MEMS MOEMS 2013; 12(3): 031105-5.
[http://dx.doi.org/10.1117/1.JMM.12.3.031105]

[9]   Yang Y, He H, Li Y, Qiu J. Using nanoimprint lithography to create robust, buoyant, superhydrophobic PVB/SiO2 coatings on wood surfaces inspired by red roses petal. Sci Rep 2019; 9(1): 9961.
[http://dx.doi.org/10.1038/s41598-019-46337-y] [PMID: 31292503]

[10]   Fernández A, Francone A, Thamdrup LH, *et al.* Hierarchical surfaces for enhanced self-cleaning applications. J Micromech Microeng 2017; 27(4): 045020.
[http://dx.doi.org/10.1088/1361-6439/aa62bb]

[11]   Boland JJ. Within touch of artificial skin. Nat Mater 2010; 9(10): 790-2.
[http://dx.doi.org/10.1038/nmat2861] [PMID: 20835233]

[12]   Tran KTM, Nguyen TD. Lithography-based methods to manufacture biomaterials at small scales. J Sci Adv Mater Devices 2017; 2(1): 1-14.
[http://dx.doi.org/10.1016/j.jsamd.2016.12.001]

[13]   Chen J, Zhou Y, Wang D, *et al.* UV-nanoimprint lithography as a tool to develop flexible microfluidic devices for electrochemical detection. Lab Chip 2015; 15(14): 3086-94.
[http://dx.doi.org/10.1039/C5LC00515A] [PMID: 26095586]

[14]   Harvey E, Ghantasala M. Nanofabrication. In: Hannink RHJ, Hill AJ, Eds. Nanostructure control of materials. Woodhead Publisher 2006; pp. 303-30.
[http://dx.doi.org/10.1533/9781845691189.303]

[15]   Tahir U, Shim YB, Kamran MA, Kim DI, Jeong MY. Nanofabrication techniques: Challenges and future prospects. J Nanosci Nanotechnol 2021; 21(10): 4981-5013.
[http://dx.doi.org/10.1166/jnn.2021.19327] [PMID: 33875085]

[16]   Huang W, Yu X, Liu Y, Qiao W, Chen L. A review of the scalable nano-manufacturing technology for flexible devices. Front Mech Eng 2017; 12(1): 99-109.
[http://dx.doi.org/10.1007/s11465-017-0416-3]

[17]   Bathe M, Chrisey LA, Herr DJC, *et al.* Roadmap on biological pathways for electronic nanofabrication and materials. Nano Futures 2019; 3(1): 012001.
[http://dx.doi.org/10.1088/2399-1984/aaf7d5]

[18]   Yang H, Jiang B, Sun Y, *et al.* Construction of polyoxometallate-based organic-inorganic hybrid nanowires for efficient oxidative desulfurization. Molecular Catalysis 2018; 448(448): 38-45.
[http://dx.doi.org/10.1016/j.mcat.2018.01.016]

[19]   Kwiat M, Cohen S, Pevzner A, Patolsky F. Large-scale ordered 1D-nanomaterials arrays: Assembly or not? Nano Today 2013; 8(6): 677-94.
[http://dx.doi.org/10.1016/j.nantod.2013.12.001]

[20]   He X, Liu P, Wu S, Liao Q, Yao J, Fu H. Multi-color perovskite nanowire lasers through kinetically controlled solution growth followed by gas-phase halide exchange. J Mater Chem C Mater Opt Electron Devices 2017; 5(48): 12707-13.
[http://dx.doi.org/10.1039/C7TC03939E]

[21]   Cui QH, Zhao YS, Yao J. Tailoring the structures and compositions of one-dimensional organic nanomaterials towards chemical sensing applications. Chem Sci (Camb) 2014; 5(1): 52-7.
[http://dx.doi.org/10.1039/C3SC51798E]

[22]   Wang S, Ding T. Buckling polystyrene beads with light. Nanoscale 2018; 10(34): 16293-7.
[http://dx.doi.org/10.1039/C8NR03697G] [PMID: 30128453]

[23]   Yan S, Zhang J, Yuan D, Li W. Hybrid microfluidics combined with active and passive approaches for continuous cell separation. Electrophoresis 2017; 38(2): 238-49.
[http://dx.doi.org/10.1002/elps.201600386] [PMID: 27718260]

[24]   Li H, Nguyen HH, Ogorzalek Loo RR, Campuzano IDG, Loo JA. An integrated native mass spectrometry and top-down proteomics method that connects sequence to structure and function of macromolecular complexes. Nat Chem 2018; 10(2): 139-48.
[http://dx.doi.org/10.1038/nchem.2908] [PMID: 29359744]

[25]   Borghei YS, Hosseini M, Ganjali MR, Hosseinkhani S. Label-free fluorescent detection of microRNA-155 based on synthesis of hairpin DNA-templated copper nanoclusters by etching (top-down approach). Sens Actuators B Chem 2017; 248(248): 133-9.
[http://dx.doi.org/10.1016/j.snb.2017.03.148]

[26]   Kroto HW, Heath JR, O'Brien SC, Curl RF, Smalley RE. C60: Buckminsterfullerene. Nature 1985; 318(6042): 162-3.
[http://dx.doi.org/10.1038/318162a0]

[27]   Iijima S, Ichihashi T. Single-shell carbon nanotubes of 1-nm diameter. Nature 1993; 363(6430): 603-5.
[http://dx.doi.org/10.1038/363603a0]

[28]   Paramasivam G, Palem VV, Sundaram T, Sundaram V, Kishore SC, Bellucci S. Nanomaterials: Synthesis and applications in theranostics. Nanomaterials (Basel) 2021; 11(12): 3228.
[http://dx.doi.org/10.3390/nano11123228] [PMID: 34947577]

[29]   Willner I, Willner B. Biomolecule-based nanomaterials and nanostructures. Nano Lett 2010; 10(10): 3805-15.
[http://dx.doi.org/10.1021/nl102083j] [PMID: 20843088]

[30]   Han J, Zhao D, Li D, Wang X, Jin Z, Zhao K. Polymer-based nanomaterials and applications for vaccines and drugs. Polymers (Basel) 2018; 10(1): 31.
[http://dx.doi.org/10.3390/polym10010031] [PMID: 30966075]

[31]   García de Arquer FP, Talapin DV, Klimov VI, Arakawa Y, Bayer M, Sargent EH. Semiconductor quantum dots: Technological progress and future challenges. Science 2021; 373(6555): eaaz8541.
[http://dx.doi.org/10.1126/science.aaz8541] [PMID: 34353926]

[32]   Gour A, Jain NK. Advances in green synthesis of nanoparticles. Artif Cells Nanomed Biotechnol 2019; 47(1): 844-51.
[http://dx.doi.org/10.1080/21691401.2019.1577878] [PMID: 30879351]

[33]   Giner-Casares JJ, Henriksen-Lacey M, Coronado-Puchau M, Liz-Marzán LM. Inorganic nanoparticles

for biomedicine: where materials scientists meet medical research. Mater Today 2016; 19(1): 19-28.
[http://dx.doi.org/10.1016/j.mattod.2015.07.004]

[34]   Guo F, Aryana S, Han Y, Jiao Y. A review of the synthesis and applications of polymer–nanoclay composites. Appl Sci (Basel) 2018; 8(9): 1696.
[http://dx.doi.org/10.3390/app8091696]

[35]   Jafari SM, McClements DJ. Nanoemulsions: formulation, applications, and characterization. Academic Press 2018; pp. 629-42.

[36]   Rafii-Tabar H, Ghavanloo E, Fazelzadeh SA. Nonlocal continuum-based modeling of mechanical characteristics of nanoscopic structures. Phys Rep 2016; 638(638): 1-97.
[http://dx.doi.org/10.1016/j.physrep.2016.05.003]

[37]   Wang ZL, Gao RP, Pan ZW, Dai ZR. Nano-scale mechanics of nanotubes, nanowires, and nanobelts. Adv Eng Mater 2001; 3(9): 657-61.
[http://dx.doi.org/10.1002/1527-2648(200109)3:9<657::AID-ADEM657>3.0.CO;2-0]

[38]   Liu J, Yu H, Wang L, *et al.* Two-dimensional metal-organic frameworks nanosheets: Synthesis strategies and applications. Inorg Chim Acta 2018; 483(16): 550-64.
[http://dx.doi.org/10.1016/j.ica.2018.09.011]

[39]   Han W, Ren L, Gong L, *et al.* Self-assembled three-dimensional graphene-based aerogel with embedded multifarious functional nanoparticles and its excellent photoelectrochemical activities. ACS Sustain Chem& Eng 2014; 2(4): 741-8.
[http://dx.doi.org/10.1021/sc400417u]

[40]   El-Khawaga AM, Zidan A, Abd El-Mageed AI. Preparation methods of different nanomaterials for various potential applications: A Review. J Mol Struct 2023; 1281(2023): 135148.

[41]   Pearce AK, Wilks TR, Arno MC, O'Reilly RK. Synthesis and applications of anisotropic nanoparticles with precisely defined dimensions. Nat Rev Chem 2020; 5(1): 21-45.
[http://dx.doi.org/10.1038/s41570-020-00232-7] [PMID: 37118104]

[42]   Prasad Yadav T, Manohar Yadav R, Pratap Singh D. Mechanical milling: a top down approach for the synthesis of nanomaterials and nanocomposites. Nanoscience and Nanotechnology 2012; 2(3): 22-48.
[http://dx.doi.org/10.5923/j.nn.20120203.01]

[43]   Li D, Xia Y. Electrospinning of nanofibers: reinventing the wheel? Adv Mater 2004; 16(14): 1151-70.
[http://dx.doi.org/10.1002/adma.200400719]

[44]   Barcelo S, Li Z. Nanoimprint lithography for nanodevice fabrication. Nano Converg 2016; 3(1): 21.
[http://dx.doi.org/10.1186/s40580-016-0081-y] [PMID: 28191431]

[45]   Yang G. Laser ablation in liquids: principles and applications in the preparation of nanomaterials. CRC Press. 2012: 1992.
[http://dx.doi.org/10.1201/b11623]

[46]   Mehta VN, Desai ML, Basu H, Singhal RK, Kailasa SK. Recent developments on fluorescent hybrid nanomaterials for metal ions sensing and bioimaging applications: A review. J Mol Liq. 2021; 333(2021): 115950.

[47]   Chaira D, Karak SK. Fabrication of nanostructured materials by mechanical milling. In: Mahmood A, Ed. Handbook of Mechanical Nanostructuring, Wiley-VCH Verlag GmbH & Co KGaA. 2015; pp. 379-416.
[http://dx.doi.org/10.1002/9783527674947.ch16]

[48]   Zhang Z, Wen G. Synthesis and characterization of carbon-encapsulated magnetite, martensite and iron nanoparticles by high-energy ball milling method. Mater Charact. 2020; 167(2020): 110502.
[http://dx.doi.org/10.1016/j.matchar.2020.110502]

[49]   Bishoyi SS, Behera SK. Synthesis and structural characterization of nanocrystalline silicon by high energy mechanical milling using Al_2O_3 media. Adv Powder Technol 2022; 33(7): 103639.

[http://dx.doi.org/10.1016/j.apt.2022.103639]

[50]   Li Y, Zhu J, Cheng H, *et al.* Developments of advanced electrospinning techniques: A critical review. Adv Mater Technol 2021; 6(11): 2100410.
[http://dx.doi.org/10.1002/admt.202100410]

[51]   Ramesh Kumar P, Khan N, Vivekanandhan S, Satyanarayana N, Mohanty AK, Misra M. Nanofibers: effective generation by electrospinning and their applications. J Nanosci Nanotechnol 2012; 12(1): 1-25.
[http://dx.doi.org/10.1166/jnn.2012.5111] [PMID: 22523944]

[52]   Guo LJ. Recent progress in nanoimprint technology and its applications. J Phys D Appl Phys 2004; 37(11): R123-41.
[http://dx.doi.org/10.1088/0022-3727/37/11/R01]

[53]   Zhou ZJ. Electron beam lithography. Handbook of microscopy for nanotechnology. 2005: 287-321.
[http://dx.doi.org/10.1007/1-4020-8006-9_10]

[54]   Zeng H, Du XW, Singh SC, *et al.* Nanomaterials *via* laser ablation/irradiation in liquid: a review. Adv Funct Mater 2012; 22(7): 1333-53.
[http://dx.doi.org/10.1002/adfm.201102295]

[55]   Shojaei TR, Soltani S, Derakhshani M. Synthesis, properties, and biomedical applications of inorganic bionanomaterials. In: Ahmed B, Jaison J, Michael KD, Eds. Fundamentals of Bionanomaterials. Elsevier 2022; pp. 139-74.
[http://dx.doi.org/10.1016/B978-0-12-824147-9.00006-6]

[56]   Kuriakose AC, Nampoori VPN, Thomas S. Facile synthesis of Au/CdS core-shell nanocomposites using laser ablation technique. Mater Sci Semicond Process 2019; 101(101): 124-30.
[http://dx.doi.org/10.1016/j.mssp.2019.05.030]

[57]   Qian C, Qi H, Gao B, *et al.* Fabrication of small diameter few-walled carbon nanotubes with enhanced field emission property. J Nanosci Nanotechnol 2006; 6(5): 1346-9.
[http://dx.doi.org/10.1166/jnn.2006.140] [PMID: 16792363]

[58]   Yao J, Yang M, Duan Y. Chemistry, biology, and medicine of fluorescent nanomaterials and related systems: new insights into biosensing, bioimaging, genomics, diagnostics, and therapy. Chem Rev 2014; 114(12): 6130-78.
[http://dx.doi.org/10.1021/cr200359p] [PMID: 24779710]

[59]   Li W, Lee J. Microwave-assisted Sol− Gel synthesis and photoluminescence characterization of LaPO4: Eu3+, Li+ nanophosphors. J Phys Chem C 2008; 112(31): 11679-84.
[http://dx.doi.org/10.1021/jp800101d]

[60]   Zhang L, Yang X, Yin Z, Sun L. A review on carbon quantum dots: Synthesis, photoluminescence mechanisms and applications. Luminescence 2022; 37(10): 1612-38.
[http://dx.doi.org/10.1002/bio.4351] [PMID: 35906748]

[61]   Tian Z, Wu W, Li ADQ. Photoswitchable fluorescent nanoparticles: preparation, properties and applications. ChemPhysChem 2009; 10(15): 2577-91.
[http://dx.doi.org/10.1002/cphc.200900492] [PMID: 19746389]

[62]   Ghaemi F, Ali M, Yunus R, Othman RN. Synthesis of carbon nanomaterials using catalytic chemical vapor deposition technique. In: Suraya AR, Raja NIRO, Mohd ZH. Synthesis, technology and applications of carbon nanomaterials. Elsevier. 2019: 1-27.
[http://dx.doi.org/10.1016/B978-0-12-815757-2.00001-2]

[63]   Singh LP, Bhattacharyya SK, Kumar R, *et al.* Sol-Gel processing of silica nanoparticles and their applications. Adv Colloid Interface Sci 2014; 214(214): 17-37.
[http://dx.doi.org/10.1016/j.cis.2014.10.007] [PMID: 25466691]

[64]   Khataee A, Mansoori GA. Nanostructured titanium dioxide materials: properties, preparation and applications. World Scientific 2011; p. 204.

[http://dx.doi.org/10.1142/8325]

[65]   Bhavani Y, Chitti Babu N, Uday Kumar K. Decolorisation of Congo red synthetic solution using Fe doped ZnO nano particles and optimization using response surface methodology. Mater Today Proc 2023; 72(72): 232-41.
[http://dx.doi.org/10.1016/j.matpr.2022.07.049]

[66]   Peeples B, Goornavar V, Peeples C, Spence D, Parker V, Bell C, *et al.* Structural, stability, magnetic, and toxicity studies of nanocrystalline iron oxide and cobalt ferrites for biomedical applications. J Nanopart Res. 2014; 16(2014): 2290.

[67]   Choudhury S, Paul S, Goswami S, Deb K. Methods for nanoparticle synthesis and drug delivery. In: Talukdar AD, Sarker SD, Patra JK, Eds. Advances in Nanotechnology-Based Drug Delivery Systems. Elsevier. 2022: 21-44.
[http://dx.doi.org/10.1016/B978-0-323-88450-1.00005-3]

[68]   Lei W, Madni I, Ren YL, Yuan CL, Luo GQ, Faraone L. Controlled vapour-phase deposition synthesis and growth mechanism of Bi2Te3 nanostructures. Appl Phys Lett. 2016; 109(2016): 083106.
[http://dx.doi.org/10.1063/1.4961632]

[69]   Jian X, Li S, Liu J, *et al.* Three-Dimensional Graphene-Foam-Supported Hierarchical Nickel Iron Phosphide Nanosheet Arrays as Efficient and Stable Bifunctional Electrocatalysts for Overall Water Splitting. ChemElectroChem 2019; 6(21): 5407-12.
[http://dx.doi.org/10.1002/celc.201901420]

[70]   Coblas DG, Fatu A, Maoui A, Hajjam M. Manufacturing textured surfaces: State of art and recent developments. P I Mech Eng J-. J Eng 2015; 229(1): 3-29.

[71]   Figueira RB, Silva CJR, Pereira EV. Organic–inorganic hybrid sol–gel coatings for metal corrosion protection: a review of recent progress. J Coat Technol Res 2015; 12(1): 1-35.
[http://dx.doi.org/10.1007/s11998-014-9595-6]

[72]   Karthik K, Pandian SK, Jaya NV. Effect of nickel doping on structural, optical and electrical properties of TiO2 nanoparticles by sol–gel method. Appl Surf Sci 2010; 256(22): 6829-33.
[http://dx.doi.org/10.1016/j.apsusc.2010.04.096]

[73]   Ramesh S, Sivasamy A, Rhee KY, Park SJ, Hui D. Preparation and characterization of maleimide–polystyrene/SiO2–Al$_2$O$_3$ hybrid nanocomposites by an in situ sol–gel process and its antimicrobial activity. Compos, Part B Eng 2015; 75(75): 167-75.
[http://dx.doi.org/10.1016/j.compositesb.2015.01.040]

[74]   Wojciechowska P. Organic–Inorganic Hybrid Materials for Active Packaging Applications. In: Shukla AS. Eds, Bio-and Nano-sensing Technologies for Food Processing and Packaging. RSC. 2022: 63-80.
[http://dx.doi.org/10.1039/9781839167966-00063]

[75]   Nagrath M, Alhalawani A, Rahimnejad Yazdi A, Towler MR. Bioactive glass fiber fabrication *via* a combination of sol-gel process with electro-spinning technique. Mater Sci Eng C 2019; 101(101): 521-38.
[http://dx.doi.org/10.1016/j.msec.2019.04.003] [PMID: 31029347]

[76]   Gan YX, Jayatissa AH, Yu Z, Chen X, Li M. Hydrothermal synthesis of nanomaterials. J Nanomater 2020; 2020: 1-3.
[http://dx.doi.org/10.1155/2020/8917013]

[77]   Feng SH, Li GH. Hydrothermal and solvothermal syntheses. In: Xu R, Xu Y, Eds. Modern inorganic synthetic chemistry. Elsevier 2017; pp. 73-104.
[http://dx.doi.org/10.1016/B978-0-444-63591-4.00004-5]

[78]   Rane AV, Kanny K, Abitha VK, Thomas S. Methods for synthesis of nanoparticles and fabrication of nanocomposites. In: Bhagyaraj SM, Ed. Synthesis of inorganic nanomaterials. Woodhead publishing 2018; pp. 121-39.
[http://dx.doi.org/10.1016/B978-0-08-101975-7.00005-1]

[79]   Asjadi F, Yaghoobi M. Characterization and dye removal capacity of green hydrothermal synthesized ZnO nanoparticles. Ceram Int 2022; 48(18): 27027-38.
[http://dx.doi.org/10.1016/j.ceramint.2022.06.015]

[80]   Isik E, Tasyurek LB, Isik I, Kilinc N. Synthesis and analysis of TiO2 nanotubes by electrochemical anodization and machine learning method for hydrogen sensors. Microelectron Eng. 2022; 262(2022): 111834.
[http://dx.doi.org/10.1016/j.mee.2022.111834]

[81]   Srivastava A, Katiyar A. Zinc oxide nanostructures. In: Misra PK, Misra RDK, Eds. Ceramic Science and Engineering. Elsevier 2022; pp. 235-62.
[http://dx.doi.org/10.1016/B978-0-323-89956-7.00012-7]

[82]   Jagadeeshan S, Parsanathan R. Nano-metal oxides for antibacterial activity. In: Naushad M, Rajendran S, Gracia F, Eds. Advanced Nanostructured Materials for Environmental Remediation. Springer Publisher 2019; pp. 59-90.
[http://dx.doi.org/10.1007/978-3-030-04477-0_3]

[83]   Syu WJ, Huang CC, Hsiao JK, *et al.* Co-precipitation synthesis of near-infrared iron oxide nanocrystals on magnetically targeted imaging and photothermal cancer therapy *via* photoablative protein denature. Nanotheranostics 2019; 3(3): 236-54.
[http://dx.doi.org/10.7150/ntno.24124] [PMID: 31263656]

[84]   Dey C, De D, Nandi M, Goswami MM. A high performance recyclable magnetic CuFe2O4 nanocatalyst for facile reduction of 4-nitrophenol. Mater Chem Phys 2020; 242(2020): 122237.

[85]   Yalcin B, Ozcelik S, Icin K, Senturk K, Ozcelik B, Arda L. Structural, optical, magnetic, photocatalytic activity and related biological effects of CoFe2O4 ferrite nanoparticles. J Mater Sci Mater Electron 2021; 32(10): 13068-80.
[http://dx.doi.org/10.1007/s10854-021-05752-6]

[86]   Sidhu AK, Verma N, Kaushal P. Role of biogenic capping agents in the synthesis of metallic nanoparticles and evaluation of their therapeutic potential. Front Nanotechnol. 2022; 3(2022): 801620.
[http://dx.doi.org/10.3389/fnano.2021.801620]

[87]   Adeyemi JO, Oriola AO, Onwudiwe DC, Oyedeji AO. Plant extracts mediated metal-based nanoparticles: Synthesis and biological applications. Biomolecules 2022; 12(5): 627.
[http://dx.doi.org/10.3390/biom12050627] [PMID: 35625555]

[88]   Sampath S, Madhavan Y, Muralidharan M, Sunderam V, Lawrance AV, Muthupandian S. A review on algal mediated synthesis of metal and metal oxide nanoparticles and their emerging biomedical potential. J Biotechnol 2022; 360(360): 92-109.
[http://dx.doi.org/10.1016/j.jbiotec.2022.10.009] [PMID: 36272578]

[89]   Michael A, Singh A, Roy A, Islam MR. [Retracted] Fungal- and Algal-Derived Synthesis of Various Nanoparticles and Their Applications. Bioinorg Chem Appl 2022; 2022(1): 3142674.
[http://dx.doi.org/10.1155/2022/3142674] [PMID: 36199747]

[90]   Das BS, Das A, Mishra A, Arakha M. Microbial cells as biological factory for nanoparticle synthesis. Front Mater Sci 2021; 15(2): 177-91.
[http://dx.doi.org/10.1007/s11706-021-0546-8]

[91]   Jain AS, Pawar PS, Sarkar A, Junnuthula V, Dyawanapelly S. Bionanofactories for green synthesis of silver nanoparticles: Toward antimicrobial applications. Int J Mol Sci 2021; 22(21): 11993.
[http://dx.doi.org/10.3390/ijms222111993] [PMID: 34769419]

[92]   Bakishzade A, Nasibova A. Future prospects of biomaterials in nanomedicine. Adv Bio Earth Sci 2023; 2023(8): 5-10.

CHAPTER 2

Recent Advances and Future Perspective in Green Synthesis of Nanomaterial for Healthcare Management

Soumya Subhashree Satapathy[1], Ruchi Bhuyan[2,3], Arun Kumar Pradhan[1], Manoranjan Arakha[1] and **Sanat Kumar Bhuyan[3,*]**

[1] *Center for Biotechnology, School of Pharmaceutical Science, Siksha 'O' Anusandhan (Deemed to be University), Bhubaneswar, Odisha 751030, India*

[2] *Department of Medical Research, IMS and SUM Hospital, Siksha 'O' Anusandhan (Deemed to be University), Bhubaneswar, Odisha 751030, India*

[3] *Institute of Dental Science, Siksha 'O' Anusandhan (Deemed to be University), Bhubaneswar, Odisha 751030, India*

Abstract: Nanotechnology, which operates on the 1 to 100 nanometer scale and has a wide range of applications in medicine and pharmaceuticals, has grown rapidly in recent years. Nanomaterials, with their distinct features, hold great promise for cancer treatment due to their low side effects and faster healing potential. However, traditional techniques for producing nanomaterials have drawbacks such as high costs, energy usage, and environmental hazards. To address these concerns, researchers have turned to green nanotechnology, which aims to reduce potential risks associated with traditional production methods. Green nanoparticles, synthesized using environmentally friendly procedures, provide a sustainable solution with reduced energy consumption and non-toxic byproducts. They can be produced using a variety of natural sources such as bacteria, fungi, algae, and plant components. The green method has shown a variety of therapeutic effects, including antiviral, antioxidant, and anticancer qualities. It promotes efficient oral cancer treatment through precise drug administration, thereby reducing side effects. Furthermore, the green synthesis process causes the production of reactive oxygen species (ROS) and free radicals, which promotes apoptosis in cancer cells. This chapter discusses an overview of recent developments in green nanomaterials, their synthesis process, and healthcare management.

Keywords: Apoptosis, Biocompatibility, Diagnostics, Environment friendly, Green synthesis, Metallic nanoparticles, Nontoxic, Nanoparticles, Therapeutics, Targeted drug delivery.

---

* **Corresponding author Sanat Kumar Bhuyan:** Institute of Dental Science, Siksha 'O' Anusandhan (Deemed to be University), Bhubaneswar, Odisha 751030, India; E-mail: sanatbhuyan@soa.ac.in

INTRODUCTION

Nanotechnology, with nanosizes ranging from 1 to 100 nm, has marked the beginning of an era in the healthcare sector, assisting in both disease detection and management. It is the most critical area of research in the early twenty-first century. Nanotechnology has improved healthcare and sparked the development of groundbreaking nanosystems for the detection, imaging, and therapy of numerous diseases, including cancer, heart disorders, and diseases linked to the central nervous system [1]. Due to their high surface-to-volume ratio, nanoparticles are ideal carriers for many clinical applications [2]. Nanomedicine, which combines modern biology and medicine, is creating a new benchmark for disease prevention and treatment in the pharmaceutical industry [3]. By building smart devices that can aid in disease detection at an early stage, nanomedicine is recognized to be efficient against *in vitro* and *in vivo* diagnostics [4]. The most cutting-edge interdisciplinary area of nanotechnology in recent years is known as regenerative medicine, which is capable of delivering medications and hormones for tissue healing [5]. Nanosensors can measure the concentration of sodium, glucose, cholesterol, cancer biomarkers, and other infectious agents. Advances in nanotechnology raise the chances of invasively and precisely delivering drugs to the desired sites [6]. Nanomaterials have unique physical, chemical, and biological capabilities due to their small size, which increases strength, reactivity, and conductivity. They are important in a variety of sectors such as they can be applied to targeted medication delivery, diagnostics, and tissue engineering, used for water purification and air filtration, and are necessary for effective energy storage devices, sensors, and renewable energy technologies such as solar and fuel cells [7]. Hepatobiliary disorders affect over 2 million people each year, and inefficient medication delivery systems typically prevent therapy. Nanotechnology provides an appropriate strategy by allowing customized medicine delivery to the liver while using its unique properties. Surface modification improves medication delivery effectiveness and reduces side effects using a variety of nanomaterials, including polymer, inorganic, and multifunctional nanoparticles. However, toxicity and stability difficulties exist, which may cause inflammation and protein adsorption. Current studies in nanotechnology have become essential for developing targeted treatments for liver ailments [8]. Nanobiomaterials have significance in cardiovascular applications since they improve therapy effectiveness and patient outcomes by delivering drugs more precisely and with fewer adverse effects. Their large surface area-to-volume ratios facilitate interactions with biological molecules, hence improving diagnostic and therapeutic capacities [9].

Nanotechnology can be synthesized using several chemical methods. However, these methods use toxic and expensive solvents, producing toxic and lethal end products. Thus, there is an urge to develop an eco-friendly approach to overcome these drawbacks [10].

Green nanotechnology, developed by biological methods, has recently gained popularity in the healthcare industry. It is an environmentally friendly and biocompatible method that leads to large-scale production with few hazardous side effects [11]. The greener method of nanoparticle synthesis is mostly used in nanomedicines and nano-drug delivery systems. Furthermore, green approaches are thought to reduce the risks of the production of noxious intermediates and by-products because they are inexpensive and have few complications. Two major aspects of green synthesized nanomaterials involve [12]:

• Developing nanomaterials that will address environmental issues.
• Developing nanomaterials to reduce any negative effects on human health.

A comprehensive and fundamental literature review was conducted to discover the recent advances and advantages of green-synthesized nanoparticles. Several online databases were evaluated, such as Scopus, PubMed, NCBI, and Google Scholar, between the years 2000 and 2022 with the keywords "nanoparticles", "green synthesis", "therapeutics", "diagnostics", and "targeted drug delivery".

SYNTHESIS OF NANOMATERIALS

Several chemical processes can be used to synthesize the nanoparticles. Some of them include:

Chemical Precipitation Method

This method involves converting a solution into a solid by either making it insoluble or extremely saturated. Chemical agents are added, and subsequently, the precipitates and solution are separated [13]. Chemical precipitation is the most common way for removing ionic metals from solutions, especially in industrial effluent contaminated with hazardous metals. Ionic metals are converted into insoluble particles *via* a chemical reaction involving soluble metal compounds and a precipitating agent. These particles are subsequently removed from the solution using filtration and settling methods [14] (Fig. **1**).

Sol-gel Method

The sol-gel method is a chemical synthesis technique for producing nanomaterials. The two major steps in this process are condensation and hydrolysis. It begins with converting a liquid "sol" into a solid "gel" phase by various chemical reactions. Initially, precursor molecules are hydrolyzed and condensed, forming a colloidal suspension (sol). This sol then polymerizes to produce a solid three-dimensional network (gel) [15]. It is a quick, affordable, and efficient way to make high-quality nanoparticles [16]. Furthermore, it enables the manufacture of a wide range of materials, including oxides, glasses, and composites, with exact control over porosity and homogeneity. Adjusting factors such as precursor concentration, pH, and temperature allows for greater freedom in modifying the properties of the final nanomaterials.

Hydrothermal Method

This technique uses a high temperature of nearly 470 °C and low pressure of fewer than 300 MPa to generate nanoparticles. Compounds that are ordinarily insoluble under normal circumstances can be diluted using this technique [17]. Hydrothermal synthesis takes place in a sealed container in which a high-pressure liquid, usually water, is heated past its boiling point, resulting in specialized conditions for material synthesis. This approach promotes the formation of aligned nanorods on substrates by controlling nucleation *via* substrate preparation. Various compounds, including metal oxides such as ZnO and TiO_2, have been successfully produced utilizing hydrothermal techniques, which use a variety of precursors, mineralizers, and surfactants [18].

Sputtering

It is a technique where high-intensity external stimuli are used to displace nanoparticles from the surface of the target material. When a high level of energy is supplied in comparison to ordinary thermal energies, successful nanoparticle ejection occurs. High-purity nanoparticles are synthesized using this process.

Electrochemical Reduction

The electrochemical approach was utilized by researchers to produce metallic nanoparticles. Electricity serves as the driving power in this technique. Electric current flows between two electrodes, which are separated by an electrolyte [19]. The metallic anode sheet was dissolved, and the metallic salt obtained was reduced by the cathode to produce metallic particles. Tetraalkylammonium salts helped to stabilize the produced metallic particles.

Photochemical Decomposition

The photochemical approach uses radiation to excite the system and produce active reducing agents such as radicals, electrons, and excited components. This approach has a considerable benefit over chemical reduction in that it removes impurities at lower temperatures. It is frequently used in noble metal production. First, water, alcohol, or organic solvents are used to produce the required salt solutions. The solution is then subjected to radiation, which causes electrons to be released and metals to be reduced [20].

Chemical Reduction Method

Bimetallic nanoparticle synthesis is carried out using this method. It mostly involves the reduction and interaction processes. Several reducing agents, including ascorbate and sodium borohydride, are employed in Tollen's reagent to generate metallic nanoparticles in a zero-valent state.

Microemulsion Method

A microemulsion is made up of three components: a surfactant, a polar component, and a non-polar component. The surfactant's job is to form a layer between the polar and non-polar components. Microemulsions are classified as water-in-oil (w/o) or oil-in-water (o/w) depending on the type of dispersed and continuous phase.

Even though there are numerous chemical methods for producing nanoparticles, this approach has many drawbacks to this approach, including high equipment costs, unsafe by-products, and the requirement for using high temperatures, which is extremely dangerous and can cause a variety of dermatological disorders. Researchers are working on a green nanoparticle synthesis method that will overcome these constraints.

GREEN SYNTHESIS OF NANOPARTICLES

Green synthesis has acquired widespread responsiveness as a sustainable and nature-friendly approach for the synthesis of several nanomaterials. Green synthesis is regarded as a key method for reducing the adverse consequences associated with the chemical methods of synthesis for nanoparticles. The green synthesis method is a one-step process that extracts metallic nanoparticles using several biological methods. The metallic nanoparticles produced by this method have a wide range of medical applications, including early diagnosis and therapeutics. Interestingly, the rising popularity of green approaches is not only confined to plants but has also sparked the synthesis of NPs utilizing a variety of sources, including plants, bacteria [21], algae, and fungi, leading to large-scale manufacturing with much less contamination [22].

Plant-based Nanoparticle Synthesis

Plants are frequently used in the synthesis of NPs because they are readily available, safe for humans, and include an array of metabolites [23]. Metabolites act as reducing, capping, and stabilizing agents, preventing MNPs from aggregating and agglomerating. The use of plants for the synthesis of nanoparticles has managed to gain significant attention in recent years due to its quick, environmentally sustainable, non-toxic, and cost-effective procedure that offers a single-step method for biological processes [24].

Plant-based NP synthesis is considered the pinnacle among green biological approaches due to its ease of extraction [25]. Surprisingly, any part of a plant, including its leaves, stem, roots, latex, and flowers, can be used to synthesize nanoparticles. Phytonanotechnology has enabled the synthesis of nearly all nanoparticles, including silver, gold, zinc oxide, copper oxide, palladium, iron, and others [26].

Microorganisms-based Nanoparticle Synthesis

Nanoparticle synthesis is not constricted to plants; bacteria, fungi, and yeast can all be used in the green synthesis of nanoparticles (Table **1**). Microbe-produced nanoparticles are thought to be non-toxic, inexpensive, and do not require a high amount of energy for their production. Microorganisms can detoxify heavy metals because of the several reductase enzymes that can convert metal salts into metal nanoparticles. In recent years, microorganisms, including fungi, bacteria, and yeasts, have been studied both extracellularly and intracellularly for the production of metallic nanoparticles [26]. Nonetheless, the extracellular strategy is preferred since it eradicates the need for downstream processing to produce NPs.

From Bacteria

Prokaryotic bacteria are key to the green synthesis method by utilizing low energy. Bacteria aid in the transformation of hazardous metals into nontoxic nanoparticles [27]. Bacteria are well known for producing inorganic materials both intracellularly and extracellularly. Some commonly used bacteria in the green synthesis approaches include Escherichia coli., Pseudomonas aeruginosa, Vibrio cholerae, Staphylococcus currens [28].

Cyanobacteria, a photosynthetic bacterium, because of the existence of their bioactive chemicals, which can stabilize and functionalize the nanoparticles, are extensively investigated to generate nanoparticles. Cyanobacteria's faster growth rate leads to the higher production of biomass to support nano synthesis [29].

From Actinomycetes

Actinomycetes have essential traits in common with fungi and prokaryotes. Despite being categorized as prokaryotes, they were formerly called ray fungi.

Because of their saprophytic nature, they produce a variety of bioactive components and extracellular enzymes. Actinomycetes are regarded as superior categories among microbes and have attracted substantial attention for the synthesis of metallic NPs [30]. Actinomycetes produce nanoparticles when metal ions undergo intracellular reduction on the surface of mycelia and cytoplasmic membranes [31]. Thermomonospora sp. and Rhodococcus sp [32] are some of the actinomycetes used in the green synthesis approach.

From Fungi and Yeast

Not only bacteria, mycosynthesis is in trend for nanoparticle synthesis. Fascinatingly, fungi are thought to be more productive than bacteria because they include a variety of bioactive substances with better bioaccumulation potential, are affordable, and have a higher tolerance rate [33]. The possible mechanisms put out to explain mycosynthesis include nitrate reductase activity, electron shuttle quinones, or a combination of the two. The synthesis can be done either in an intracellular manner or through an extracellular manner. Extracellular biosynthesis is involved in the utilization of fungus extracts [34, 35]. Fungal enzymes, specifically nitrate reductase from Penicillium species and Fusarium oxysporum, play a vital role in the production of nanoparticles [36].

It is presumed that yeast is a viable carrier for NP production because of its simple encapsulation method and an excellent source of dietary supplements [37].

From Algae

Algae are thought to be the best candidate for nanoparticle biogenesis because they can accumulate metal and reduce metal ions. Furthermore, algae have several advantages, including the ability to synthesize at relatively low temperatures, resulting in increased energy efficiency and less environmental risk and toxic effects (Table **1**). Anticancer, antibacterial, wound-healing, and antifungal activities are among the potential applications of algal nanoparticles [38].

From Plant Virus

Plant viruses have been studied in recent years for their ability to produce nanoparticles. Because of their structural and biochemical stability, plant viruses are an effective method of producing nanoconjugates and nanocomposites with metal nanoparticles, which are critical components in the delivery of drugs and treatment for cancer [39].

Fig. (**1**) describes the methods involved in the extraction of nanoparticles using the chemical method and green approach and explains the pitfalls of chemical methods.

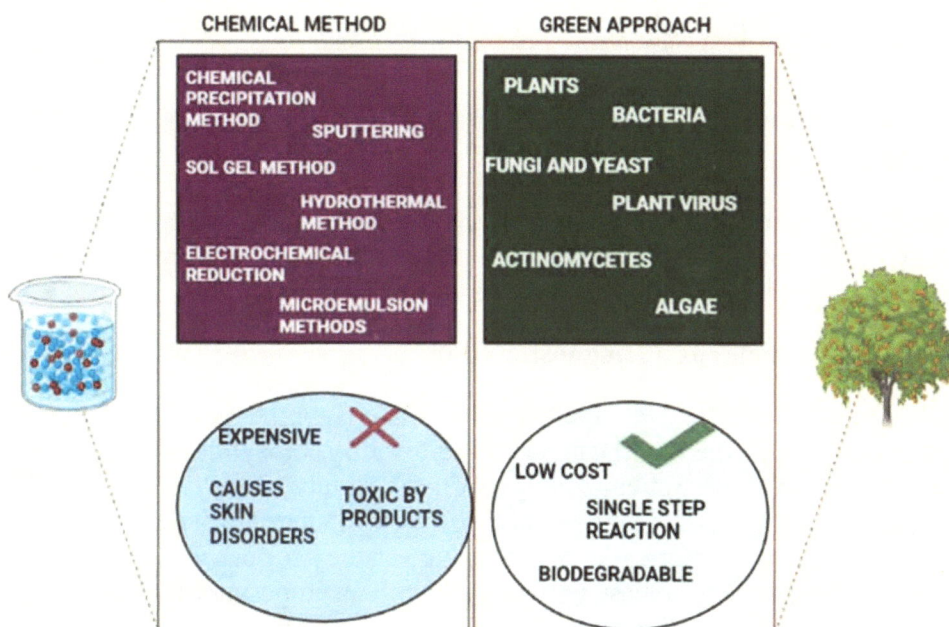

Fig. (1). Describes the methods involved in the extraction of nanoparticles using the chemical method and green approach and explains the pitfalls of chemical methods.

Table 1. describes the sources from where NPs were extracted along with their nano size.

Serial No.	Sources		NPs	Size of NPs	References
01	Plants	*Ginkgo biloba*	Copper	15 – 20nm	[40]
		Nigella sativa	Silver	15nm	[41]
		Camellia sinensis	Zinc	9 – 17.5nm	[42]
02	Bacteria	*Bhargavaea indica*	Silver and gold	30 – 100nm	[43]
		Weissella oryzae	Silver	10 -30nm	[44]
03	Actinomycetes	*Streptomyces* sp. LK3	Silver	5nm	[45]
04	Fungi	*Neurospora crassa*	Silver, gold, bimetallic silver, and gold	> 100nm	[46]

(Table 1) cont.....

Serial No.	Sources		NPs	Size of NPs	References
05	Yeast	Extremophilic yeast	Silver and gold	20nm (silver), 30 -100 nm (gold)	[47]
		Rhodosporidium diobovatum	Lead	2 -5nm	[48]
06	Algae	*Spirulina platensis*	Gold	5nm	[49]

RECENT ADVANCES IN NANOMATERIALS IN THE HEALTHCARE SECTOR

With the emergence of revolutionary nanosystems for the diagnosis, imaging, and therapy of multiple diseases, the domain of nanotechnology has recently attracted significant interest in various healthcare industries (Fig. **2**).

Green Nanoparticles as a Diagnostic Tool

Green nanomaterials, which have recently been synthesized, have enormous potential for use in the healthcare sector. Currently, biosynthesized NPs are the focus of extensive research and analysis for the timely identification of a variety of diseases. NPs have been pivotal in multifunctional molecular imaging. Nanotechnology has emerged as a promising strategy for cancer diagnosis due to its nano size, controlled release patterns, and enhanced permeability and retention effect [50]. Here, we will discuss the role of some of the nanomaterials that are used as diagnostic tools.

Role of Zinc Oxide in Diagnostics

ZnO nanoparticles have grown in popularity in diagnostics due to their unique characteristics, which include semiconducting properties and wide band gap semiconductors with a high exciton binding energy, resulting in an efficient source of excitonic blue and near-UV emissions [51, 52]. ZnO NPs are ideal for bio-imaging due to their excitonic blue and near-UV emissions. They are also attractive nanomaterials for biosensors because of their large surface area, excellent biocompatibility/stability, low toxicity, and high electron transfer capability [53]. Zno-based biosensors can detect phenol, glucose, H_2O_2, urea, and other small molecule analytes. An elevated level of urea can cause many disorders, such as gastrointestinal hemorrhage, renal failure, and urinary tract obstruction. Thus, it is essential to measure the urea level in blood for the diagnosis of multiple diseases. Recently, zinc oxide nanowires have been created on gold-coated materials for urea biosensors, which immobilize urease through physical adsorption [54].

Role of Silver Oxide and Gold Nanomaterials in Diagnostics

For centuries, the antifungal, antibacterial, antiviral, anticancer, wound-healing, and bone-repairing capabilities of silver oxide have been extensively observed. AgNPs can also be employed as inexpensive surface-enhanced Raman scattering substrates. AgNP-containing nanoparticles can serve as biosensors to find infections, enzymes, blood glucose, molecular markers of tumor cells, *etc.* Due to their distinctive optical and electrical characteristics, silver nanoclusters are a suitable material for synthetic probes. Certain mutations causing sickle cell anemia were identified by the fluorescent silver nanoclusters [55]. Due to their remarkable physicochemical properties, gold nanoparticles have engrossed the attention of many researchers in recent times. In general, Au NPs display size- and shape-dependent optical and electrical characteristics and biocompatibility. Au NPs have exceptional properties that exemplify their enormous potential for use in multiple biomedical fields [56]. AuNPs have sparked the most interest as an X-ray contrast agent due to their high X-ray absorption coefficient, nontoxicity, surface functionalization for colloidal stability, and precise delivery. Gold nanoparticles are employed for the identification of microbial cells and their metabolites, the bio-imaging of tumor cells, the identification of receptors on their surface, and the observation of endocytosis [57].

Some other Nanomaterials used in Diagnostics

Palladium NPs are an excellent possibility for imaging enhancement in early diagnosis due to their ultra-thin, high photothermal efficiency, and catalytic characteristics [58]. Pd NPs were used to generate extremely sensitive, unlabelled electrochemical biosensors for the detection of microRNA (miRNA) [59]. Iron oxide nanomaterial is a promising candidate for a magnetic resonance imaging (MRI) contrast agent that enables the timely detection or monitoring of a variety of ailments, including inflammatory disorders and cancer [60]. With several remarkable advantages, including low cost, non-toxicity, and significant electrochemical activity, copper is a noble metal substitute that may be used in biosensors for the early identification of illness [61]. To identify the cervical cancer-causing human papillomavirus type 16 (HPV16), CuO was used as colored tags in lateral flow strip biosensors [62].

Green Synthesized Nanomaterials in Therapeutics

Nanomaterials play a major role in therapeutics due to their nano size, which makes them ideal for crossing the barriers such as the blood-brain barrier and ensures minimal toxicity with enhanced permeability and a higher retention effect that helps to release the drug in a precise and dose-dependent manner to the targeted location. In addition, the green approach is cost-effective, and its ease of availability makes it suitable for therapeutics.

In Gene Therapy

Gene therapy involves the replacement of a defective gene with a healthy one. Vectors, particularly viral vectors, are used to insert a normal gene into a stem cell [63]. Unfortunately, the use of viral vectors is linked with multiple risk factors, including the occurrence of off-target changes in the host body, immune response, and inflammation. Due to the drawbacks, these methods are not encouraged anymore, and the globe is currently shifting toward the use of nanomaterials for gene therapy. Non-viral nanostructures are used in gene therapy, which is considered safe, less carcinogenic, and rarely triggers the immune system. The ability of nanoparticles to penetrate the membrane deeply and their high surface-to-volume ratio makes them ideal vectors for gene therapy [64].

In Targeted Drug Delivery

Drug targeting to a specific site has become a significant challenge in recent years. This is because the major difficulties in conventional drug delivery include nonspecific drug delivery that affects healthy tissues and cells, along with drug degradation. In recent times, thanks to the drug delivery system, which helped to conquer the hurdles associated with the traditional drug delivery method, the DDS assists the drug in precisely reaching its target in a dose-dependent manner. Furthermore, this novel DDS shields the drug from photodegradation [65]. The use of nanoparticles is the most approachable method of drug delivery because of their nano size, the ability to penetrate deep into the tumor cells, the distribution of drugs only to the desired target, not disrupting the surrounding sites, achieving higher therapeutic efficiency as they can load multiple drugs, increasing the drug circulation time in bloodstream, and suppressing of fast renal excretion of the drug. Gold NPs have proven to be the best among several metallic NPs. The reason for selecting Au NPs is that they have proven to have a high capacity for delivering drugs and their ease of functionalization with other molecules, biocompatibility, and high stability make them an ideal drug delivery carrier. The palladium NPs are also thought to be an excellent vehicle for delivering anticancer drugs [66].

In the Treatment of Cardiovascular Diseases

Worldwide, cardiovascular disorders are responsible for millions of fatalities [67]. Several treatments have improved patient survival, but none of them have completely restored the heart, particularly in those who have had a myocardial infarction [68]. Several conventional medications can help reduce CVD. In addition to these traditional methods, novel therapeutic models are widely employed to treat CVD. Indeed, in recent years, nanomedicine has grown in popularity as a potent, pivotal, and safe platform that can be employed in the management of ischemic, inflammatory, angiogenic, and metabolic disorders such as hypertension, hyperlipidemia, and atherosclerosis [69]. For treating CVDs, iron oxide nanoparticles are highly desirable. Iron oxide nanoparticles are essential for the magnetically assisted transport of mesenchymal stem cells to the infarcted myocardium. Furthermore, it appears that using these magnetic elements seems to benefit various aspects of stem cell therapy. Moreover, it increases the availability of stem cells near the injury site, improving their capacity for regeneration. The use of iron oxide NPs in the treatment of heart failure will be a successful and appealing strategy since it will increase the myocardium's stem cell supply while reducing the migration of these cells to other organs [70].

Besides this, AuNPs also possess important cardioprotective features. This is mostly attributable to their effects on oxidative stress and hypertrophy, which are brought about by the downregulation of beta-adrenoceptors and consequent reduction in the ERK1/2-mediated hypertrophic pathway. Inhibiting beta-adrenoceptors is crucial for reducing myocardial infarction morbidity since they are the main mediators of myocardial hypertrophy [71].

Ag NPs have been used as a coating on pacemakers and cardiac stents to lower the likelihood of infections caused by foreign bodies. AgNPs' antibacterial properties are responsible for this beneficial lowering of infections [72].

In the Treatment of Brain Disorders

The brain is the body's most complex organ that helps regulate cognitive, behavioral, and emotional activities. Brain disorders mostly include disability and ultimately lead to death. Moreover, the brain is a target for countless disorders and diseases ranging from injuries to cancers. For many years, researchers have faced difficulties in developing a drug that can cross the blood-brain barrier and other mechanical barriers. However, in recent years, nanoparticles have raised hopes for the cure of brain disorders. Owing to their small size, minimal toxic effect, and high bioavailability, nanoparticles can easily cross and penetrate the blood-brain barrier [73]. Nanomaterials help in precise drug delivery by enhancing the cytotoxic effect of the conventional drug at the affected site. The large surface-t--volume area of nanomaterials enhances the solubility and bioavailability of the drug so that the drug can easily dissolve and can reach systemic blood circulation and afterward to the desired site [74, 75].

In the Treatment of Ocular Diseases

Ocular diseases have a direct impact on human vision [76]. There has been significant progress in the understanding of pathological mechanisms and the treatment of ocular disease [77]. Despite this, diagnoses and treatments for these diseases can be ineffective and lack specificity due to the unique anatomical structures and physiological barriers of the visual system. Current therapeutic methods aid in the restoration of total vision loss. The unique properties of nanomaterials, such as their improved bio adhesiveness, stealth capabilities, and stimuli-responsive release, are intended to treat eye ailments.

Gold NPs are frequently used in ophthalmology since they have higher stability, antioxidant, biocompatibility, anti-angiogenic, and modifiability properties and can easily pass the blood-retinal barrier without harming the retina [78]. According to research, conjugating the combination of Au NPs and hyaluronic acid enables them to penetrate through the eye's physiological barrier and gain entry into the retina. Upon entering the human retina, the hyaluronic acid-gold NPs can deliver relevant drugs, significantly increasing the drug's bioavailability and possible therapeutic impact [79].

Inflammation is a major cause of ophthalmic disorders, and gold, silver, and titanium dioxide nanoparticles all have anti-inflammatory properties [80]. Another *in vitro* study discovered that selenium nanoparticles could suppress the inflammatory responses caused by hypoxia in RPE cells [81]. ZnO-NPs are effective antibacterial agents for reducing the risk of infection when wearing contact lenses [82].

In the Treatment of Oral Diseases

Nanotechnology presents novel therapeutic strategies for the prevention of oral disorders, including dental caries and periodontitis [83]. To more quickly recognize and eliminate the harmful bacteria that can cause gingivitis and periodontitis, nanorobots are being integrated into mouthwash [84]. Dental caries is mostly carried on by bacteria that stick to the tooth's surface and generate acid, which induces demineralization of the tooth surfaces and phosphate and calcium loss. Hopefully, there are several nanomaterials, such as zinc oxide, that have antibacterial properties that can be incorporated to prevent the growth of bacteria. The nanomaterials inhibit bacterial growth using mechanisms like the production of reactive oxygen species [85], disruptions of the bacterial cell membrane [86], and disruption of the electron transport across the bacterial membrane [87]. Moreover, nanoparticles help in the selective killing of oral cancer cells *via* inducing apoptosis of the cancer cell and are known to be a vowing approach for drug delivery to a specific location. The nanomaterials are used to develop novel adhesive films that help cure oral ulcers and induce salivary stimulation. Nanoparticles act as a sealer to reduce root canal disinfection.

In Treating Cancer

The multidisciplinary branch of nanotechnology is among the most promising areas for cancer therapy. Recently, there have been a lot of accents on the use of nanomaterials in cancer treatment. It provides a distinct strategy against cancer through early detection, prediction, prevention, and therapy [88]. Several metallic nanoparticles, including silver, gold, and zinc oxide, have shown excellent anticancer properties against different cell lines. The nanoparticles aid in the *in vivo* tumor imaging and diagnosis of tumor cells. It induces apoptosis in a dose-dependent manner. The nanomaterials help block the signalling of VEGF and EDGR angiogenic factors [89]. Gold NPs are known to promote apoptosis initiation and cell cycle arrest at the G2/M transition [90].

Nano Antibiotics

Antibiotic resistance is considered the major consequence, where bacteria become resistant to antimicrobial treatments, hastening the evolution of the bacterial genome. Nanoparticles have proven effective in treating multidrug-resistant bacteria [91]. Three processes are primarily involved in the antimicrobial properties of metallic nanoparticles: ROS production, the ion release mechanism, and the interaction of NPs with cell membranes [92].

Fig. (**2**) explains the various applications of nanomaterials along with their examples that are associated with healthcare sectors.

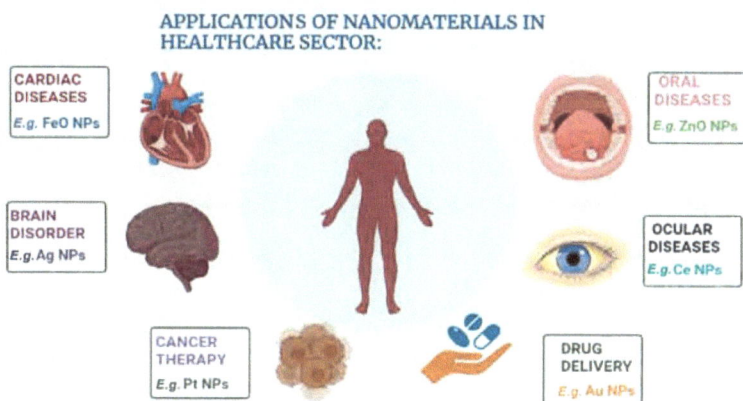

APPLICATIONS OF NANOMATERIALS IN HEALTHCARE SECTOR:

CARDIAC DISEASES
E.g. FeO NPs

ORAL DISEASES
E.g. ZnO NPs

BRAIN DISORDER
E.g. Ag NPs

OCULAR DISEASES
E.g. Ce NPs

CANCER THERAPY
E.g. Pt NPs

DRUG DELIVERY
E.g. Au NPs

Fig. (2). Explains the various applications of nanomaterials along with their examples that are associated with healthcare sectors.

CONCLUSION AND FUTURE PERSPECTIVE

This article briefly summarizes the limitations of the chemical techniques of synthesis as well as the green synthesis methods of nanomaterials. Recently, a lot of researchers have attempted to develop a metal NP-based early disease diagnosis and therapeutics. Metal NP-based technologies offer greater sensitivity and selectivity, or they offer unique innovations that are not achievable with conventional procedures. Metal NPs' distinctive and adaptable optical characteristics have shown great promise for the development of biosensing and bioimaging methods.

The exigence for the green approach of nanoparticle extraction has sparked the advancement of various green synthesis procedures that use plants, microorganisms, and other natural resources to produce nanomaterials. Focus has been placed on the ecologically friendly synthesis of NPs using this method. Much study has been done on green synthesized extract-mediated NPs generation and their possible applications in a broad range of sectors because of their affordability, nontoxicity, simplicity of availability, and eco-friendliness. Although it is becoming extremely prevalent and is anticipated to increase massively in the coming years, protracted effects on humans and animals, alongside the accumulation of the nanomaterials in the environment and their sway, must be considered.

ACKNOWLEDGEMENTS

The authors would like to thank the president of Siksha 'O' Anusandhan (deemed to be university), India, for providing the necessary resources to carry out the study.

AUTHOR CONTRIBUTION

SSS did the literature review and wrote the manuscript. SKB designed the manuscript and gave the concept. RB helped in making diagrams, and AKP and MA edited the manuscript.

LIST OF ABBREVIATIONS

Ag Silver

Au Gold

CVD Cardiovascular Diseases

NPs Nanoparticles

ROS Reactive Oxygen Species

ZnO Zinc oxide

REFERENCES

[1]   Saul JM, Annapragada AV, Bellamkonda RV. A dual-ligand approach for enhancing targeting selectivity of therapeutic nanocarriers. J Control Release 2006; 114(3): 277-87.
[http://dx.doi.org/10.1016/j.jconrel.2006.05.028] [PMID: 16904220]

[2]   Nath D, Banerjee P. Green nanotechnology – A new hope for medical biology. Environ Toxicol Pharmacol 2013; 36(3): 997-1014.
[http://dx.doi.org/10.1016/j.etap.2013.09.002] [PMID: 24095717]

[3]   Elayan H, Shubair RM, Almoosa N. Revolutionizing the healthcare of the future through nanomedicine: Opportunities and challenges. In 2016 12th International Conference on Innovations in Information Technology (IIT) 2016; 1-5

[4]   Chen G, Roy I, Yang C, Prasad PN. Nanochemistry and nanomedicine for nanoparticle-based diagnostics and therapy. Chem Rev 2016; 116(5): 2826-85.
[http://dx.doi.org/10.1021/acs.chemrev.5b00148] [PMID: 26799741]

[5]   Singh S, Singh A. Current status of nanomedicine and nanosurgery. Anesth Essays Res 2013; 7(2): 237-42.
[http://dx.doi.org/10.4103/0259-1162.118976] [PMID: 25885840]

[6]   Sarkar S, Misra S. From Micro to Nano: The Evolution of Wireless Sensor-Based Health Care. IEEE Pulse 2016; 7(1): 21-5.
[http://dx.doi.org/10.1109/MPUL.2015.2498498] [PMID: 26799723]

[7]   Karadağ M, Omarova S. USE OF Prunus armeniaca L. Seed Oil and pulp in health and cosmetic products. Adv Bio Earth Sci. 2024 Jan 2; 9.
[http://dx.doi.org/10.62476/abess105]

[8]   Huseynov E, Khalilov R, Mohamed AJ. Novel nanomaterials for hepatobiliary diseases treatment and future perspectives. Adv Bio Earth Sci. 2024 Jan 2; 9.
[http://dx.doi.org/10.62476/abes9s81]

[9]   Erdil N. Cardiovascular disease, signalling, gene/cell therapy and advanced nanobiomaterials. Adv Bio Earth Sci. Jan 2; 9.
[http://dx.doi.org/10.62476/abes9s58]

[10]   Srivastava S. Green Nanotechnology. J Nanotechnol Mater Sci 2016; 3(1): 1-7.
[http://dx.doi.org/10.15436/2377-1372.16.022]

[11]   Ahmed S, Ahmad M, Swami BL, Ikram S. A review on plants extract mediated synthesis of silver nanoparticles for antimicrobial applications: A green expertise. J Adv Res 2016; 7(1): 17-28.
[http://dx.doi.org/10.1016/j.jare.2015.02.007] [PMID: 26843966]

[12]   Karn BP, Bergeson LL. Green nanotechnology: straddling promise and uncertainty. Nat Resour Envt 2009; 24(2): 24-9.

[13]   Sharma G, Gupta VK, Agarwal S, Kumar A, Thakur S, Pathania D. Fabrication and characterization of Fe@MoPO nanoparticles: Ion exchange behavior and photocatalytic activity against malachite green. J Mol Liq 2016; 219: 1137-43.
[http://dx.doi.org/10.1016/j.molliq.2016.04.046]

[14]   Dahman Y, Deonanan K, Dontsos T, Iammatteo A. Nanopolymers. Nanotechnology and functional materials for engineers. 2017; 121-44.
[http://dx.doi.org/10.1016/B978-0-323-51256-5.00006-X]

[15]   Suroshe JS. Chemical methods for the synthesis of nanomaterials. Nanochemistry. CRC Press 2023; pp. 77-87.
[http://dx.doi.org/10.1201/9781003081944-5]

[16]   Sharma G, Kumar A, Naushad M, Pathania D, Sillanpää M. Polyacrylamide@Zr(IV) vanadophosphate nanocomposite: Ion exchange properties, antibacterial activity, and photocatalytic behavior. J Ind Eng

Chem 2016; 33: 201-8.
[http://dx.doi.org/10.1016/j.jiec.2015.10.011]

[17]   Sharma G, Kumar A, Sharma S, *et al.* Novel development of nanoparticles to bimetallic nanoparticles and their composites: A review. J King Saud Univ Sci 2019; 31(2): 257-69.
[http://dx.doi.org/10.1016/j.jksus.2017.06.012]

[18]   Djurisic AB, Xi YY, Hsu YF, Chan WK. Hydrothermal synthesis of nanostructures. Recent Pat Nanotechnol 2007; 1(2): 121-8.
[http://dx.doi.org/10.2174/187221007780859591] [PMID: 19076026]

[19]   Katwal R, Kaur H, Sharma G, Naushad M, Pathania D. Electrochemical synthesized copper oxide nanoparticles for enhanced photocatalytic and antimicrobial activity. J Ind Eng Chem 2015; 31: 173-84.
[http://dx.doi.org/10.1016/j.jiec.2015.06.021]

[20]   Ghorbani HR. Chemical Synthesis of Copper Nanoparticles. Orient J Chem 2014; 30(2): 803-6.
[http://dx.doi.org/10.13005/ojc/300254]

[21]   Daphne J, Francis A, Mohanty R, Ojha N, Das N. Green synthesis of antibacterial silver nanoparticles using yeast isolates and its characterization. Research Journal of Pharmacy and Technology 2018; 11(1): 83-92.
[http://dx.doi.org/10.5958/0974-360X.2018.00016.1]

[22]   Ahmad S, Munir S, Zeb N, *et al.* Green nanotechnology: a review on green synthesis of silver nanoparticles — an ecofriendly approach. Int J Nanomedicine 2019; 14: 5087-107.
[http://dx.doi.org/10.2147/IJN.S200254] [PMID: 31371949]

[23]   Nath D, Banerjee P. Green nanotechnology – A new hope for medical biology. Environ Toxicol Pharmacol 2013; 36(3): 997-1014.
[http://dx.doi.org/10.1016/j.etap.2013.09.002] [PMID: 24095717]

[24]   Li S, Shen Y, Xie A, *et al.* Rapid, room-temperature synthesis of amorphous selenium/protein composites using *Capsicum annuum L* extract. Nanotechnology 2007; 18(40): 405101.
[http://dx.doi.org/10.1088/0957-4484/18/40/405101]

[25]   Hano C, Abbasi BH. Plant-Based Green Synthesis of Nanoparticles: Production, Characterization and Applications. Biomolecules 2021; 12(1): 31.
[http://dx.doi.org/10.3390/biom12010031] [PMID: 35053179]

[26]   Singh P, Kim YJ, Zhang D, Yang DC. Biological Synthesis of Nanoparticles from Plants and Microorganisms. Trends Biotechnol 2016; 34(7): 588-99.
[http://dx.doi.org/10.1016/j.tibtech.2016.02.006] [PMID: 26944794]

[27]   Garole DJ, Choudhary BC, Paul D, Borse AU. Sorption and recovery of platinum from simulated spent catalyst solution and refinery wastewater using chemically modified biomass as a novel sorbent. Environ Sci Pollut Res Int 2018; 25(11): 10911-25.
[http://dx.doi.org/10.1007/s11356-018-1351-5] [PMID: 29397510]

[28]   Khursigara CM, Koval SF, Moyles DM, Harris RJ. Inroads through the bacterial cell envelope: seeing is believing. Can J Microbiol 2018; 64(9): 601-17.
[http://dx.doi.org/10.1139/cjm-2018-0091] [PMID: 30169124]

[29]   Ghosh S, Ahmad R, Zeyaullah M, Khare SK. Microbial Nano-Factories: Synthesis and Biomedical Applications. Front Chem 2021; 9: 626834.
[http://dx.doi.org/10.3389/fchem.2021.626834] [PMID: 33937188]

[30]   Kumar SA, Peter YA, Nadeau JL. Facile biosynthesis, separation and conjugation of gold nanoparticles to doxorubicin. Nanotechnology 2008; 19(49): 495101.
[http://dx.doi.org/10.1088/0957-4484/19/49/495101] [PMID: 21730661]

[31]   Hulkoti NI, Taranath TC. Biosynthesis of nanoparticles using microbes—A review. Colloids Surf B Biointerfaces 2014; 121: 474-83.

[http://dx.doi.org/10.1016/j.colsurfb.2014.05.027] [PMID: 25001188]

[32]   Ahmad A, Mukherjee P, Senapati S, *et al.* Extracellular biosynthesis of silver nanoparticles using the fungus Fusarium oxysporum. Colloids Surf B Biointerfaces 2003; 28(4): 313-8.
[http://dx.doi.org/10.1016/S0927-7765(02)00174-1]

[33]   Alghuthaymi MA, Almoammar H, Rai M, Said-Galiev E, Abd-Elsalam KA. Myconanoparticles: synthesis and their role in phytopathogens management. Biotechnol Biotechnol Equip 2015; 29(2): 221-36.
[http://dx.doi.org/10.1080/13102818.2015.1008194] [PMID: 26019636]

[34]   Zhang L, Gu FX, Chan JM, Wang AZ, Langer RS, Farokhzad OC. Nanoparticles in medicine: therapeutic applications and developments. Clin Pharmacol Ther 2008; 83(5): 761-9.
[http://dx.doi.org/10.1038/sj.clpt.6100400] [PMID: 17957183]

[35]   Molnár Z, Bódai V, Szakacs G, *et al.* Green synthesis of gold nanoparticles by thermophilic filamentous fungi. Sci Rep 2018; 8(1): 3943.
[http://dx.doi.org/10.1038/s41598-018-22112-3] [PMID: 29500365]

[36]   Anil Kumar S, Abyaneh MK, Gosavi SW, *et al.* Nitrate reductase-mediated synthesis of silver nanoparticles from AgNO3. Biotechnol Lett 2007; 29(3): 439-45.
[http://dx.doi.org/10.1007/s10529-006-9256-7] [PMID: 17237973]

[37]   Roychoudhury A. Yeast-mediated Green Synthesis of Nanoparticles for Biological Applications. Indian J Pharm Biol Res 2020; 8(3): 26-31.

[38]   LewisOscar F, Vismaya S, Arunkumar M, Thajuddin N, Dhanasekaran D, Nithya C. Algal nanoparticles: synthesis and biotechnological potentialsAlgae-Organisms Imminent Biotechnol.. 2016 Jun 29; 7: 157-82.
[http://dx.doi.org/10.5772/62909]

[39]   Chandrakala V, Aruna V, Angajala G. Review on metal nanoparticles as nanocarriers: current challenges and perspectives in drug delivery systems. Emergent Materials 2022; 5(6): 1593-615.
[http://dx.doi.org/10.1007/s42247-021-00335-x] [PMID: 35005431]

[40]   Nasrollahzadeh M, Mohammad Sajadi S. Green synthesis of copper nanoparticles using Ginkgo biloba L. leaf extract and their catalytic activity for the Huisgen [3 + 2] cycloaddition of azides and alkynes at room temperature. J Colloid Interface Sci 2015; 457: 141-7.
[http://dx.doi.org/10.1016/j.jcis.2015.07.004] [PMID: 26164245]

[41]   Amooaghaie R, Saeri MR, Azizi M. Synthesis, characterization and biocompatibility of silver nanoparticles synthesized from Nigella sativa leaf extract in comparison with chemical silver nanoparticles. Ecotoxicol Environ Saf 2015; 120: 400-8.
[http://dx.doi.org/10.1016/j.ecoenv.2015.06.025] [PMID: 26122733]

[42]   Zahran M, El-Kemary M, Khalifa S, El-Seedi H. Spectral studies of silver nanoparticles biosynthesized by Origanum majorana. Green Processing and Synthesis 2018; 7(2): 100-5.
[http://dx.doi.org/10.1515/gps-2016-0183]

[43]   Singh P, Kim YJ, Singh H, Mathiyalagan R, Wang C, Yang DC. Biosynthesis of anisotropic silver nanoparticles by Bhargavaea indica and their synergistic effect with antibiotics against pathogenic microorganisms. J Nanomater 2015; 2015(1): 234741.
[http://dx.doi.org/10.1155/2015/234741]

[44]   Singh P, Kim YJ, Wang C, Mathiyalagan R, Yang DC. *Weissella oryzae* DC6-facilitated green synthesis of silver nanoparticles and their antimicrobial potential. Artif Cells Nanomed Biotechnol 2016; 44(6): 1569-75.
[http://dx.doi.org/10.3109/21691401.2015.1064937] [PMID: 26212222]

[45]   Karthik L, Kumar G, Kirthi AV, Rahuman AA, Bhaskara Rao KV. Streptomyces sp. LK3 mediated synthesis of silver nanoparticles and its biomedical application. Bioprocess Biosyst Eng 2014; 37(2): 261-7.

[http://dx.doi.org/10.1007/s00449-013-0994-3] [PMID: 23771163]

[46]   Castro-Longoria E, Vilchis-Nestor AR, Avalos-Borja M. Biosynthesis of silver, gold and bimetallic nanoparticles using the filamentous fungus Neurospora crassa. Colloids Surf B Biointerfaces 2011; 83(1): 42-8.
[http://dx.doi.org/10.1016/j.colsurfb.2010.10.035] [PMID: 21087843]

[47]   Mourato A, Gadanho M, Lino AR, Tenreiro R. Biosynthesis of crystalline silver and gold nanoparticles by extremophilic yeasts. Bioinorg Chem Appl. 2011; 2011.
[http://dx.doi.org/10.1155/2011/546074]

[48]   Seshadri S, Saranya K, Kowshik M. Green synthesis of lead sulfide nanoparticles by the lead resistant marine yeast, *Rhodosporidium diobovatum*. Biotechnol Prog 2011; 27(5): 1464-9.
[http://dx.doi.org/10.1002/btpr.651] [PMID: 21710608]

[49]   Naveena BE, Prakash S. Biological synthesis of gold nanoparticles using marine algae Gracilaria corticata and its application as a potent antimicrobial and antioxidant agent. Asian J Pharm Clin Res 2013; 6(2): 179-82.

[50]   Samuel MS, Ravikumar M, John J A, *et al.* A review on green synthesis of nanoparticles and their diverse biomedical and environmental applications. Catalysts 2022; 12(5): 459.
[http://dx.doi.org/10.3390/catal12050459]

[51]   Yang P, Yan R, Fardy M. Semiconductor nanowire: what's next? Nano Lett 2010; 10(5): 1529-36.
[http://dx.doi.org/10.1021/nl100665r] [PMID: 20394412]

[52]   Wang ZL. Splendid one-dimensional nanostructures of zinc oxide: a new nanomaterial family for nanotechnology. ACS Nano 2008; 2(10): 1987-92.
[http://dx.doi.org/10.1021/nn800631r] [PMID: 19206442]

[53]   Kumar SA, Chen SM. Nanostructured zinc oxide particles in chemically modified electrodes for biosensor applications. Anal Lett 2008; 41(2): 141-58.
[http://dx.doi.org/10.1080/00032710701792612]

[54]   Solanki PR, Kaushik A, Ansari AA, Sumana G, Malhotra BD. Zinc oxide-chitosan nanobiocomposite for urea sensor. Appl Phys Lett 2008; 93(16): 163903.
[http://dx.doi.org/10.1063/1.2980448]

[55]   Xu L, Wang YY, Huang J, Chen CY, Wang ZX, Xie H. Silver nanoparticles: Synthesis, medical applications and biosafety. Theranostics 2020; 10(20): 8996-9031.
[http://dx.doi.org/10.7150/thno.45413] [PMID: 32802176]

[56]   Bansal SA, Kumar V, Karimi J, Singh AP, Kumar S. Role of gold nanoparticles in advanced biomedical applications. Nanoscale Adv 2020; 2(9): 3764-87.
[http://dx.doi.org/10.1039/D0NA00472C] [PMID: 36132791]

[57]   Dykman LA, Khlebtsov NG. Gold nanoparticles in biology and medicine: recent advances and prospects. Acta Nat (Engl Ed) 2011; 3(2): 34-55.
[http://dx.doi.org/10.32607/20758251-2011-3-2-34-55] [PMID: 22649683]

[58]   Dumas A, Couvreur P. Palladium: a future key player in the nanomedical field? Chem Sci (Camb) 2015; 6(4): 2153-7.
[http://dx.doi.org/10.1039/C5SC00070J] [PMID: 28694948]

[59]   Zhang C, Li D, Li D, Wen K, Yang X, Zhu Y. Rolling circle amplification-mediated *in situ* synthesis of palladium nanoparticles for the ultrasensitive electrochemical detection of microRNA. Analyst (Lond) 2019; 144(12): 3817-25.
[http://dx.doi.org/10.1039/C9AN00427K] [PMID: 31086898]

[60]   Jouyandeh M, Sajadi SM, Seidi F, *et al.* Metal nanoparticles-assisted early diagnosis of diseases. OpenNano 2022; 8: 100104.
[http://dx.doi.org/10.1016/j.onano.2022.100104]

[61]   Santos FCU, Paim LL, Luiz da Silva J, Stradiotto NR. Electrochemical determination of total reducing sugars from bioethanol production using glassy carbon electrode modified with graphene oxide containing copper nanoparticles. Fuel 2016; 163: 112-21.
[http://dx.doi.org/10.1016/j.fuel.2015.09.046]

[62]   Yang Z, Yi C, Lv S, *et al.* Development of a lateral flow strip biosensor based on copper oxide nanoparticles for rapid and sensitive detection of HPV16 DNA. Sens Actuators B Chem 2019; 285: 326-32.
[http://dx.doi.org/10.1016/j.snb.2019.01.056]

[63]   Gonçalves GAR, Paiva RMA. Gene therapy: advances, challenges and perspectives. Einstein (Sao Paulo) 2017; 15(3): 369-75.
[http://dx.doi.org/10.1590/s1679-45082017rb4024] [PMID: 29091160]

[64]   Singh M, Briones M, Ott G, O'Hagan D. Cationic microparticles: A potent delivery system for DNA vaccines. Proc Natl Acad Sci USA 2000; 97(2): 811-6.
[http://dx.doi.org/10.1073/pnas.97.2.811] [PMID: 10639162]

[65]   Martinho N, Damgé C, Reis CP. Recent advances in drug delivery systems. J Biomater Nanobiotechnol 2011; 2(5): 510-26.
[http://dx.doi.org/10.4236/jbnb.2011.225062]

[66]   Klębowski B, Depciuch J, Parlińska-Wojtan M, Baran J. Applications of Noble Metal-Based Nanoparticles in Medicine. Int J Mol Sci 2018; 19(12): 4031.
[http://dx.doi.org/10.3390/ijms19124031] [PMID: 30551592]

[67]   Prajnamitra RP, Chen HC, Lin CJ, Chen LL, Hsieh PCH. Nanotechnology approaches in tackling cardiovascular diseases. Molecules 2019; 24(10): 2017.
[http://dx.doi.org/10.3390/molecules24102017] [PMID: 31137787]

[68]   Calin M, Stan D, Simion V. Stem cell regenerative potential combined with nanotechnology and tissue engineering for myocardial regeneration. Curr Stem Cell Res Ther 2013; 8(4): 292-303.
[http://dx.doi.org/10.2174/1574888X11308040005] [PMID: 23547964]

[69]   Younis NK, Ghoubaira JA, Bassil EP, Tantawi HN, Eid AH. Metal-based nanoparticles: Promising tools for the management of cardiovascular diseases. Nanomedicine 2021; 36: 102433.
[http://dx.doi.org/10.1016/j.nano.2021.102433] [PMID: 34171467]

[70]   Jasmin , Souza GT, Andrade Louzada R, Rosado-de-Castro PH, Mendez-Otero R, Carvalho ACC. Tracking stem cells with superparamagnetic iron oxide nanoparticles: perspectives and considerations. Int J Nanomedicine 2017; 12: 779-93.
[http://dx.doi.org/10.2147/IJN.S126530]

[71]   Qiao Y, Zhu B, Tian A, Li Z. PEG-coated gold nanoparticles attenuate β-adrenergic receptor-mediated cardiac hypertrophy. Int J Nanomedicine 2017; 12: 4709-19.
[http://dx.doi.org/10.2147/IJN.S130951] [PMID: 28740379]

[72]   Talapko J, Matijević T, Juzbašić M, Antolović-Požgain A, Škrlec I. Antibacterial activity of silver and its application in dentistry, cardiology and dermatology. Microorganisms 2020; 8(9): 1400.
[http://dx.doi.org/10.3390/microorganisms8091400] [PMID: 32932967]

[73]   Ngowi EE, Wang YZ, Qian L, *et al.* The application of nanotechnology for the diagnosis and treatment of brain diseases and disorders. Front Bioeng Biotechnol 2021; 9: 629832.
[http://dx.doi.org/10.3389/fbioe.2021.629832] [PMID: 33738278]

[74]   Chow SC. Bioavailability and bioequivalence in drug development. Wiley Interdiscip Rev Comput Stat 2014; 6(4): 304-12.
[http://dx.doi.org/10.1002/wics.1310] [PMID: 25215170]

[75]   Alany R. Solid state characterization, solid dispersions, solubility enhancement, drug dissolution and drug release. Pharm Dev Technol 2017; 22(1): 1-1.
[http://dx.doi.org/10.1080/10837450.2017.1275305] [PMID: 28117622]

[76] Weng Y, Liu J, Jin S, Guo W, Liang X, Hu Z. Nanotechnology-based strategies for treatment of ocular disease. Acta Pharm Sin B 2017; 7(3): 281-91.
[http://dx.doi.org/10.1016/j.apsb.2016.09.001] [PMID: 28540165]

[77] Schoenfeld ER, Greene JM, Wu SY, Leske MC. Patterns of adherence to diabetes vision care guidelines. Ophthalmology 2001; 108(3): 563-71.
[http://dx.doi.org/10.1016/S0161-6420(00)00600-X] [PMID: 11237912]

[78] Yang C, Yang J, Lu A, *et al*. Nanoparticles in ocular applications and their potential toxicity. Front Mol Biosci 2022; 9: 931759.
[http://dx.doi.org/10.3389/fmolb.2022.931759] [PMID: 35911959]

[79] Laradji A, Karakocak BB, Kolesnikov AV, Kefalov VJ, Ravi N. Hyaluronic acid-based gold nanoparticles for the topical delivery of therapeutics to the retina and the retinal pigment epithelium. Polymers (Basel) 2021; 13(19): 3324.
[http://dx.doi.org/10.3390/polym13193324] [PMID: 34641139]

[80] Tsirouki T, Dastiridou A, Symeonidis C, *et al*. A focus on the epidemiology of uveitis. Ocul Immunol Inflamm 2018; 26(1): 2-16.
[http://dx.doi.org/10.1080/09273948.2016.1196713] [PMID: 27467180]

[81] Özkaya D, Nazıroğlu M, Vanyorek L, Muhamad S. Involvement of TRPM2 channel on hypoxia-induced oxidative injury, inflammation, and cell death in retinal pigment epithelial cells: modulator action of selenium nanoparticles. Biol Trace Elem Res 2021; 199(4): 1356-69.
[http://dx.doi.org/10.1007/s12011-020-02556-3] [PMID: 33389617]

[82] Shayani Rad M, Sabeti Z, Mohajeri SA, Fazly Bazzaz BS. Preparation, characterization, and evaluation of zinc oxide nanoparticles suspension as an antimicrobial media for daily use soft contact lenses. Curr Eye Res 2020; 45(8): 931-9.
[http://dx.doi.org/10.1080/02713683.2019.1705492] [PMID: 31847595]

[83] Neel A, Bozec L, Perez RA, Kim HW, Knowles JC. Nanotechnology in dentistry: prevention, diagnosis, and therapy. Int J Nanomedicine 2015; 10: 6371-94.
[http://dx.doi.org/10.2147/IJN.S86033] [PMID: 26504385]

[84] Vaishali S, Nashra K. Nanotechnology in Periodontics: A Review. J Res Med Dent Sci 2021; 9: 339-44.

[85] Foldbjerg R, Olesen P, Hougaard M, Dang DA, Hoffmann HJ, Autrup H. PVP-coated silver nanoparticles and silver ions induce reactive oxygen species, apoptosis and necrosis in THP-1 monocytes. Toxicol Lett 2009; 190(2): 156-62.
[http://dx.doi.org/10.1016/j.toxlet.2009.07.009] [PMID: 19607894]

[86] Kawabata N, Nishiguchi M. Antibacterial activity of soluble pyridinium-type polymers. Appl Environ Microbiol 1988; 54(10): 2532-5.
[http://dx.doi.org/10.1128/aem.54.10.2532-2535.1988] [PMID: 3202632]

[87] Li Q, Mahendra S, Lyon DY, *et al*. Antimicrobial nanomaterials for water disinfection and microbial control: Potential applications and implications. Water Res 2008; 42(18): 4591-602.
[http://dx.doi.org/10.1016/j.watres.2008.08.015] [PMID: 18804836]

[88] Jin C, Wang K, Oppong-Gyebi A, Hu J. Application of Nanotechnology in Cancer Diagnosis and Therapy - A Mini-Review. Int J Med Sci 2020; 17(18): 2964-73.
[http://dx.doi.org/10.7150/ijms.49801] [PMID: 33173417]

[89] Gurunathan S, Lee KJ, Kalishwaralal K, Sheikpranbabu S, Vaidyanathan R, Eom SH. Antiangiogenic properties of silver nanoparticles. Biomaterials 2009; 30(31): 6341-50.
[http://dx.doi.org/10.1016/j.biomaterials.2009.08.008] [PMID: 19698986]

[90] Al-Dulimi AG, Al-Saffar AZ, Sulaiman GM, *et al*. Immobilization of l-asparaginase on gold nanoparticles for novel drug delivery approach as anti-cancer agent against human breast carcinoma cells. J Mater Res Technol 2020; 9(6): 15394-411.

[http://dx.doi.org/10.1016/j.jmrt.2020.10.021]

[91]   Nadeem M, Tungmunnithum D, Hano C, *et al.* The current trends in the green syntheses of titanium oxide nanoparticles and their applications. Green Chem Lett Rev 2018; 11(4): 492-502.
[http://dx.doi.org/10.1080/17518253.2018.1538430]

[92]   Nisar P, Ali N, Rahman L, Ali M, Shinwari ZK. Antimicrobial activities of biologically synthesized metal nanoparticles: an insight into the mechanism of action. J Biol Inorg Chem 2019; 24(7): 929-41.
[http://dx.doi.org/10.1007/s00775-019-01717-7] [PMID: 31515623]

CHAPTER 3

Biosynthesis of Selenium Nanoparticles and their Biological and Pharmaceutical Applications

Satabdi Mishra[1], Banishree Sahoo[1,*] and Manoranjan Arakha[1]

[1] Center for Biotechnology, Siksha 'O' Anusandhan (Deemed to be University), Bhubaneswar, Odisha 751030, India

Abstract: Numerous methods have been established to fabricate nanoparticles with varying physico-chemical properties to suit the applications. However, because of eco-friendly, non-toxic, and cost-effective approaches, fabrication methods have attracted the interest of many research groups to synthesize different nanoparticles, especially for biological and pharmaceutical applications. In this context, microorganisms and plant extracts are widely utilized for the bio-fabrication of nanoparticles. Hence, this chapter intends to discuss the bio-fabrication of selenium nanoparticles and their biological applications. In this context, SeNPs have antioxidative, antimicrobial, and anticancer activities. Hence, SeNP has a broad range of therapeutic effects and is often used to treat various diseases, including diabetes, cancer, arthritis, *etc.* Selenium nanoparticles can be used for numerous pharmaceutical applications, such as gene and drug delivery. Additionally, SeNPs are used for environmental remediation and nano-biosensors, which can be used to detect pathogens, nucleic acids, antibodies, *etc.* To this end, the chapter will explore the potential of biological agents for the bio-fabrication of nanoparticles and their application in different pharmaceutical industries.

Keywords: Antimicrobial activity, Anticancer activity, Bio-fabrication, Nanoparticles, Selenium.

INTRODUCTION

In the twenty-first century, nanotechnology has become an emerging field of science. In this context, the synthesis of nanoparticles using natural products is gaining importance, and they already have a variety of beneficial properties. There are various techniques to synthesize nanoparticles, such as physical, chemical, and biological approaches.

However, using chemical and physical methods to fabricate nanoparticles has disadvantages, such as the use of harmful substances that are not eco-friendly.

* **Corresponding author Banishree Sahoo:** Center for Biotechnology, Siksha 'O' Anusandhan (Deemed to be University), Bhubaneswar, Odisha 751030, India; E-mail: sahoo.banishree1@gmail.com

Manoranjan Arakha & Arun Kumar Pradhan (Eds.)

Hence, biological/green synthetic approaches for nanoparticle synthesis have piqued the interest of several research groups since they are cleaner, have fewer side effects on the human body on application, and are non–toxic and eco-friendly in nature. Among the essential micronutrients necessary for humans and animals, selenium (Se) plays an important role. Although selenium comes in two forms—organic and inorganic—it has been shown that organic selenium compounds are more resilient than inorganic ones. It has been reported in various studies that selenium nanoparticles are a little toxic and highly bio-active; hence, they possess promising biomedical applications [1]. This chapter will briefly discuss the biological approach to synthesizing selenium nanoparticles and their pharmaceutical applications. Even though there are numerous techniques for the fabrication of selenium nanoparticles, biological methods using different biological agents, such as microbes, plant extracts, *etc.*, are attracting the interest of different scientific groups. The use of microorganisms for nanoparticle synthesis involves certain risks, such as pathogenicity, and necessitates the maintenance of massive cultures. As a result, the production of nanoparticles using plants and plant extracts is gaining popularity [2]. Furthermore, manufacturing of nanoparticles utilizing plants/some plant extracts has benefits over microorganisms such as bacteria, algae, fungus, and so on. Hence, green synthetic approaches are used for the synthesis of nanoparticles using plant products that are much simpler, eco-friendly, easily available, cost-effective, efficient, and can be scaled up easily for larger operations [2, 3]. Different nanoparticles, for example, zinc oxide, silver, gold, iron, and palladium, have been successfully synthesized using green synthetic approaches [2].

NANOPARTICLES

Richard Feynman is known as the father of modern nanotechnology [4]. However, in 1925, the idea of a 'nanometer' was suggested by Nobel Laureate Richard Zsigmondy. In 1959, Richard Feynman presented a lecture named 'There's Plenty of Room at the Bottom', in which he addressed the concept of manipulating matter at the atomic level [4]. The word 'nanoparticle' was coined for the first time by Norio Taniguchi in 1974. A nanoparticle is usually described as a particle of matter sized between 1 and 100 nanometres in diameter [5]. Nanoscience is the study that helps in the construction, formation, composition, shape, and design of particle sizes between 1-100 nanometers [6]. When there is a comparison between the bulk or the large size of the same material and the nanoparticle, the nanoparticles show some advanced physicochemical features, such as tiny size and higher surface area to volume ratio, compared to bulk material [7]. Hence, nanoparticles are considered a bridge between bulk and small-sized materials onthe nanoscale. Nanoparticles usually comprise three layers: the core, surface layer, and the shell layer. Among them, the surface layer usually comprises a diversity of molecules named polymers, metal ions, and surfactants [8]. Nanoparticles are either an assembly of one material or a combination of more than one material [9]. Nanoparticles can either remain as suspensions, colloids, or dispersed aerosols mainly because of their electromagnetic properties and chemical composition [10]. In the past decades, the research on nanoparticles has opened many different and new opportunities for their use in the treatment of many complicated diseases. Due to their unique and extraordinary features, such as the tiny dimensions, higher surface area to volume ratio, surface chemistry, solubility, and multi-functionality, nanoparticles are marvelously special [11]. Nanoparticles manifest themselves as good candidates for drug transporters, achieving huge triumphs in delivering therapeutic molecules to specific target sites without any or fewer side effects. Nanoparticles synthesized using biological methods are being applied in different biomedical engineering fields, which is only due to their advanced physico-chemical properties [12]. Additionally, nanoparticle systems seem a favorable alternative to oral administration drug-delivering systems and nutritional supplementation.

Selenium Nanoparticles

Selenium or Se is one of many vital microelements in the body because of its pro-oxidative and anti-oxidative effects. In 1817, Swedish scientist Jons Jacob Berzelius first discovered the element selenium [13]. The letter 'Se' in the word selenium was adopted from the word 'selene', which is the other name of the moon in the Greek language. It is presented as group 6 of the modern periodic table and holds an atomic number of 34 [13]. It is an indirect elemental semiconductor and has always attracted the interest of researchers due to its remarkable properties, such as photoelectrical properties, catalytic activities towards organic hydration and oxidation reactions, optical properties, *etc.*, and greater future application in different fields of biological sciences [14]. Hence, selenium is a partly solid non-metal, mostly seen as red in powder form, black in vitreous form, and metallic gray in crystalline form [15]. SeNP exhibits better biological activity when it enters into the circulation of the human body compared to other organic and inorganic selenium products [11]. However, selenium deficiency in our body can result in the degeneration of many organs and tissues, such as disorders related to heart muscle and joints, prostate cancer, thyroiditis, or asthma [16].

Synthesis of Nanoparticles

Nanoparticles are of two types: naturally found and manmade. Nanoparticles are naturally available in volcanic ashes, ocean sprays, tiny sand particles, dust, and biological living organisms like viruses [17]. Manmade nanoparticles are also called anthropogenic or synthetic nanoparticles, which are made by humans or, more precisely, by researchers in the laboratory. Synthetic or manmade nanoparticles can be further classified into incidental and engineered nanoparticles. Incidental nanoparticles, as the name suggests, are produced without proper attention and are believed to be the result of some unknown human activities. They are of undesired shape, sizes, and characteristics. Human activities like using diesel engines, extracting minerals through mining, and igniting a fire can even make nanoparticles that are missed through our naked eyes [18]. However, engineered nanoparticles are precisely planned and intentionally manufactured by human beings with proper care and attention because a slight change in the synthesis process can affect the properties of nanoparticles [19]. Now, the engineered nanoparticles can be obtained by various methods like chemical, physical, and biological techniques, *etc.* Among these methods, the chemical and physical procedures used for formulating nanoparticles sometimes need high temperatures in comparison to nanoparticles produced through any other procedure and contain many perilous chemicals or ingredients; moreover, the required pH is more acidic. When all of these add up, it becomes noxious and harmful for both humans and the surrounding environment [20].

The physical method for nanoparticle synthesis applies mechanical pressure, high-energy radiations, melting, thermal energy, or electrical energy to cause material abrasion, evaporation, or condensation to formulate nanoparticles (Fig. **1**) [21]. Physical procedures mainly utilize a top-down approach, and they are even more favorable as they are excluded from solvent adulterants and formulate uniform nanoparticles [22]. The ample amount of leftover produced during the synthesis time is the reason for physical processes to become less reasonable, which is a disadvantage of the physical process of nanoparticle synthesis. Inert gas condensation, high energy ball milling, physical vapour deposition, laser ablation, flash spray pyrolysis, laser pyrolysis, melt mixing, and electro spraying, are some of the most often used physical methods to synthesize nanoparticles (Fig. **1**) [23, 24]. Next comes the chemical method to synthesize nanoparticles; it includes hydrothermal synthesis, micro-emulsion technique, polymer synthesis, sol-gel method, plasma-enhanced chemical vapor deposition process, or chemical vapor synthesis, which are frequently used methods for nanoparticle production [25]. As the name suggests, chemicals are the fundamental basis of nanoparticle formulation; however, in the case of biological processes, the fundamental aspects are the biological agents like pants and microbes (Fig. **1**). The chemical methods are simple, rapid, and cost-effective, but the employment of harmful and unsafe chemicals and stabilizing agents makes them hazardous to health [26].

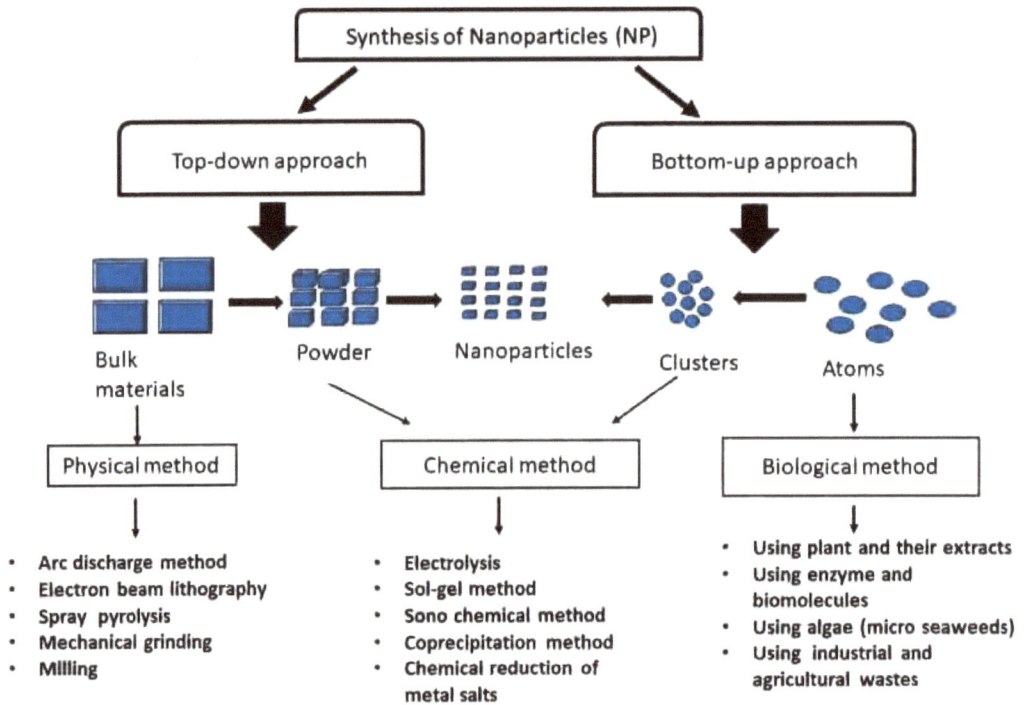

Fig. (1). Various methods for synthesizing nanoparticles.

Green Synthetic Approach for the Synthesis of Selenium Nanoparticles

Green synthesis, as the name suggests, is a method of nanoparticle synthesis using biotic agents like plants and microbes since these are considered safe and eco-friendly and have insignificant side effects. The use of plants has been investigated efficiently earlier for rapid biogenesis of several metal nanoparticles such as silver, gold, selenium, MgO, CuO, and ZnO nanoparticles because of the presence of phytochemicals, which aids in the formation of nanoparticles with less toxicity [27]. The plants used for the synthesis of nanoparticles might be scientifically proven compared to microbes by getting rid of the excessive steps taken for maintaining the culture media [28]. The following points describe a few reasons why the researchers are leaning toward the biological synthesis of nanoparticles using various biological agents:

• The amount of time spent in the biosynthesis of nanoparticles is less when compared with the physical and chemical processes of production of nanoparticles [12].
• Biologically synthesized nanoparticles are advantageous because biologically active elements get adhered to the exterior of the nanoparticles, which are

fabricated from biological sources. In the case of medicinal plants, they have ample metabolites loaded with therapeutic values that are supposed to be affixed to the synthesized nanoparticles, which provides an extra advantage by increasing the efficacy of the nanoparticles [29].

• Nanoparticles produced by biosynthesis/green synthesis methods are excluded from toxic contamination.

• Cost-effective, eco-friendly.

• Green synthesis methods do not need any stabilizing agents since the plants and microbes comprise their own self to act as capping and stabilizing agents [30].

In this context, Sudhasree *et al.* reported that the nanoparticles obtained by biological methods using *Desmodium gangeticum* are uniformly dispersed and exhibit antioxidant, antimicrobial, and biocompatible properties as compared to chemically synthesized nickel nanoparticles [31]. Mukherjee *et al.* showed greater effectiveness of biologically produced silver nanoparticles, which were obtained from *Olax scandens* leaf for its antineoplastic actions, biological compatibility for the delivery of drugs, and imaging compared with chemically synthesized silver nanoparticles [32]. The SeNP, synthesized from *Allium sativum* bud extract, has excellent antimicrobial propensity against *Staphylococcus aureus* and *Bacillus subtilis* [33]. It is reported that SeNP, biosynthesized from aloe vera extract, has excellent antifungal and antimicrobial activity against both Gram-negative and Gram-positive bacteria [34]. According to Sujatha *et al.*, SeNP that was synthesized from *Diospyros Montana* leave extract has antibacterial, antioxidant, and anticancer activity [35]. The following Table **1** shows the synthesis of selenium nanoparticles of different sizes using various plant extracts:

Table **1.** Describes synthesis of SeNPs of various sizes by using different plant extracts and their applications.

Sl. No.	Type of Se Nanoparticles	Biological Agent Used	Size of Np (nm)	Applications	References
1	SeNP	Leave extracts of *withania Somnifera* (ashwagandha, Indian ginseng)	45-90	SeNPs produced have anticancer activity against lung cancer cells	[36]
2	SeNP	Fenugreek seed extracts	50-150	SeNP has anticancer activity against breast cancer.	[2]
3	SeNP	The Hawthorn fruit extracts	Average size of 113	HE-SeNPs exhibited significant antitumor activity against HepG2 cells	[37]
4	SeNP	Broccoli	50-15	Anticancer and antioxidant properties	[38]

(Table 1) cont.....

Sl. No.	Type of Se Nanoparticles	Biological Agent Used	Size of Np (nm)	Applications	References
5	SeNP	*Emblica officinalis*	15-40	Exhibited lesser toxicity; hence, biocompatible	[29]
6	SeNPs	Dried *Vitis vinifera* (raisin) extract	3-18	After conjugation of SeNP with antibiotics, it enhances anticancer activity. It can be used as a nano biosensor and helps in environmental remediation.	[39]
7	SENPs	*Allium sativum*	7-45	It shows anti-angiogenic and antiproliferative activity against cervical cancer cells.	[40]
8	SeNPs	*Clausena dentate* plant extract	46.32-78.8	Selenium nanoparticles synthesized from *Clausena dentate* were investigated for larvicidal effectiveness against Anopheles stephensi, Aedes aegypti, and Culex quinquefasciatus fourth-instar larvae.	[41]
9	SeNP	*Aloe vera* leaf extract	121-243	SeNP shows antimicrobial and antifungal activity against pathogenic bacteria and spoilage fungi.	[34]
10	SeNP	Drumstick	200-400	It shows anticancer and photocatalytic activity towards the degradation of sunset yellow azo dye.	[42]
11	SeNPs	Leaf extract of *Cassia Auriculata* (matura tea tree)	10-20	*Cassia Auriculata* synthesizes selenium nanoparticles and their potential activity against HL-60 cells.	[43]

Application of Selenium Nanoparticles

Cancers begin as localized illnesses, but they commonly spread and become incurable. Current treatments—radiation, chemotherapy, and surgery—are based on clinical and pathological staging as assessed by histology and conventional radiography. Despite technological advances, early diagnosis and effective treatment remain challenging. However, nanomedicine offers the potential to improve cancer therapy by addressing the limitations of traditional treatments [44]. Although selenium nanoparticles are used in different branches of science and technology, the anticancer propensity of selenium nanoparticles has piqued the interest of researchers to use these nanoparticles for the treatment of cancer. In this context, selenium nanoparticles are used against prostate cancer. Generally, prostate cancer is the second most widespread and fifth most common cause of cancer-related death of humans. It is reported that selenium nanoparticles have strong anticancer activity against prostate cancer cells, and the presence of high selenium levels in the body of prostate cancer patients may show a better prognosis [45]. Additionally, it is reported that the conjugation of selenium and silver nanoparticles demonstrated antitumor activity against Dalton's lymphoma cells [11]. When SeNP is conjugated with organic molecules and drugs, it avoids the aggregation of nanoparticles and increases anticancer efficacy [46, 47]. When SeNP conjugates with spirulina polysaccharide, it suppresses tumor development by inducing apoptosis, and it is recognized by an increase in the sub-G1 cell population, chromatin condensation, and DNA fragmentation. These conjugates aid in the delivery of nanoparticles to the target cancer cells for recovery [48]. It has been reported that selenium nanoparticles do not show any harmful effect on normal cells, whereas theyshow cytotoxic effects against cancer cells (Fig. **2**) [11].

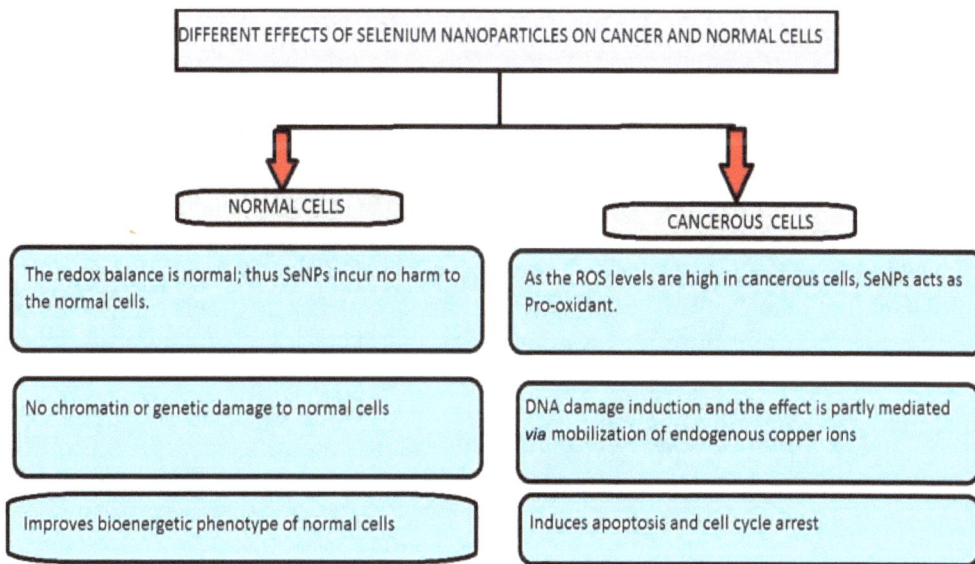

Fig. (2). Effects of selenium nanoparticles on cancer cells and normal cells.

Selenium nanoparticles can also be used for the treatment of diabetes. Selenium nanoparticles possess anti-diabetic properties either in combination with other agents or individually. It has been shown that SeNP enhances β-cell proliferation and insulin release. The suggested methods, which include the capacity to scavenge ROS and modify the proteins SIRT1 and HSP70, aid in the battle against diabetes and its associated issues [49]. Phytonanotechnology is the agronomic application of nanotechnology having advantages like biocompatibility, scalability, etc, which helps in controlling the release of agrochemicals (*e.g.*, pesticides, fertilizers, herbicides) and has the power to change traditional plant production systems [11]. Selenium nanoparticles are important nanomaterials that show a promising future in phytonanotechnology [17]. Selenium nanoparticles can be utilized to increase the productivity of the soil, which helps in the growth and development of crops. It is a trace element that helps in the functioning of biotic organisms. Generally, selenium is found in soil, water, crops, food, *etc.* It is reported that Se zero-valent NP can be used as a fertilizer and antioxidant [50].

Selenium nanoparticles are also used in the food industry. Selenium in nano form is biocompatible and non-toxic in nature than the inorganic or organic forms; hence, nano selenium acts as a better food supplement. It is used as a food additive, more precisely in individuals with selenium deficiency [51]. Selenium nanoparticles are also used as food packaging material [52]. Selenium nanoparticles are also used as a diet supplement. Generally, selenium is soluble in nature, and it is absorbed in the lower section of the small intestine as amino acid selenocysteine [53]. Selenite is absorbed by the body by the passive diffusion method, and it reacts non-enzymatically with glutathione to generate seleno-glutathine (GS-Se-SG) [54]. Selenate is also absorbed by the same method and reduced to selenite in the presence of NADPH [55]. Generally, selenium is acquired from food; however, in some geographical areas, there is a lack of selenium content in food, and some patients are advised to take selenium daily to prevent some diseases. In the case of HIV [56], Crohn's disease [57], cardiovascular disease [58], thyroid disease [59], and others, patients are prescribed to take selenium as a food supplement. The doses of selenium depend on various factors *i.e.*, physiological properties, chemical composition of Se, solubility, and others [60]. Overdose of selenium exhibits various adverse effects on the human body *i.e.*, garlickyodor in the breath, metallic taste in the mouth, and selenosis, which is characterized by nail and hair loss, fatigue, brittleness, lesions on the skin, diarrhea, nausea, and abnormalities of the nervous system [61].

SeNP, which is synthesized from biological processes, has significant antimicrobial activity against fungi, yeast, and bacteria due to different concentrations and sizes [62]. SeNP shows equally significant antimicrobial activity against both Gram-negative and Gram-positive bacteria [62]. *S. aureus* is the main causative agent of many diseases as it produces biofilm and also an antibiotic-resistant bacteria. Hence, SeNP is used to treat *S. aureus* infection [63]. It is reported that SeNP synthesized from *Prunus amygdalus* nuts extract shows antimicrobial activity against *Pseudomonas aeruginosa, Staphylococcus aureus,*

Proteus vulgaris, Bacillus subtilis, and *Escherichia coli* by using growth kinetics, well diffusion, and disc diffusion methods [64]. SeNP has antifungal activity by inhibiting spore germination of *Malassezia sympodialis* and *Malassezia furfur* [65].

Production of reactive oxygen species (ROS) that leads to oxidative stress has been reported by many researchers as one of the most significant reasons behind many diseases. It has been reported that small selenium nanoparticles have higher free radical scavenging potential than larger nanoparticles [66]. According to Zhang *et al.*, the production of SeNPs from tea extract exhibits higher antioxidant activity in the ABTS and DPPH free radical scavenging assays [67].

SeNP is used as a nanocarrier due to its lower toxicity, good availability, and drug delivery efficiency properties. In the multivalent surface of SeNP, different chemical drugs and biomacromolecules bind by covalent and non-covalent interactions. For drug delivery, the first strategy is to check the size of SeNP, and the second is flexibility for drug loading. SeNP can be modified with functional ligands in which nanoparticles can easily bind and deliver the drug to the target site.

Selenium nanoparticles have significant impacts on the environment. Reduction of the toxic form of selenite and selenate into non-toxic selenium form results in pollution control of the environment [68, 69]. Microorganisms found in contaminated sludge water reduce toxic oxyanions to amorphous elemental Se [70]. In wastewater, amorphous SeNP is transformed into crystalline form by the treatment of heat. The biogenic amorphous SeNP subsequently reduces mercury (Hg^0) and, hence, is utilized to remove mercury from polluted water [71].

CONCLUSION

Biological synthesis techniques are a promising way to produce selenium nanoparticles (SeNPs) with medicinal applications. These approaches use bacteria, fungi, and plants, each having benefits such as scalability, therapeutic functions, and reduced toxicity. Capping agents play an essential role in regulating SeNP characteristics, improving stability, and minimizing negative effects. Biogenic synthesis of selenium nanoparticles (SeNPs) is preferred for its safety and eco-friendliness. SeNPs display antibacterial properties, hinder biofilm formation, and act as drug carriers. These SeNPs, particularly those in the 50-200 nm range, show promise in combating diseases such as cancer and diabetes, further augmented by capping compounds with anti-tumor properties. Selenium is essential for human health, although deficiency can trigger various diseases. Controlled dosages are required since it functions as an antioxidant in optimum quantities but may be pro-oxidant in excess. Furthermore, SeNPs exhibit protective actions against inflammation, oxidative damage, and toxicity induced by various substances, suggesting prospective uses in liver injury, kidney, psoriasis, epilepsy, Alzheimer's disease, and wound healing. SeNPs have a wide range of therapeutic potential and are low in toxicity, making them promising for future medicinal applications such as antiviral agents and supplements. Biological synthesis approaches provide benefits in producing biocompatible and biologically active nanoparticles with a wide array of beneficial features.

REFERENCES

[1]   Pal G, Rai P, Pandey A. Green synthesis of nanoparticles: A greener approach for a cleaner future Green synthesis, characterization and applications of nanoparticles. Elsevier 2019; pp. 1-26.

[2]   Ramamurthy C, Sampath KS, Arunkumar P, *et al.* Green synthesis and characterization of selenium nanoparticles and its augmented cytotoxicity with doxorubicin on cancer cells. Bioprocess Biosyst Eng 2013; 36(8): 1131-9.
       [http://dx.doi.org/10.1007/s00449-012-0867-1] [PMID: 23446776]

[3]   Rahmani AH, Alsahli MA, Almatroodi SA. Active constituents of pomegranates (Punica granatum) as potential candidates in the management of health through modulation of biological activities. Pharmacogn J 2017; 9(5): 689-95.
       [http://dx.doi.org/10.5530/pj.2017.5.109]

[4]   Hulla J, Sahu S, Hayes A. Nanotechnology: History and future. Human Expt Toxicol. 2015; 34(12), 1318-21.
       [http://dx.doi.org/10.1177/0960327115603588]

[5]   Niroumand H, Zain MFM. The Role of nanomaterials in nanoarchitecture. Procedia Soc Behav Sci 2013; 89: 27-30.
       [http://dx.doi.org/10.1016/j.sbspro.2013.08.804]

[6]   Jeevanandam J, Barhoum A, Chan YS, Dufresne A, Danquah MK. Review on nanoparticles and nanostructured materials: history, sources, toxicity and regulations. Beilstein J Nanotechnol 2018; 9(1): 1050-74.
       [http://dx.doi.org/10.3762/bjnano.9.98] [PMID: 29719757]

[7]   Chenthamara D, Subramaniam S, Ramakrishnan SG, *et al.* Therapeutic efficacy of nanoparticles and routes of administration. Biomater Res 2019; 23(1): 20.
       [http://dx.doi.org/10.1186/s40824-019-0166-x] [PMID: 31832232]

[8]   Khan I, Saeed K, Khan I. Nanoparticles: Properties, applications and toxicities. Arab J Chem 2019; 12(7): 908-31.
       [http://dx.doi.org/10.1016/j.arabjc.2017.05.011]

[9]   Charitidis CA, Georgiou P, Koklioti MA, Trompeta AF, Markakis V. Manufacturing nanomaterials: from research to industry. Manuf Rev (Les Ulis) 2014; 1: 11.
       [http://dx.doi.org/10.1051/mfreview/2014009]

[10]   Buzea C, Pacheco II, Robbie K. Nanomaterials and nanoparticles: Sources and toxicity. Biointerphases 2007; 2(4): MR17-71.
       [http://dx.doi.org/10.1116/1.2815690] [PMID: 20419892]

[11]   Khurana A, Tekula S, Saifi MA, Venkatesh P, Godugu C. Therapeutic applications of selenium nanoparticles. Biomed Pharmacother 2019; 111: 802-12.
       [http://dx.doi.org/10.1016/j.biopha.2018.12.146] [PMID: 30616079]

[12]   Singh P, Kim YJ, Zhang D, Yang DC. Biological Synthesis of Nanoparticles from Plants and Microorganisms. Trends Biotechnol 2016; 34(7): 588-99.
       [http://dx.doi.org/10.1016/j.tibtech.2016.02.006] [PMID: 26944794]

[13]   Hariharan S, Dharmaraj S. Selenium and selenoproteins: it's role in regulation of inflammation. Inflammopharmacol 2020; pp. 1-29.
[http://dx.doi.org/10.1007/s10787-020-00690-x]

[14]   Zhang Q, Li H, Ma Y, Zhai T. ZnSe Nanostructures: Synthesis, Properties and Applications. Prog Mater Sci 2016; 83: 472-535.
[http://dx.doi.org/10.1016/j.pmatsci.2016.07.005]

[15]   El-Ramady HR, Domokos-Szabolcsy É, Shalaby TA, Prokisch J, Fári M. Selenium in agriculture: water, air, soil, plants, food, animals and nanoselenium CO2 sequestration, biofuels and depollution. Springer 2015; pp. 153-232.

[16]   Mates I, Antoniac I, Vasile L. Vicaş, Selenium nanoparticles: Production, characterization and possible applications in biomedicine and food science. *UPB Sci. Bull.* Series B: Chem Mater Sci 2019; 81: 205-16.

[17]   Griffin S, Masood M, Nasim M, *et al.* Natural nanoparticles: A particular matter inspired by nature. Antioxidants 2017; 7(1): 3.
[http://dx.doi.org/10.3390/antiox7010003] [PMID: 29286304]

[18]   Wagner S, Gondikas A, Neubauer E, Hofmann T, von der Kammer F. Spot the difference: engineered and natural nanoparticles in the environment--release, behavior, and fate. Angew Chem Int Ed 2014; 53(46): 12398-419.
[http://dx.doi.org/10.1002/anie.201405050] [PMID: 25348500]

[19]   Jeyaraj M, Gurunathan S, Qasim M, Kang MH, Kim JH. A comprehensive review on the synthesis, characterization, and biomedical application of platinum nanoparticles. Nanomaterials (Basel) 2019; 9(12): 1719.
[http://dx.doi.org/10.3390/nano9121719] [PMID: 31810256]

[20]   Wadhwani SA, Shedbalkar UU, Singh R, Chopade BA. Biogenic selenium nanoparticles: current status and future prospects. Appl Microbiol Biotechnol 2016; 100(6): 2555-66.
[http://dx.doi.org/10.1007/s00253-016-7300-7] [PMID: 26801915]

[21]   Wang Y, Xia Y. Bottom-up and top-down approaches to the synthesis of monodispersed spherical colloids of low melting-point metals. Nano Lett 2004; 4(10): 2047-50.
[http://dx.doi.org/10.1021/nl048689j]

[22]   Odularu AT. Metal nanoparticles: thermal decomposition, biomedicinal applications to cancer treatment, and future perspectives. Bioinorg. Chem. Applications., 2018; 2018.
[http://dx.doi.org/10.1155/2018/9354708]

[23]   Tissue BM. Synthesis and luminescence of lanthanide ions in nanoscale insulating hosts. Chem Mater 1998; 10(10): 2837-45.
[http://dx.doi.org/10.1021/cm9802245]

[24]   Dhand C, Dwivedi N, Loh XJ, *et al.* Methods and strategies for the synthesis of diverse nanoparticles and their applications: a comprehensive overview. RSC Advances 2015; 5(127): 105003-37.
[http://dx.doi.org/10.1039/C5RA19388E]

[25]   Iravani S, Korbekandi H, Mirmohammadi SV, Zolfaghari B. Synthesis of silver nanoparticles: chemical, physical and biological methods. Res Pharm Sci 2014; 9(6): 385-406.
[PMID: 26339255]

[26]   Sharma D, Kanchi S, Bisetty K. Biogenic synthesis of nanoparticles: A review. Arab J Chem 2019; 12(8): 3576-600.
[http://dx.doi.org/10.1016/j.arabjc.2015.11.002]

[27]   Korde P, Ghotekar S, Pagar T. Plant Extract Assisted Eco-benevolent Synthesis of Selenium Nanoparticles—A Review on Plant Parts Involved, Characterization, and Their Recent Applications. J Chem Rev 2020; 2: 157-68.

[28]   Gunti L, Dass RS, Kalagatur NK. Phytofabrication of selenium nanoparticles from Emblica officinalis fruit extract and exploring its biopotential applications: antioxidant, antimicrobial, and biocompatibility. Front Microbiol 2019; 10: 931.
[http://dx.doi.org/10.3389/fmicb.2019.00931] [PMID: 31114564]

[29]   Sintubin L, Verstraete W, Boon N. Biologically produced nanosilver: Current state and future perspectives. Biotechnol Bioeng 2012; 109(10): 2422-36.
[http://dx.doi.org/10.1002/bit.24570] [PMID: 22674445]

[30]   Makarov V, Love A. Sinitsyna, O. "Green" nanotechnologies: synthesis of metal nanoparticles using plants. Acta Naturae 2014; 6: 20.
[http://dx.doi.org/10.32607/20758251-2014-6-1-35-44]

[31]   Sudhasree S. Shakila, Banu.; A, Brindha P.; Kurian, GA. Synthesis of nickel nanoparticles by chemical and green route and their comparison in respect to biological effect and toxicity. *Toxicol. &amp*. Environ Chem 2014; 96(5): 743-54.
[http://dx.doi.org/10.1080/02772248.2014.924907]

[32]   Mukherjee S, Chowdhury D, Kotcherlakota R, *et al.* Potential theranostics application of bio-synthesized silver nanoparticles (4-in-1 system). Theranostics 2014; 4(3): 316-35.
[http://dx.doi.org/10.7150/thno.7819] [PMID: 24505239]

[33]   Rana JVS. Synthesis of selenium nanoparticles using Allium sativum extract and analysis of their antimicrobial proprty against gram positive bacteria. Pharma Innov 2018; 7(9): 262-6.
[http://dx.doi.org/10.7897/2277-7695.079262]

[34]   Fardsadegh B, Jafarizadeh-Malmiri H. Aloe vera leaf extract mediated green synthesis of selenium nanoparticles and assessment of their *In vitro* antimicrobial activity against spoilage fungi and pathogenic bacteria strains. Green Processing and Synthesis 2019; 8(1): 399-407.
[http://dx.doi.org/10.1515/gps-2019-0007]

[35]   Kokila K, Elavarasan N, Sujatha V. Diospyros montana leaf extract-mediated synthesis of selenium nanoparticles and their biological applications. New J Chem 2017; 41(15): 7481-90.
[http://dx.doi.org/10.1039/C7NJ01124E]

[36]   Tripathi RM, Hameed P, Rao RP. Biosynthesis of Highly Stable Fluorescent Selenium Nanoparticles and the Evaluation of Their Photocatalytic Degradation of Dye. Bio Nano Sci 2020; pp. 1-8.
[http://dx.doi.org/10.1007/s12995-020-00256-0]

[37]   Cui D, Liang T, Sun L, *et al.* Green synthesis of selenium nanoparticles with extract of hawthorn fruit induced HepG2 cells apoptosis. Pharm Biol 2018; 56(1): 528-34.
[http://dx.doi.org/10.1080/13880209.2018.1510974] [PMID: 30387372]

[38]   Kapur M, Soni K, Kohli K. Green synthesis of selenium nanoparticles from broccoli, characterization, application and toxicity. Advanced Techniques in Biology & Medicine 2017; 5(1): 2379-1764.
[http://dx.doi.org/10.4172/2379-1764.1000198]

[39]   Sharma G, Sharma A, Bhavesh R, *et al.* Biomolecule-mediated synthesis of selenium nanoparticles using dried Vitis vinifera (raisin) extract. Molecules 2014; 19(3): 2761-70.
[http://dx.doi.org/10.3390/molecules19032761] [PMID: 24583881]

[40]   Vyas J, Rana S. Antioxidant activity and green synthesis of selenium nanoparticles using allium sativum extract. Int J Phytomed 2017; 9(4): 634.
[http://dx.doi.org/10.5138/09750185.2185]

[41]   Sowndarya P, Ramkumar G, Shivakumar MS. Green synthesis of selenium nanoparticles conjugated *Clausena dentata* plant leaf extract and their insecticidal potential against mosquito vectors. Artif Cells Nanomed Biotechnol 2017; 45(8): 1490-5.
[http://dx.doi.org/10.1080/21691401.2016.1252383] [PMID: 27832715]

[42]   Hassanien R, Abed□Elmageed AA. Husein, DZ. Eco□Friendly Approach to Synthesize Selenium Nanoparticles: Photocatalytic Degradation of Sunset Yellow Azo Dye and Anticancer Activity. Chem

Select 2019; 4(31): 9018-26.

[43]   Anu K, Devanesan S, Prasanth R, AlSalhi MS. Biogenesis of selenium nanoparticles and their anti-leukemia activity. J. King Saud University-Sci 2020.
[http://dx.doi.org/10.1016/j.jksus.2020.04.018]

[44]   Rosic G. Selakovic, D.; & Omarova, S. Cancer signaling, cell/gene therapy, diagnosis and role of nanobiomaterials. Adv Biol & Earth Sci 2024; 9: 81-91.

[45]   Liao G, Tang J, Wang D, *et al.* Selenium nanoparticles (SeNPs) have potent antitumor activity against prostate cancer cells through the upregulation of miR-16. World J Surg Oncol 2020; 18(1): 81.
[http://dx.doi.org/10.1186/s12957-020-01850-7] [PMID: 32357938]

[46]   Ahmad MS, Yasser MM, Sholkamy EN, Ali AM, Mehanni MM. Anticancer activity of biostabilized selenium nanorods synthesized by Streptomyces bikiniensis strain Ess_amA-1. Int J Nanomedicine 2015; 10: 3389-401.
[PMID: 26005349]

[47]   Hossein Y. M Mahdavi, M. Setayesh, N. Selenium nanoparticle-enriched Lactobacillus brevis causes more efficient immune responses *in vivo* and reduces the liver metastasis in metastatic form of mouse breast cancer. Daru 2013; 21: 4.
[http://dx.doi.org/10.1186/2008-2231-21-4]

[48]   Yang F, Tang Q, Zhong X, *et al.* Surface decoration by Spirulina polysaccharide enhances the cellular uptake and anticancer efficacy of selenium nanoparticles. Int J Nanomedicine 2012; 7: 835-44.
[PMID: 22359460]

[49]   Kumar GS, Kulkarni A, Khurana A, Kaur J, Tikoo K. Selenium nanoparticles involve HSP-70 and SIRT1 in preventing the progression of type 1 diabetic nephropathy. Chemico Biol Intr 2012; 223:125-33.
[http://dx.doi.org/10.1016/j.cbi.2014.10.004]

[50]   Gudkov SV, Shafeev GA, Glinushkin AP, *et al.* Production and Use of Selenium Nanoparticles as Fertilizers. ACS Omega 2020; 5(28): 17767-74.
[http://dx.doi.org/10.1021/acsomega.0c02448] [PMID: 32715263]

[51]   Hosnedlova B, Kepinska M, Skalickova S, *et al.* Nano-selenium and its nanomedicine applications: a critical review. Int J Nanomedicine 2018; 13: 2107-28.
[http://dx.doi.org/10.2147/IJN.S157541] [PMID: 29692609]

[52]   Vera P, Echegoyen Y, Canellas E, *et al.* Nano selenium as antioxidant agent in a multilayer food packaging material. Anal Bioanal Chem 2016; 408(24): 6659-70.
[http://dx.doi.org/10.1007/s00216-016-9780-9] [PMID: 27497969]

[53]   Tapiero H, Townsend DM, Tew KD. The antioxidant role of selenium and seleno-compounds. Biomed Pharmacother 2003; 57(3-4): 134-44.
[http://dx.doi.org/10.1016/S0753-3322(03)00035-0] [PMID: 12818475]

[54]   Weekley CM, Aitken JB, Vogt S, *et al.* Metabolism of selenite in human lung cancer cells: X-ray absorption and fluorescence studies. J Am Chem Soc 2011; 133(45): 18272-9.
[http://dx.doi.org/10.1021/ja206203c] [PMID: 21957893]

[55]   Hsieh HS, Ganther HE. Acid-volatile selenium formation catalyzed by glutathione reductase. Biochemistry 1975; 14(8): 1632-6.
[http://dx.doi.org/10.1021/bi00679a014] [PMID: 235962]

[56]   Stone CA, Kawai K, Kupka R, Fawzi WW. Role of selenium in HIV infection. Nutr Rev 2010; 68(11): 671-81.
[http://dx.doi.org/10.1111/j.1753-4887.2010.00337.x] [PMID: 20961297]

[57]   Kuroki F, Matsumoto T, Iida M. Selenium is depleted in Crohn's disease on enteral nutrition. Dig Dis 2003; 21(3): 266-70.
[http://dx.doi.org/10.1159/000073346] [PMID: 14571102]

[58]   Benstoem C, Goetzenich A, Kraemer S, *et al.* Selenium and its supplementation in cardiovascular disease--what do we know? Nutrients 2015; 7(5): 3094-118.
[http://dx.doi.org/10.3390/nu7053094] [PMID: 25923656]

[59]   Wu Q, Rayman MP, Lv H, *et al.* Low population selenium status is associated with increased prevalence of thyroid disease. *The J. Clinic. Endocrinol. &amp.* J Clin Endocrinol Metab 2015; 100(11): 4037-47.
[http://dx.doi.org/10.1210/jc.2015-2222] [PMID: 26305620]

[60]   Thomson CD. Selenium speciation in human body fluids. Analyst (Lond) 1998; 123(5): 827-31.
[http://dx.doi.org/10.1039/a707292i] [PMID: 9709477]

[61]   Nuttall KL. Evaluating selenium poisoning. Annals Clinic. Laboratory Sci 2006; 36(4): 409-20.
[http://dx.doi.org/10.1177/014662160603600406]

[62]   Hariharan H, Al-Harbi N, Karuppiah P, Rajaram S. Microbial synthesis of selenium nanocomposite using Saccharomyces cerevisiae and its antimicrobial activity against pathogens causing nosocomial infection. Chalcogenide Lett 2012; 9(12): 509-15.

[63]   Chudobova D, Cihalova K, Dostalova S, *et al.* Comparison of the effects of silver phosphate and selenium nanoparticles on *Staphylococcus aureus* growth reveals potential for selenium particles to prevent infection. FEMS Microbiol Lett 2014; 351(2): 195-201.
[http://dx.doi.org/10.1111/1574-6968.12353] [PMID: 24313683]

[64]   Sadalage PS, Nimbalkar MS, Sharma KKK, Patil PS, Pawar KD. Sustainable approach to almond skin mediated synthesis of tunable selenium microstructures for coating cotton fabric to impart specific antibacterial activity. J Colloid Inter Sci 2020; 569: 346-57.
[http://dx.doi.org/10.1016/j.jcis.2020.02.094]

[65]   Kazempour ZB, Yazdi MH, Rafii F, Shahverdi AR. Sub-inhibitory concentration of biogenic selenium nanoparticles lacks post antifungal effect for Aspergillus niger and Candida albicans and stimulates the growth of Aspergillus niger. Iran J Microbiol 2013; 5(1): 81-5.
[PMID: 23466957]

[66]   Torres SK, Campos VL, León CG, *et al.* Biosynthesis of selenium nanoparticles by Pantoea agglomerans and their antioxidant activity. J Nanopart Res 2012; 14(11): 1236.
[http://dx.doi.org/10.1007/s11051-012-1236-3]

[67]   Zhang W, Zhang J, Ding D, *et al.* Synthesis and antioxidant properties of Lycium barbarum polysaccharides capped selenium nanoparticles using tea extract. Artif Cells Nanomed Biotechnol 2018; 46(7): 1463-70.
[http://dx.doi.org/10.1080/21691401.2017.1373657] [PMID: 28880681]

[68]   Garbisu C, Ishii T, Leighton T, Buchanan BB. Bacterial reduction of selenite to elemental selenium. Chem Geol 1996; 132(1-4): 199-204.
[http://dx.doi.org/10.1016/S0009-2541(96)00056-3]

[69]   Pickett TM, Ma Y, Sonstegard J. Selenium removal using chemical oxidation and biological reduction. Google Patents 2013.

[70]   Lenz, M.; Van, Aelst A.; Smit, M.editors. Biological production of selenium nanoparticles from waste waters. Adv. Mater. Res., 2009.

[71]   Jiang S, Ho CT, Lee JH, Duong HV, Han S, Hur HG. Mercury capture into biogenic amorphous selenium nanospheres produced by mercury resistant Shewanella putrefaciens 200. Chemosphere 2012; 87(6): 621-4.
[http://dx.doi.org/10.1016/j.chemosphere.2011.12.083] [PMID: 22386108]

CHAPTER 4

The Roles of Nanomaterials in Disease Diagnosis

Akhlash P. Singh[1,*] and **Shivani Devi**[1]

[1] *Department of Biochemistry, GGDSD College (Panjab University), Chandigarh, India*

Abstract: Nanotechnology is an interdisciplinary field of science that has revolutionized different fields of science and technology, agriculture, and medicine. Current clinical diagnostic methods are less sensitive, are unable to detect multiple analytes, and have adverse effects on the human body. Hence, there is a need for a diagnostic method that can detect the early onset of disease, conduct complete health checks, and offer a reliable pretext for effective treatment. Currently, many nanomaterials are produced to prevent, diagnose, and treat various diseases. There are a few nanomaterials, such as nanowires, nanotubes, nanocrystals, cantilevers, dendrimers, quantum dots, and liposomes. These materials are quite suitable and effective for imaging technologies for the highly specific detection of DNA and proteins. Nanomaterials are used to diagnose and treat a wide range of diseases, including communicable and non-communicable diseases. The main objectives of the current chapter are to introduce various nanomaterials and their applications in the diagnostics of different diseases that affect human life.

Keywords: Carbon nanotubes, Nanoparticle, Nanotechnology, Quantum dots.

INTRODUCTION

Recently, a large number of nanotechnologies and nanomaterials have been discovered and utilized for numerous scientific applications owing to their outstanding physio-chemical properties. However, the applications of nanomaterials are very wide and include electronics, agriculture, biosciences, and science & technology, but health care and diagnostics are the most predominant areas where nano-materials are commonly used [1]. Nanomaterials possess highly diverse structural properties and, hence, perform versatile functions. Therefore, nanomaterials are multi-functional and provide a highly sensitive signal that is highly suitable for the imaging process. Nanomaterials are suitable candidates for use in diverse types of nanodevices used for qualitative and quantitative diagnostic purposes in humans and animals. Chemically, there are mainly three

* **Corresponding author Akhlash P. Singh:** Department of Biochemistry, GGDSD College (Panjab University), Chandigarh, India; E-mail: akhlash@ggdsd.ac.in

Manoranjan Arakha & Arun Kumar Pradhan (Eds.)

types of nanomaterials, namely, inorganic, organic, and hybrid nano-materials, which have been extensively studied and exploited in diagnostics and healthcare [1].

Carbon nanotubes, nanocrystals, liposomes, dendrimers, micelles, hyper-branched organic polymers, molecularly imprinted nanostructures, and polymeric hydrogel nanoparticles are the best examples of organic nanomaterials that have been widely used as imaging and therapeutic agents [2]. Inorganic nanomaterials such as quantum dots, superparamagnetic iron oxide nanoparticles, metallic nanoparticles, and metal oxides have also received considerable interest in healthcare diagnostics, particularly in biosensing and biosensor construction. Nanomaterials are exploited in diverse types of imaging technologies such as magnetic resonance imaging, positron emission tomography, computed tomography, single-photon emission tomography, optical imaging, and ultrasound imaging [3].

Nanomaterials possess special and exceptional properties that allow for a wide range of diagnostic approaches. Nanomaterials, for instance, may be used to alter surfaces in order to provide more binding sites for immobilized binding receptor molecules, thereby improving diagnostic effectiveness. Furthermore, illness biomarkers may be labeled with nanoparticles to enhance the sensitivity and specificity of bio-detection techniques used in diagnosis. Nanomaterials have revolutionized molecular diagnostics, point-of-care diagnosis, disease therapeutics, and personalized medicine. The incorporation of biomarker discovery into nanodevices, in addition to the modification and application of nanomaterials, has significantly improved clinical outcomes [4].

More recently, a wide range of biomarkers have been discovered to identify various diseases at the early stage of their onset or early occurrence in humans and animals. However, highly minute amounts of biomarkers from the range of nano to atto levels are present, and their detection in biological samples is an uphill task. Hence, highly sensitive nanomaterials that have very strong binding sites for immobilized binding receptor molecules and the early appearance of biomarkers in biological fluids are required. Consequently, biomarker discoveries made by using nanodevices and nanomaterial alterations will improve the detection of cancer, cardiovascular, infectious, and neurological illnesses. Nanomaterials offer help in the development of advanced molecular diagnostics, point-of-care diagnosis, disease care with treatments, and customized medicine. Hence, healthcare diagnostics will improve dramatically and lead to synergetic medicinal solutions using nanomaterial engineering in the near future [3].

Nano-materials and their Properties and Applications

There are nanomaterials, for example, nanoparticles, nanotubes, nanowires, and graphene, that have different physical, chemical, and biological characteristics. These materials are used in various fields of science and technology, including electronics, energy, medicine, and environmental clean-up; moreover, nanoparticles can also help in various chemical processes, such as water purification, catalysts, sensing, and medication delivery. Recently, many nanomaterials have been used in the production of nanodevices and biosensors to detect biomarkers, which are biochemicals, metabolites, and other responsible agents produced in a specific type of disease.

Nanotubes

In recent years, interest has emerged in various kinds of nanotubes. The nanotubes contain one dimension between 1 and 100 nanometers and possess entirely different physical, chemical, and biological characteristics in comparison to bulk materials. Although many nanomaterials are already mentioned above, carbon nanotubes (CNTs) are predominantly used for many purposes [5], such as in electronics, energy, medicine, and environmental clean-up. Moreover, CNTs are also applied to target body cells *via* catalysts, sensors, and medication delivery systems. Graphene, another form of carbon, has also been studied for use in electronics and energy storage, but the current paragraph is mainly focused on CNTs and their applications in diagnostic processes. Although CNTs offer attractive possibilities for many applications in their environmental and health cures [5]. Currently, nanotechnology has become so advanced that it can manipulate nanomaterials at the level of individual molecules and atoms. Therefore, several types of nanotubes were constructed and utilized in diagnostics, healthcare, food, electronics, *etc.*

Structural Features

Nanotubes are nanosized cylindrical structures with nanoscale diameters and lengths on the micrometer scale. Carbon nanotubes are constructed from carbon atoms and are also known as **carbon nanotubes (CNTs)**. Apart from carbon, nanotubes are also produced from other carbon sources like methane, iron, nickel, cobalt, silicon, quartz, sapphire, and alumina [6].

Advantages

A wider range of applications for CNTs are attributed to their physico-chemical properties, such as mechanical strength, electrical conductivity, thermal conductivity, and surface area. However, there are other characteristics that make nanotubes highly suitable for diagnostic applications. These are high levels of light absorption, good photonics, sensing, and biocompatibility [5]. In brief, the mechanical strength, high level of stiffness, thermal conductivity, and high electrical conductivity make nanotubes suitable for exploitation since CNTs are very conductive and used as electrical transducers in biosensors [7]. Additionally, because CNTs have a huge surface area-to-volume ratio, they offer more binding sites for biomolecules, which act as biomarkers. There are a few characteristics of nanotubes, like high light absorption, strong signals, and biocompatibility, that make them suitable for photonics, sensing applications, medication delivery, and tissue creation in biomedical applications. Moreover, nanotubes, due to their tiny size, also react to slight changes in the environment, produce excellent sensitivity, and can be used for biosensor applications [8].

Disadvantages

There are several drawbacks that must also be addressed before CNTs can be extensively used. For instance, nanotubes' stability in biological contexts is unknown, and CNTs are prone to deterioration over time. Moreover, the interactions of nanotubes with biological systems are still poorly known, and there are worries about nanotube toxicity [7]. While nanotubes have great electrical conductivity, their tiny size makes it difficult to detect the minuscule signals created in biosensing applications. Another major concern is the high cost of nanotube production, which prevents their widespread use in biosensors [9]. Additionally, nanotube manufacturing is a difficult job, and the reproducibility of their properties is also not yet reliable in biosensors [10]. Despite this, research in the nanotechnology field continues, and substantial progress has been made in recent years, providing optimism for the future development of nanotube-based biosensors.

Applications

Due to their distinct physical, chemical, and biological characteristics, carbon nanotubes (CNTs) are the best candidates for biomedical diagnostics [8]. So, CNTs are used to make biosensors, drug delivery systems, and imaging devices, all of which are important parts of diagnostic science in the 21st century. In fact, biosensors are analytical nanodevices that combine a biological component with a physico-chemical transducer to detect or quantify [10] biomolecules, metabolites, or ions used as biomarkers. Recently, many biosensors, including optical biosensors, electrochemical biosensors, piezoelectric biosensors, and thermal biosensors, have been developed with high sensitivity, selectivity, and specificity, real-time monitoring, and the ability to operate in complex environments [7]. Therefore, modern biosensors are very tiny or miniaturized integrated electronics and are available in the form of portables, which are ideal for point-of-care testing and other on-site applications [11]. CNTs can detect cancer and infectious disorders in the near future, even before their full onset.

CNT-based biosensors have been used in the detection of a wide range of communicable and non-communicable diseases, such as COVID-19 and malaria. Many applications of nanomaterials are given in Table **1**. Another major application of biosensors is in imaging, such as MRI and CT, because nanotubes act as contrast agents in the imaging process [12]. In most cases, researchers put targeting chemicals on nanotubes to mark tissues or cells for imaging in the treatment of many diseases, including cancer [13]. Due to nanotubes' high surface area and aspect ratio, they are suitable for drug-loading, and their functionalization enables pinpoint delivery. In addition, nanotubes can also be used to explore living cells. Researchers may investigate mitochondria or cell membranes by functionalizing nanotubes with targeted chemicals and receptors [13].

Table 1. Applications of various nanomaterials in the diagnosis of communicable diseases.

Nanomaterial	Diseases	Application
Carbon nanotube	Tuberculosis	The detection of TB-causing *Mycobacterium tuberculosis* in sputum samples [14]
Multiwalled carbon nanotubes (MWCNTs)	Human papillomavirus (HPV)	Detection of HPV antigens and antibodies [15]
Silver nanoparticles	Typhoid fever	Detection of *Salmonella Typhi* antigens in blood samples [16]
Gold nanoparticle	Rotavirus	Detection of rotavirus antigens in stool samples [17]
Gold nanoparticle-based lateral flow assays	Norovirus	Detection of norovirus, a cause of gastroenteritis, in stool samples [18].
Gold nanoparticles and gadolinium oxide	Cholera	Used to detect *Vibrio cholerae*, which causes cholera, in environmental samples [19].
Magnetic nanoparticle-based immunoassays	Dengue	Used to detect anti-dengue virus antibodies [20]
Nanocrystal-based fluorescence assays	Influenza	Rapid detection of influenza virus in respiratory samples [21]
Liposomes	Lyme Disease	Detection of Borrelia burgdorferi in blood samples [22]
Liposome-based ELISA	Chlamydia	Detection of *Chlamydia trachomatis* antigen in genital swab samples [23]

(Table 1) cont.....

Nanomaterial	Diseases	Application
Dendrimers	Norovirus	Detection of norovirus in stool and environmental samples [24]
Dendrimer-based fluorescent biosensors	Ebola virus	Detection of the Ebola virus in blood samples [25]
Quantum Dots	Chlamydia	Detection of *Chlamydia trachomatis* antigen in genital swab samples [26]
Quantum dot-based fluorescence immunoassays	Syphilis	Detection of Treponema pallidum antibodies in blood samples [27]
Micelle-based surface-enhanced Raman scattering (SERS) biosensors	Hepatitis B	Quantitative detection of hepatitis B virus DNA in serum samples [28]
Hyper-branched organic polymers	Influenza	Detection of influenza virus antigens in respiratory samples [29]
Polymeric hydrogel nanoparticle	Tuberculosis	Detection of *Mycobacterium tuberculosis* antigens in sputum and urine samples [30]
Hydrogel-based biosensors	COVID-19	Detection of SARS-CoV-2 RNA in saliva and nasal swab samples [31]
Titanium dioxide nanoparticles	Methicillin-resistant *Staphylococcus aureus* (MRSA)	Detects MRSA in biological samples [32]
Graphene oxide	Chikungunya	Detection of chikungunya virus RNA in serum samples [33]

Dendrimer

Dendrimers are nanoscale synthetic polymers with a clear and well-defined structure. They have a core, an outer shell made up of repeating units, and many functional groups attached to the surface [34]. These cores are internally structured by the network of dendrimer branches, or dendrons, that radiate out from their center. The overall chemical and physical properties of dendrimers are altered by substituting various substituents for the free functional groups. They are used for a variety of applications, including medication administration, gene therapy, and imaging, due to their ability to precisely regulate their size, shape, and surface functions [35].

Structural Features

Dendrimers are usually constructed of polymers like polyamidoamine (PAMAM), polypropyleneimine (PPI), or polyester, depending on their purpose [36]. A dendrimer's central core may be formed of metals, quantum dots, or organic compounds like polyethylene glycol (PEG) or polyethyleneimine (PEI) [37]. The central core scaffolds dendrimer development and affects its physical and chemical characteristics. Monomer units are carefully added to dendrimers' repeating outer shells during synthesis. Adjusting the outer shell composition helps fine-tune dendrimer characteristics. Drugs, targeting moieties, imaging agents, and other functional groups may change dendrimer characteristics [38].

Advantages

Due to their highly suitable properties and features, dendrimers have been exploited in the areas of drug delivery, gene therapy, and imaging. These are the main qualities, for example, exact size and structure, which make dendrimers appropriate for uniformity and reproducibility, hence making them ideal for medicinal and biological applications [39]. Dendrimers have a high surface-to-volume ratio that allows them to efficiently attach functional groups like medicines, targeting moieties, and imaging agents [36]. Moreover, the versatility associated with dendrimers offers a great level of surface functions that may be changed to fine-tune the dendrimer. Low toxicity and biocompatibility make them appropriate for biological applications; hence, dendrimers stabilize visiting molecules *in vitro* and *in vivo*. Dendrimers also act as good solubility enhancers; for instance, PAMAM dendrimers capture hydrophobic, water-soluble compounds to increase solubility. Even in biological fluids, they retain their structure and characteristics. Dendrimers also comprise a variety of functional groups, with several groups that target genes, or help create imaging agents, particularly in cells or tissues; therefore, they can help in targeted medication delivery that enhances the effectiveness, toxicity, and specificity of diagnostic methods over standard techniques [40].

Disadvantages

These include the cost of dendrimer synthesis, which requires costly raw ingredients and is time-consuming. Sometimes, dendrimers limit their usage in large-scale applications like medication administration and imaging [41]. Biodegradation and excretion are two problems with dendrimers. Another drawback is that the aggregations of dendrimers, especially in biological fluids, reduce their effectiveness and toxicity. Currently, there is inadequate knowledge of dendrimer processes related to cell and tissue uptake, which might alter delivery specificity [42].

Applications

Dendrimers, extremely branched, tree-like polymers, offer promising biological uses, including diagnostics, biosensing, and biosensors [43]. Dendrimers are used in imaging, such as medical imaging, including MRI, CT, and PET, due to their contrasting properties. These dendrimers offer enormous surface area and the ability to transport numerous imaging agents to improve medical imaging sensitivity and specificity [44]. Dendrimers can detect proteins and DNA in bodily fluids with accuracy and enhanced selectivity and sensitivity. Dendrimers are also used to deliver medications to particular cells or tissues by targeting ligands that attach to cell surface receptors that may transport the medicine directly to the target region. Currently, molecular diagnostics that include DNA sequencing, PCR, and gene therapy involve dendrimers. Therefore, dendrimers can identify nucleic acid sequences using DNA or RNA probes. Dendrimers have significant potential to improve illness detection accuracy and sensitivity, and researchers are exploring their uses [45].

Dendrimers are widely used in a variety of areas, including the modulation of gene expression, which is responsible for the development of many diseases. In this direction, dendrimers are used in the transport of genes and medications to cure serious ailments; for example, dendrimers carry siRNA for gene silencing or medicines to particular cells or organs [45]. They are also used in drug delivery, targeted medication, vaccine developments, and diagnostic techniques.

There are several diseases linked to tissue damage that can be overcome by using regenerative medicine. In the case of regenerative medicine, a large number of growth factors and genes, as well as cell and tissue growth, are involved. Hence, dendrimers can provide growth factors or therapeutic genes to repair and regenerate tissue. This can help cure tissue disorders and imaging disorders, including neurodegenerative and cardiovascular ailments [46]. Recently, dendrimers combined with antibodies helped in the detection of blood cancer and prostate-specific antigen (PSA); therefore, in the near future, biomarkers will revolutionize illness identification and personalized medication [47].

Quantum Dots

The nanoscale semiconductor quantum dots exhibit unique optical and electrical features. Quantum confinement gives these nanoscale structures unique energy levels based on their size and form [48]. Quantum dots can be made from cadmium, lead, and indium arsenide and used in solar cells, LED displays, and biological imaging. Their unique optical qualities enable quantum dots to be part of more sensitive sensors for detecting minute changes in temperature, pressure, and other environmental conditions [49].

Structural Features

Nanoscale semiconductor quantum dots' compositions vary, which determines their features and functioning. There are a few metals that are used in quantum dot production, for example, gold, zinc oxide, titanium dioxide, iron oxide, silver, or platinum [50]. These metals provide nanoparticles with distinct electrical and optical characteristics; for instance, gold nanoparticles are used in biomedical imaging and medication delivery [51].

In order to explain the general structure of a quantum dot, a core-shell system is proposed [52]. According to this model, quantum dots have a core typically made of semiconductor material, with a diameter ranging from a few nanometers to tens of nanometers. Generally, the core-shell structure of many quantum dots has a semiconductor core with a distinct material shell; for example, zinc sulfide is used to make a core of quantum dots. The second component is a shell that controls quantum dot stability and emission wavelength. The third component, *i.e.*, the surface of the quantum dot, is typically passivated with organic molecules to prevent the surface from oxidizing and to make the quantum dot more stable [53]. The passivation, or outer layer, also allows for the attachment of functional groups that can be used to link the quantum dot to other materials or biomolecules. On the surface of quantum dots, many functional groups, such as thiols, amines, and carboxylic acids, are used to regulate quantum dot solubility and stability, which are used in biomedical imaging to target cancer cells. However, overall, the precise structure of a quantum dot can vary depending on the specific materials used and the intended application, but the core-shell structure is a common feature [54].

Advantages

Quantum dots, which have many benefits due to their unique properties of nanoscale size, confinement, wide spectrum light absorption, composition, and pliable structure, can be used in lighting, displays, solar cells, and bioimaging ties [55]. Recently, displays, lights, and bioimaging have been developed using quantum dots' size-dependent features. Bioimaging is an inseparable part of the diagnostic process; for example, multi-colour labeling uses quantum dots. Quantum dots can produce light at a specified wavelength. This allows them to create vivid, high-quality colors in displays. These are used as bright and photo-stable bioimaging probes [56]. They remain stable for a long time. This makes them valuable for lighting, displays, and bioimaging due to their ability to retain brightness and emission qualities over time. The surfaces of quantum dots are crucial for bioimaging and medicine delivery benefits, which can be attributed to their high surface area-to-volume ratio. Moreover, the presence of ligands on the surface of quantum dots is easily manipulated, which affects solubility, stability, and biocompatibility [57].

In addition to bioimaging, biosensors are crucial in modern diagnostics and health care because they detect and analyze biological substances by using quantum dots. Therefore, quantum dots can detect and measure biomolecules, including proteins, DNA, and viruses, as fluorescent markers due to their unique optical characteristics [58].

Disadvantages

Currently, quantum dots have numerous severe downsides, including toxicity. Most QDs are manufactured with hazardous heavy metals like cadmium. QDs may pollute land and water if discarded improperly. Cost is another concern in large-scale quantum dot production. This limits their usage in large-scale solar energy generation [59, 60].

Applications

QDs are small particles with unique optical and electrical capabilities that are employed to diagnose and investigate non-communicable and communicable illnesses and their tracking. Human cancers are diagnosed using quantum dots. Quantum dots target cancer cells and generate light signals when they bind to malignant tissue. This makes cancer cell identification very sensitive and specific, thereby helping in early diagnosis and therapeutic efforts. Cardiovascular diseases are a major concern and are responsible for a huge number of deaths [61]. Quantum dots can be used in the monitoring of cardiovascular illnesses like atherosclerosis by targeting and binding to biomarkers. This helps doctors assess therapy efficacy and make better patient care choices. Moreover, neurological diseases are also diagnosed by using quantum dots that can target and see brain neuronal structures, helping diagnose and cure neurological illnesses, including Alzheimer's and Parkinson's. Due to their optical and electrical features, quantum dots offer great disease diagnostic potential [62]. Nevertheless, further study is required to improve their clinical usage, safety, and effectiveness. (Table **1**).

Infectious illness is another major cause of mortality, and its early detection can prevent substantial numbers of deaths due to new emerging infectious agents such as COVID-19 [63]. Currently, new pathogens are emerging, so there is an urgent need to detect germs and viruses in patient samples. In this quest, quantum dots can play a very significant role. Thus, QDsDs are designed in such a way to target and bind to pathogen biomolecules for precise diagnostics (Table **2**).

Table 2. Applications of various nanomaterials in the diagnosis of non-communicable diseases.

Nanomaterial	Disease	Application
Carbon nanotube and gold nanoparticle-based immunoassays	Cancer	Detection of prostate-specific antigen (PSA) in prostate cancer patients [64]
Carbon nanotube	Parkinson's disease	Detect dopamine in biological samples, which is a biomarker for Parkinson's disease [65].
Carbon nanotube	Chronic pain	Detection of neuropeptides and calcitonin gene-related peptide (CGRP) in cerebrospinal fluid samples [66]
Quantum dots, barium titanate (BTO)	Cardiovascular disease	Detection of high-sensitivity C-reactive protein (hs-CRP) [67].
Quantum dots and copper sulfide	Diabetes	Detection of glucose and haemoglobin A1c (HbA1c) in diabetics [68]
Quantum dot	Multiple sclerosis	To detect auto-antibodies in blood samples [69]
Quantum dot	Inflammatory bowel disease	Detection of mucosal addressin cell adhesion molecule 1 (MAdCAM-1) as a reliable marker for IBD [70]
Gold nanoparticle	Osteoporosis	Detection of osteoprotegerin (OPG) and receptor activator in blood samples [71]
Magnetic nanoparticle	Kidney disease	Detection of creatinine, cystatin C, and kidney injury molecule-1 (KIM-1 in kidney disease [72].
Gold Nanoclusters	Liver disease	BSA-stabilized gold nanoclusters (AuNCs) were used as probes for biorecognition [73].
Gold nanoparticles	Rheumatoid arthritis	Detection of miR-93 and miR-223 in the blood [74]
Graphene oxide	Chronic kidney disease	Determine the concentration of human serum albumin in urine [75].

Solid Lipid Nanoparticles (SLNs)

Surfactant-stabilized solid lipid nanoparticles (SLNs) are nanoscale devices that are used as drug delivery devices. The solid lipid core of SLN disperses a lipophilic medication, while its surfactant coating on nanoparticles inhibits aggregation and stabilizes them [76]. Currently, SLNs are employed in drug delivery, gene therapy, and imaging due to their being biocompatible, biodegradable, and non-toxic. Therefore, SLNs are applied to deliver medications to target cells or tissues with great effectiveness and reduced adverse effects ascribed to their tiny size and high surface area to volume ratio [39].

Structural Characteristics

Lipids, surfactants, and co-solvents generate stable solid lipid nanoparticles (SLNs). SLNs' lipid cores are natural or manufactured and produced by triglycerides, mono- and diglycerides, fatty acids, and waxes. The choice of lipid is based on its capacity to form a stable particle, biocompatibility, biodegradability, and melting temperature [77]. The surfactant coating on the lipid core is usually one or more surfactants that stabilize the particle and prevent aggregation. Polysorbate 80, Tween 80, cationic, and anionic surfactants are employed in SLNs. Co-solvents may be added to the SLN formulation to improve medication solubility and lipid matrix incorporation. Ethanol, propylene glycol, and glycerine are SLN co-solvents [78]. To target specific cells and tissues, SLNs may be functionalized by targeting ligands like antibodies or peptides. Covalently binding the targeted ligand to the particle surface does this. SLNs may be chemically altered to improve stability, drug loading, and targeting. SLN composition is being optimized for biological purposes [79].

Advantages

Currently, SLNs are found to be extremely appealing for drug delivery systems due to their biocompatibility, high surface area to volume ratio, loading of lipophilic medicines, regulated release of medicine, and enhanced therapeutic effectiveness [80]. These all-important properties increase drug solubility, stability, and bioavailability. Moreover, targeting of SLNs inside cells and tissues can be customized using antibodies or peptides, and because they are non-toxic and biodegradable, they are metabolized by the body and do not accumulate in organs or tissues [78].

Disadvantages

Solid lipid nanoparticles (SLNs) offer several benefits as a drug delivery technology, but they also have certain downsides, such as restricted drug loading capacity, which may require higher doses or more frequent administration [81]. Physical instability is another concern because agglomeration or fusion alters particle size and changes the release kinetics of a specific medication and imaging process. The difficult nature of their manufacturing process, which requires specialized equipment and processes, may raise production costs and complexity. Further, regulatory challenges still persist, and there is a need to maintain strict safety and effectiveness standards for SLNs [19].

Applications

Recently, SLNs have been used in drug delivery, cosmetics, and nutraceuticals and are also used to cure cancer, inflammatory illnesses, and neurological problems. However, from a diagnostic point of view, bioimaging is highly significant. SLNs are generally incorporated as imaging agents and replaced with magnetic nanoparticles or fluorescent dyes, making them highly contrasting agents. This provides non-invasive imaging of particular tissues or cells, for example, in the study of the diagnosis of neurodegenerative diseases and brain cancers [82].

Magnetic Nanoparticles (MNPs)

Magnetic nanoparticles range in size from 1 to 100 nanometers, generally comprising an iron, cobalt, or nickel magnetic core with a biocompatible covering like silica or dextran [83]. Due to their capacity to be functionalized and controlled by a magnetic field, engineered magnetic nanoparticles (MNPs) are a cutting-edge medical technology that helps in magnetic resonance imaging (MRI), pharmaceuticals, gene delivery, magnetic hyperthermia, cancer care, tissue engineering, cell tracking, and bio- separations. These fields have all greatly benefited from their use.

Structural Characteristics

Currently, to construct the magnetic core of MNPs and a surface coating, iron oxide (Fe_3O_4), magnetite, or maghemite (-Fe_2O_3) are the most common magnetic core materials used; however, cobalt and nickel are also utilized. There are a few biocompatible materials, such as silica, dextran, or polyethylene glycol-coated magnetic nanoparticles (PEG), that are also predominantly exploited in the production of MNPs [84]. The coating stabilizes the magnetic core, prevents agglomeration, and improves biocompatibility. To target particular cells or tissues, the surface coating is functionalized using antibodies or peptides.

Advantages

Magnetic nanoparticles have unique properties such as magnetic properties, biocompatibility, nanometre size, targeting, stability, and multimodal imaging. They can be manipulated by magnetic fields, coated with biocompatible ingredients, targeted with biomolecules, and synthesized in large numbers [85].

Magnetic nanoparticles have several biological and therapeutic uses. Magnetic nanoparticles may be functionalized with targeting molecules to deliver medications to particular cells or regions. Magnetic targeting increases medication accumulation in specific areas, decreasing adverse effects and enhancing therapy. Magnetic nanoparticles are used in cancer treatment by exploiting the mechanism of magnetic hyperthermia, which is an alternating heating method used by MNPs. This non-invasive cancer treatment uses this feature [86]. In addition, MNPs are used as contrast agents in MRI imaging, which is also a non-invasive method. The suitability of magnetic nanoparticles in imaging is attributed to their better relativity than traditional contrast agents, hence making them more sensitive and improving MRI contrast [87]. Biomolecular sensing is another field where MNPs are conjugated with antibodies or aptamers to detect biomolecules, which is a significant part of diagnostics. This property can be applied to diagnoses of non-communicable and communicable diseases, for instance, cancer and infectious disorders [88].

Disadvantages

Apart from the various benefits and applications of MNPs, they also have some downsides, such as toxicity. Magnetic nanoparticles are dangerous for living cells and tissues, although most magnetic nanoparticles are biocompatible and non-toxic. But they can harm cells by producing oxidative stress. The next limitation is their agglomeration and stability. The process of forming agglomerates in physiological fluids reduces efficacy and toxicity. Recently, it has been shown that surface coatings or functionalization can reduce toxicity and enhance stability, but it is difficult [89]. Furthermore, magnetic nanoparticles may accumulate in numerous organs and tissues, affecting their clearance and biodistribution.

Applications

MNPs have tremendous promise for many uses, but their potential toxicity and other issues must be addressed to guarantee their safe and successful usage in clinical practice. Recently, due to their huge surface area, superparamagnetic, and biocompatibility, magnetic nanoparticles (MNPs) have generated great interest for their applications, especially in biological applications that include the diagnosis of many diseases. A few examples of magnetic nanoparticles in human disease diagnosis are given in the Table **3** [90]. MNPs are used as contrast agents in MRI to better see soft tissues and malignancies [91]. They target particular cells or tissues with specific ligands for more precise imaging. Biosensors are very significant for diseases and their onset in humans and animals [89]. Enhancement of the effectiveness or efficacy of MNPs functionalized with antibodies, peptides, or nucleic acids is needed to construct disease diagnostic biosensors [92]. Generally, biomolecules preferentially bind to disease indicators for sensitive and specific detection. Magnetic particle imaging (MPI) is another application, which is an innovative imaging method that leverages MNPs' magnetic characteristics to produce high-resolution pictures of biological tissues, such as in the case of cancer imaging and cardiovascular disease diagnostics, and other uses are possible [93].

Liposomes

Liposomes are spherical nanoparticles with phospholipid bilayer membranes. The central part of liposomes is generally filled with aqueous solutions, medicines, or DNA. Liposomes are compatible with biological systems due to their phospholipid bilayer membrane structure. The size of biodegradable and biocompatible liposomes ranges from tens of nanometers to several micrometers [94]. Currently, liposomes are used in medicine delivery, gene therapy, and cosmetics due to their unique structure and transport capabilities.

Structural Features

Liposome formation is based on the principle that the amphipathic phospholipid molecules stabilize the lipid bilayer structure and prevent cargo leakage. To improve the target accuracy of liposomes, they can also be coated or functionalized with specific types of receptor molecules that substantially increase their stability and targeting [95]. Customizing the interior cargo of liposomes enables their use for therapeutic or diagnostic purposes.

Liposomes possess a few unique features that are very helpful in many applications, such as biocompatibility, non-toxicity, and pliability for targeted processes, which make liposomes appropriate for medicinal and biological applications. Generally, liposomes are known for their targeting capability, which can be further enhanced by coating or functionalization to target certain tissues or cell types; therefore, drug delivery and diagnostic imaging processes make them more efficient [96].

Advantages

In the case of biological and industrial applications, liposome nanoparticles are frequently used as medication delivery vehicles because they encapsulate and target pharmaceuticals to particular tissues or cells. Therefore, liposomes boost therapeutic effectiveness and minimize toxicity, finally improving patient outcomes. So far, liposomes are also used in cosmetics, food industries, environmental sciences, and agriculture, but imaging is very significant from a diagnostic point of view because liposomes can carry fluorescent dyes or magnetic nanoparticles for diagnostic imaging [97].

Disadvantages

There are other limitations of liposomes, where drug delivery suffers substantially, such as the transport of a limited quantity of medication. Liposomes are not suitable for poorly soluble medicines and dye molecules. The short circulation time of liposomes limits their imaging potential. Additionally, the immune system can eliminate them very quickly, and liver and spleen accumulation complicates the diagnostic process [98]. Liposome manufacturing is a very complicated and time-consuming process that affects its economic viability.

Applications

Liposomes prove excellent imaging agents for tumor-specific antibodies or peptides that bind to cancer cells or are absorbed by cancer cells. Imaging techniques such as MRI, CT, and PET scans are used to identify various types of cancer by using liposomes as imaging agents within the cell, thereby aiding cancer detection.

Cardiac diseases are another non-communicable ailment that is responsible for a huge number of modalities worldwide. Currently, liposomes loaded with contrast agents are targeted to atherosclerosis or thrombosis locations to show arterial walls and inflammation, pre-existing conditions for cardiovascular diseases. Moreover, various diagnostic tests, such as MRI, CT, and PET scans, are used to investigate heart conditions such as valves and blockages of coronary arteries. Various heart disease conditions, such as cardiomyopathy, can be detected by imaging technologies; thus, pre-existing technologies can be made more efficient by using liposomes as imaging agents. It is well known that liposomes carry nucleic acid probes or CRISPR/Cas's gene editing tools to target the expression of particular genes [99]. Apart from this, many diseases, such as cystic fibrosis, sickle cell anemia, and Huntington's disease, are also diagnosed and treated by using liposomes.

Table 3. List of biomarkers developed by using different types of nanomaterials.

Name of the biomarker	Disease Diagnostic/ Infected Location	Reason	Biosensor Techniques
CA 125	Ovarian cancer	CA125 is a protein that is produced by ovarian cancer cells.	Electrochemical biosensors, piezoelectric biosensors [100]

(Table 3) cont.....

Name of the biomarker	Disease Diagnostic/ Infected Location	Reason	Biosensor Techniques
Alpha-fetoprotein (AFP)	Liver cancer	AFP is a protein that cancerous liver cells produce.	Electrochemical biosensors, optical biosensors [101]
hFABP, GPBB, S100, PAPP-A, RP, TNF, and IL6	Acute myocardial infarction (AMI)	These biomarkers are produced at various stages of heart disease.	Enzymatic and nonenzymatic [102]
Thyroglobulin	Thyroid disease	Treatment monitoring	Surface plasmon resonance (SPR) [103]
CA 19-9	Pancreatic cancer	CA 19-9 is a protein that is produced by pancreatic cancer cells.	Ultrasensitive immunosensor [104]
Troponin	Heart attack	Troponin is a protein that is released into the bloodstream when heart muscle cells are damaged,	Optical biosensors, immunosensors [105]
H. pylori antibodies	Stomach ulcers	H. pylori-specific antibodies are detected in ulcers.	Electrochemical Immunosensor [106]
Amyloid beta	Alzheimer's disease	Amyloid beta protein in the brain	Hydrogel biosensors and nano-biosensors [107]
C-reactive protein (CRP)	Inflammation	CRP is the best biomarker for inflammation.	Electrochemical biosensors, optical biosensors [108]
CA 15-3	Breast cancer	CA 15-3 is a protein used to detect breast cancer.	Electrochemistry-Assisted Surface Plasmon Resonance Biosensor [109]
Glucose	Diabetes	monitor of glucose to diagnose and monitor diabetes,	Enzymatic and non-enzymatic nanozyme-based colorimetric biosensor [110]
Human chorionic gonadotropin (HCG)	Pregnancy	HCG is a hormone that is produced by the placenta during pregnancy and is used as a biomarker.	Flow Immunoassay [111]
Cancer-Testis (CT) Antigens	Prostate, liver, lung, bladder, skin	Diagnosis and prognosis of various cancers	Electrochemical biosensors [112]
Immunoglobulin E (IgE)	Allergies	IgE is an antibody that is produced by the immune system in response to allergens.	Fluorescent biosensors, optoelectrical biosensors [113]

(Table 3) cont.....

Name of the biomarker	Disease Diagnostic/ Infected Location	Reason	Biosensor Techniques
Brain natriuretic peptide (BNP)	Heart failure	BNP is a hormone that is produced by heart disease.	Electrochemical immunoassay [114]
Alpha-fetoprotein (AFP)	Liver cancer	AFP is a protein that is produced by liver cancer cells.	Photoelectrochemical biosensor, electrochemical biosensor [115]
Troponin	Heart attack	Troponin is a protein that is released into the bloodstream and is found in heart muscles.	Electrochemiluminescence (ECL) immunosensor [116]
Prostate-specific antigen (PSA)	Prostate cancer	The prostate gland produces a protein known as PSA.	Impedimetric biosensors, surface plasmon resonance biosensors [117]
Creatinine	Kidney disease	Creatinine is a biomarker to diagnose and monitor kidney disease.	Amperometric creatinine biosensor, electrochemical biosensor [118]
Human papillomavirus (HPV) DNA	Cervical cancer	HPV DNA biomarker to screen.	electrochemical DNA biosensor [119]
Alpha-1 antitrypsin (AAT)	Lung disease	AAT is a protein that is produced by the liver and is a biomarker to diagnose and monitor lung disease.	Electrochemical and carbon nanotubes [120]
D-dimer	Blood clots	D-dimer is a protein fragment produced in blood clots; it is used as a biomarker of clotting-related disease.	Electrochemical bioanalytical devices [121]

Nanocrystals

Traditional organic fluorophores have limitations such as low photostability, narrow absorption spectra, and broad emission spectra; however, nanocrystals provide a potential alternative. In addition to being extremely photo-stable, they feature broad absorption spectra and narrow, size-tuneable emission spectra [122]. Recent advancements in the synthesis of these materials have resulted in the creation of biocompatible semiconductor fluorophores that are brilliant, sensitive, and extremely photo-stable. These nanocrystals are now commercially accessible and may be employed in a wide range of novel biological research, such as multiplexed cellular imaging, long-term *in vitro* and *in vivo* labeling of deep tissue structure mapping, and single particle analysis of dynamic cellular processes

[122]. Nanocrystals are one of the first examples of nanotechnology that enables imaging systems to work more efficiently for biological applications.

Structural Features

Nanocrystals, also known as nano-suspensions, are actually nanoparticle formulations made of drug molecules or other active substances ranging in size from 10 to 1000 nanometers and suspended in a liquid medium. In order to enhance the stability of nanocrystals, surfactants, polymers, or lipids are used to stabilize the nanocrystals, which are crystalline or partly crystalline [123]. There are several types of nanocrystals, such as zero-dimensional (0-D) nanocrystals, 1-D nanowires and nanorods, hollow structures, and superlattices, that have been modified to target and control drug release, and their tiny size improves medication absorption and distribution [124]. Nanocrystals are used in the processes of nanofabrication, nanodevices, nanobiology, and nanocatalysis. are

Advantages

There are many benefits associated with nanocrystals. Due to their tiny size and large surface area, nanocrystals can be efficiently used as active ingredients. Moreover, polymer coatings or matrix materials can regulate their functions. Nanocrystals can target a specific tissue or cell in diagnostic methods by connecting to cell surface receptors *via* ligands or antibodies, which enable tailored delivery [125].

Disadvantages

There are also many significant limitations that must be considered while creating and employing them in diverse applications. These disadvantages include low solubility, which restricts their applicability, for example, with restricted drug delivery. Agglomeration of nanocrystals is another problem that affects their performance [126]. This makes controlling their size and stability difficult, which limits their usage in several applications. Nanocrystals' stability degrades over time, affecting shelf life and efficacy; consequently, long-term storage and usage are difficult [127]. Nanocrystal fabrication is complicated and time-consuming, which limits its availability and cost. This makes biological and environmental applications difficult for nanocrystals.

Applications

Due to their tiny size, vast surface area, and adjustable optical and magnetic characteristics, nanocrystals are promising in human disease diagnostics. Disease diagnosis using nanocrystals includes cancer diagnosis, for example, because engineered nanocrystals can target cancer cells. Researchers have observed cancer cells in tissue samples or *in vivo* by conjugating nanocrystals with malignancy-specific biomarkers (Table **3**). Nanocrystals may also be employed as contrast agents in MRI and CT to identify malignancies [128]. Nanocrystals can also detect infectious illnesses. Currently, two-photon photoluminescence is used to make high-biocompatibility monodispersed perovskite-in-silica nanocrystals. This method has been used to detect cancer very early. To make this nanoprobe, single CsPbBr3 perovskite nanocrystals (NCs) were put inside a silica nanoshell to stop Pb2+ ions from leaking out of them [129].

In the recent past, gold nanorods were used to detect influenza viruses by changing color. Moreover, quantum dots conjugated with bacterial cell wall antibodies may also detect microorganisms. Genetic illness diagnosis also got a boost due to nanocrystals that identify certain genetic mutations. Researchers can identify DNA mutations by conjugating nanocrystals with complementary DNA. This method detects cystic fibrosis and sickle cell anemia mutations. Nanocrystals can detect cardiovascular disease biomarkers (Table **3**); for instance, gold nanoparticles containing troponin antibodies can detect cardiac attacks [130]. In summary, nanocrystals possess unique characteristics that make them attractive disease diagnostic tools. Further research is required to maximize their therapeutic utility.

CONCLUSION AND FUTURE PERSPECTIVE

Nanomaterials enhance illness assessment, providing substantial promise in the early detection and optimization of therapy for diseases including cancer, Alzheimer's, and heart ailments. The heightened sensitivity of nanomaterials enables them to detect minuscule quantities of metabolites, proteins, enzymes, and ions in biofluids, resulting in earlier and more precise diagnoses. Anticipated advancements in the future are expected to generate nanomaterials with enhanced sensitivity, enabling accurate identification of diseases and customized medical treatments by assessing biomarkers particular to each patient (Table **3**). This individualized strategy guarantees more efficient therapies with reduced adverse effects. In addition, nanomaterials will facilitate painless and non-invasive diagnostic techniques, such as non-invasive blood samples for cancer detection. Additionally, they might be employed in sensors for immediate health monitoring, such as glucose monitoring in individuals with diabetes, enabling remote disease tracking and modifications to care plans. In general, the utilization of nanomaterials in the field of disease detection has significant potential.

REFERENCES

[1]   Ghosh S, Das P. Nanomaterials in healthcare diagnostics and treatment: A critical review. J Drug Deliv Sci Technol 2019; 49: 298-316.
[http://dx.doi.org/10.1016/j.jddst.2018.12.019]

[2]   Nile SH, Baskar V, Selvaraj D, Nile A, Xiao J, Kai G. Nanotechnologies in food science: applications, recent trends, and future perspectives. Nano-Micro Lett 2020; 12(1): 45.
[http://dx.doi.org/10.1007/s40820-020-0383-9] [PMID: 34138283]

[3]   Pandey CM, Pal P. Nanomaterials for biosensors: A review. J Nanosci Nanotechnol 2021; 21(7): 3799-819.
[http://dx.doi.org/10.1166/jnn.2021.19262]

[4]   Milanova Sertova N. Contribution of nanotechnology in animal and human health care. Adv Mater Lett 2020; 11(9): 1-7.
[http://dx.doi.org/10.5185/amlett.2020.091552]

[5]   Dong X, Li Y. Carbon nanotube-based biosensors: a review. Anal Methods 2019; 11(20): 2588-601.
[http://dx.doi.org/10.1039/C9AY00552H]

[6]   Kumar S, Rani R, Dilbaghi N, Tankeshwar K, Kim KH. Carbon nanotubes: a novel material for multifaceted applications in human healthcare. Chem Soc Rev 2017; 46(1): 158-96.
[http://dx.doi.org/10.1039/C6CS00517A] [PMID: 27841412]

[7]   Wang C, Tao H, Cheng L, Liu Z, Liu Y. Emerging advances in carbon nanotube-based biosensors for healthcare applications. J Mater Chem B Mater Biol Med 2019; 7(16): 2539-61.
[PMID: 32255001]

[8]   Karimi M, Solati-Hashjin M, Alizadeh E. Carbon nanotubes as a novel drug delivery system in medical applications. J Biomed Sci 2016; 23(1): 1-14.
[http://dx.doi.org/10.1186/s12929-016-0234-2]

[9]   Ngo YH, Li D, Simon GP, Garnier G. Carbon nanotube membranes: synthesis, properties, and future filtration applications. J Membr Sci 2016; 499: 480-98.
[http://dx.doi.org/10.3390/nano7050099]

[10]   Sui X, Yan Q, Lv L. Carbon nanotubes-based biosensors: Current status and future prospects. Anal Chim Acta 2021; 1172: 338572.
[http://dx.doi.org/10.1016/j.aca.2021.338572]

[11]   Bardhan NM, Jansen P, Belcher AM. Graphene, carbon nanotube and plasmonic nanosensors for detection of viral pathogens: opportunities for rapid testing in pandemics like COVID-19. Frontiers in Nanotechnology 2021; 3: 733126.
[http://dx.doi.org/10.3389/fnano.2021.733126]

[12]   Misra R, Acharya S, Sushmitha N. Nanobiosensor-based diagnostic tools in viral infections: Special emphasis on Covid-19. Rev Med Virol 2022; 32(2): e2267.
[http://dx.doi.org/10.1002/rmv.2267] [PMID: 34164867]

[13]   Pumera M. Carbon nanotubes for biomedical applications. J Mater Chem 2011; 21(40): 15862-70.
[http://dx.doi.org/10.1039/C1JM13011K]

[14]   Sheikhpour M, Delorme V, Kasaeian A, *et al.* An effective nano drug delivery and combination therapy for the treatment of Tuberculosis. Sci Rep 2022; 12(1): 9591.
[http://dx.doi.org/10.1038/s41598-022-13682-4] [PMID: 35688860]

[15]   Mahmoodi P, Rezayi M, Rasouli E, *et al.* Early-stage cervical cancer diagnosis based on an ultra-sensitive electrochemical DNA nanobiosensor for HPV-18 detection in real samples. J Nanobiotechnology 2020; 18(1): 11.
[http://dx.doi.org/10.1186/s12951-020-0577-9] [PMID: 31931815]

[16]   Baker A, Iram S, Syed A, *et al.* Fruit derived potentially bioactive bioengineered silver nanoparticles.

Int J Nanomedicine 2021; 16: 7711-26.
[http://dx.doi.org/10.2147/IJN.S330763] [PMID: 34848956]

[17]   Biswas S, Devi YD, Sarma D, Namsa ND, Nath P. Gold nanoparticle decorated blu-ray digital versatile disc as a highly reproducible surface-enhanced Raman scattering substrate for detection and analysis of rotavirus RNA in laboratory environment. J Biophotonics 2022; 15(11): e202200138.
[http://dx.doi.org/10.1002/jbio.202200138] [PMID: 36054627]

[18]   Wang N, Pan G, Guan S, *et al.* A Broad-Range Disposable Electrochemical Biosensor Based on Screen-Printed Carbon Electrodes for Detection of Human Noroviruses. Front Bioeng Biotechnol 2022; 10: 845660.
[http://dx.doi.org/10.3389/fbioe.2022.845660] [PMID: 35402404]

[19]   Kumar A, Sarkar T, Solanki PR. Amine Functionalized Gadolinium Oxide Nanoparticles-Based Electrochemical Immunosensor for Cholera. Biosensors (Basel) 2023; 13(2): 177.
[http://dx.doi.org/10.3390/bios13020177] [PMID: 36831943]

[20]   Alejo-Cancho I, Navero-Castillejos J, Peiró-Mestres A, *et al.* Evaluation of a novel microfluidic immuno-magnetic agglutination assay method for detection of dengue virus NS1 antigen. PLoS Negl Trop Dis 2020; 14(2): e0008082.
[http://dx.doi.org/10.1371/journal.pntd.0008082] [PMID: 32069280]

[21]   Farka Z, Juřík T, Kovář D, Trnková L, Skládal P. Nanoparticle-based immunochemical biosensors and assays: recent advances and challenges. Chem Rev 2017; 117(15): 9973-10042.
[http://dx.doi.org/10.1021/acs.chemrev.7b00037] [PMID: 28753280]

[22]   Federizon J, Frye A, Huang WC, *et al.* Immunogenicity of the Lyme disease antigen OspA, particleized by cobalt porphyrin-phospholipid liposomes. Vaccine 2020; 38(4): 942-50.
[http://dx.doi.org/10.1016/j.vaccine.2019.10.073] [PMID: 31727504]

[23]   Christensen D, Agger EM, Andreasen LV, Kirby D, Andersen P, Perrie Y. Liposome-based cationic adjuvant formulations (CAF): Past, present, and future. J Liposome Res 2009; 19(1): 2-11.
[http://dx.doi.org/10.1080/08982100902726820] [PMID: 19515003]

[24]   Gao J, Zhang L, Xue L, *et al.* Development of a High-Efficiency Immunomagnetic Enrichment Method for Detection of Human Norovirus *via* PAMAM Dendrimer/SA-Biotin Mediated Cascade-Amplification. Front Microbiol 2021; 12: 673872.
[http://dx.doi.org/10.3389/fmicb.2021.673872] [PMID: 34354679]

[25]   Wang C, Liu M, Wang Z, Li S, Deng Y, He N. Point-of-care diagnostics for infectious diseases: From methods to devices. Nano Today 2021; 37: 101092.
[http://dx.doi.org/10.1016/j.nantod.2021.101092] [PMID: 33584847]

[26]   Zhao J, Zhu ZZ, Huang X, Hu X, Chen H. Magnetic gold nanocomposite and aptamer assisted triple recognition electrochemical immunoassay for determination of brain natriuretic peptide. Mikrochim Acta 2020; 187(4): 231.
[http://dx.doi.org/10.1007/s00604-020-4221-z] [PMID: 32180025]

[27]   Ahmad Najib M, Selvam K, Khalid MF, Ozsoz M, Aziah I. Quantum dot-based lateral flow immunoassay as point-of-care testing for infectious diseases: A narrative review of its principle and performance. Diagnostics (Basel) 2022; 12(9): 2158.
[http://dx.doi.org/10.3390/diagnostics12092158] [PMID: 36140559]

[28]   Gholizadeh O, Yasamineh S, Amini P, *et al.* Therapeutic and diagnostic applications of nanoparticles in the management of COVID-19: a comprehensive overview. Virol J 2022; 19(1): 206.
[http://dx.doi.org/10.1186/s12985-022-01935-7] [PMID: 36463213]

[29]   Francese R, Cecone C, Costantino M, *et al.* Identification of a βCD-Based Hyper-Branched Negatively Charged Polymer as HSV-2 and RSV Inhibitor. Int J Mol Sci 2022; 23(15): 8701.
[http://dx.doi.org/10.3390/ijms23158701] [PMID: 35955832]

[30]   Abdelghany S, Alkhawaldeh M, AlKhatib HS. Carrageenan-stabilized chitosan alginate nanoparticles

loaded with ethionamide for the treatment of tuberculosis. J Drug Deliv Sci Technol 2017; 39: 442-9.
[http://dx.doi.org/10.1016/j.jddst.2017.04.034]

[31] Mavrikou S, Tsekouras V, Hatziagapiou K, *et al.* Clinical application of the novel cell-based biosensor for the ultra-rapid detection of the SARS-CoV-2 S1 spike protein antigen: a practical approach. Biosensors (Basel) 2021; 11(7): 224.
[http://dx.doi.org/10.3390/bios11070224] [PMID: 34356695]

[32] Mostafa Abdalhamed A, Zeedan GSG, Ahmed Arafa A, Shafeek Ibrahim E, Sedky D, Abdel nabey Hafez A. Abdel nabey Hafez A. Detection of methicillin-resistant Staphylococcus aureus in clinical and subclinical mastitis in ruminants and studying the effect of novel green synthetized nanoparticles as one of the alternative treatments. Vet Med Int 2022; 2022: 1-14.
[http://dx.doi.org/10.1155/2022/6309984] [PMID: 36457891]

[33] Yupapin P, Trabelsi Y, Vigneswaran D, Taya SA, Daher MG, Colak I. Ultra-high-sensitive sensor based on surface plasmon resonance structure having Si and graphene layers for the detection of chikungunya virus. Plasmonics 2022; 17(3): 1315-21.
[http://dx.doi.org/10.1007/s11468-022-01631-w]

[34] Esfand R, Tomalia DA. Dendrimers in biomedical applications. Curr Opin Biotechnol 2001; 12(5): 468-77.
[PMID: 11301287]

[35] Lee BE, Kang T, Jenkins D, Li Y, Wall MM, Jun S. A single-walled carbon nanotubes-based electrochemical impedance immunosensor for on-site detection of *Listeria monocytogenes*. J Food Sci 2022; 87(1): 280-8.
[http://dx.doi.org/10.1111/1750-3841.15996] [PMID: 34935132]

[36] Chauhan AS. Dendrimers for Drug Delivery. Molecules 2018; 23(4): 938.
[http://dx.doi.org/10.3390/molecules23040938] [PMID: 29670005]

[37] Filipczak N, Yalamarty SSK, Li X, Parveen F, Torchilin V. Developments in treatment methodologies using dendrimers for infectious diseases. Molecules 2021; 26(11): 3304.
[http://dx.doi.org/10.3390/molecules26113304] [PMID: 34072765]

[38] Majoros IJ, Thomas TP, Mehta CB, Baker JR Jr. Poly(amidoamine) dendrimer-based multifunctional engineered nanodevice for cancer therapy. J Med Chem 2005; 48(19): 5892-9.
[http://dx.doi.org/10.1021/jm0401863] [PMID: 16161993]

[39] Singh U, Dar M, Hashmi A. Dendrimers: synthetic strategies, properties and applications. Orient J Chem 2014; 30(3): 911-22.
[http://dx.doi.org/10.13005/ojc/300301]

[40] Klajnert B, Bryszewska M. Dendrimers: properties and applications. Acta Biochim Pol 2001; 48(1): 199-208.
[http://dx.doi.org/10.18388/abp.2001_5127] [PMID: 11440170]

[41] Pandita D, Madaan K, Kumar S, Poonia N, Lather V. Dendrimers in drug delivery and targeting: Drug-dendrimer interactions and toxicity issues. J Pharm Bioallied Sci 2014; 6(3): 139-50.
[http://dx.doi.org/10.4103/0975-7406.130965] [PMID: 25035633]

[42] Kesharwani P, Jain K, Jain NK. Dendrimer as nanocarrier for drug delivery. Prog Polym Sci 2014; 39(2): 268-307.
[http://dx.doi.org/10.1016/j.progpolymsci.2013.07.005]

[43] Esfand R, Tomalia DA. Dendrimers in biomedical applications. Curr Opin Biotechnol 2001; 12(5): 468-77.
[PMID: 11301287]

[44] Qiu Z, Huang J, Liu L, Li C, Cohen Stuart MA, Wang J. Effects of pH on the Formation of PIC Micelles from PAMAM Dendrimers. Langmuir 2020; 36(29): 8367-74.
[http://dx.doi.org/10.1021/acs.langmuir.0c00598] [PMID: 32610910]

[45] Kesharwani P, Xie L, Banerjee SK. Dendrimer nanotechnology: A futuristic tool for drug delivery, diagnostics and imaging. Nanomedicine 2018; 14(3): 937-53.
[http://dx.doi.org/10.2217/nnm-2017-0371]

[46] Wu P, Malkoch M, Hunt JN. Dendrimers in Medicine: Therapeutic Concepts and Pharmaceutical Challenges. Bioconjug Chem 2018; 29(4): 839-42.
[PMID: 25654320]

[47] Svenson S, Tomalia DA. Dendrimers in biomedical applications—reflections on the field. Adv Drug Deliv Rev 2012; 64: 102-15.
[http://dx.doi.org/10.1016/j.addr.2012.09.030] [PMID: 16305813]

[48] Yamada K, Kim CT, Kim JH, Chung JH, Lee HG, Jun S. Single walled carbon nanotube-based junction biosensor for detection of Escherichia coli. PLoS One 2014; 18;9(9): e105767.
[http://dx.doi.org/10.1371/journal.pone.0105767] [PMID: 25233366]

[49] Gan L, Li Z, Lv Q, Huang W. Rabies virus glycoprotein (RVG29)-linked microRNA-124-loaded polymeric nanoparticles inhibit neuroinflammation in a Parkinson's disease model. Int J Pharm 2019; 567: 118449.
[http://dx.doi.org/10.1016/j.ijpharm.2019.118449] [PMID: 31226473]

[50] Sui X, Yan Q, Lv L. Carbon nanotubes-based biosensors: Current status and future prospects. Anal Chim Acta 2021; 1172: 338572.
[http://dx.doi.org/10.1016/j.aca.2021.338572]

[51] Valizadeh A, Mikaeili H, Samiei M, *et al.* Quantum dots: synthesis, bioapplications, and toxicity. Nanoscale Res Lett 2012; 7(1): 480.
[http://dx.doi.org/10.1186/1556-276X-7-480] [PMID: 22929008]

[52] Hussain S, Al-Naib IA. Nanomaterials in biosensors and their applications in healthcare. Biosens Bioelectron 2021; 178: 113008.
[http://dx.doi.org/10.1016/j.bios.2021.113008]

[53] Ahmad Najib M, Selvam K, Khalid MF, Ozsoz M, Aziah I. Quantum dot-based lateral flow immunoassay as point-of-care testing for infectious diseases: A narrative review of its principle and performance. Diagnostics (Basel) 2022; 12(9): 2158.
[http://dx.doi.org/10.3390/diagnostics12092158] [PMID: 36140559]

[54] Truffi M, Sevieri M, Morelli L, *et al.* Anti-MAdCAM-1-conjugated nanocarriers delivering quantum dots enable specific imaging of inflammatory bowel disease. Int J Nanomedicine 2020; 15: 8537-52.
[http://dx.doi.org/10.2147/IJN.S264513] [PMID: 33173291]

[55] Salman HD, Kadhim MM. Serum Levels of Receptor Activator of Nuclear Factor-κβ Ligand, Osteoprotegerin, Interlukin-17 and association with Receptor Activator of Nuclear Factor Kappa B/Osteoprotegerin Ratio in Patients with Osteoporosis. Ann Rom Soc Cell Biol 2021; 2(1): 3805-21.
[http://dx.doi.org/10.5281/zenodo.4567890]

[56] Lagos-Arevalo P, Palijan A, Vertullo L, *et al.* Cystatin C in acute kidney injury diagnosis: early biomarker or alternative to serum creatinine? Pediatr Nephrol 2015; 30(4): 665-76.
[http://dx.doi.org/10.1007/s00467-014-2987-0] [PMID: 25475610]

[57] Sharma M, Mehta I. Surface stabilized atorvastatin nanocrystals with improved bioavailability, safety and antihyperlipidemic potential. Sci Rep 2019; 9(1): 16105.
[http://dx.doi.org/10.1038/s41598-019-52645-0] [PMID: 31695118]

[58] Pitou M, Papi RM, Tzavellas AN, Choli-Papadopoulou T. ssDNA-Modified Gold Nanoparticles as a Tool to Detect miRNA Biomarkers in Osteoarthritis. ACS Omega 2023; 8(8): 7529-35.
[http://dx.doi.org/10.1021/acsomega.2c04806] [PMID: 36873033]

[59] Chawjiraphan W, Apiwat C, Segkhoonthod K, *et al.* Albuminuria detection using graphene oxide-mediated fluorescence quenching aptasensor. MethodsX 2020; 7: 101114.
[http://dx.doi.org/10.1016/j.mex.2020.101114] [PMID: 33194563]

[60]   Mowbray DJ, Skolnick MS. New physics and devices based on self-assembled semiconductor quantum dots. J Phys D Appl Phys 2005; 38(13): 2059-76.
[http://dx.doi.org/10.1088/0022-3727/38/13/002]

[61]   Bruchez MP. Quantum dots find their stride in single molecule tracking. Curr Opin Chem Biol 2011; 15(6): 775-80.
[http://dx.doi.org/10.1016/j.cbpa.2011.10.011] [PMID: 22055494]

[62]   Shao L, Gao Y, Yan F. Semiconductor quantum dots for biomedicial applications. Sensors (Basel) 2011; 11(12): 11736-51.
[http://dx.doi.org/10.3390/s111211736] [PMID: 22247690]

[63]   Peng CW, Li Y. Application of quantum dots-based biotechnology in cancer diagnosis: current status and future perspectives. J Nanomater 2010; 2010(1): 676839.
[http://dx.doi.org/10.1155/2010/676839]

[64]   Parak WJ, Manna L, Simmel FC, Gerion D, Alivisatos P. Quantum dots. Nanoparticles: from theory to application. 2010 Aug 25:3-47.
[http://dx.doi.org/10.1002/9783527631544.ch2]

[65]   Bera D, Qian L, Tseng TK, Holloway PH. Quantum dots and their multimodal applications: a review. Materials (Basel) 2010; 3(4): 2260-345.
[http://dx.doi.org/10.3390/ma3042260]

[66]   Jin Y, Gao X. Plasmonic fluorescent quantum dots. Nat Nanotechnol 2009; 4(9): 571-6.
[http://dx.doi.org/10.1038/nnano.2009.193] [PMID: 19734929]

[67]   Xu L, Yuan S, Zeng H, Song J. A comprehensive review of doping in perovskite nanocrystals/quantum dots: evolution of structure, electronics, optics, and light-emitting diodes. Materials Today Nano 2019; 6: 100036.
[http://dx.doi.org/10.1016/j.mtnano.2019.100036]

[68]   Mohamed WAA, Abd El-Gawad H, Mekkey S, *et al.* Quantum dots synthetization and future prospect applications. Nanotechnol Rev 2021; 10(1): 1926-40.
[http://dx.doi.org/10.1515/ntrev-2021-0118]

[69]   Valizadeh A, Mikaeili H, Samiei M, *et al.* Quantum dots: synthesis, bioapplications, and toxicity. Nanoscale Res Lett 2012; 7(1): 480.
[http://dx.doi.org/10.1186/1556-276X-7-480] [PMID: 22929008]

[70]   Jasieniak J, Smith L, van Embden J, Mulvaney P. Califonia–Australia light alliance (CALA) meeting on optical materials: Quantum dot light-emitting diodes: The path to solid-state lighting. Mater Today 2011; 14(11): 526-33.
[http://dx.doi.org/10.1016/j.mattod.2011.09.003]

[71]   Klimov VI, Baker TA, Lim J, Velizhanin KA, McDaniel H. Quality factor of luminescent solar concentrators and practical concentration limits attainable with semiconductor quantum dots. ACS Photonics 2016; 3(6): 1138-48.
[http://dx.doi.org/10.1021/acsphotonics.6b00307]

[72]   Rahmani H, Mansouri Majd S, Salimi A, Ghasemi F. Ultrasensitive immunosensor for monitoring of CA 19-9 pancreatic cancer marker using electrolyte-gated TiS_3 nanoribbons field-effect transistor. Talanta 2023; 257: 124336.
[http://dx.doi.org/10.1016/j.talanta.2023.124336] [PMID: 36863296]

[73]   Bera D, Qian L, Tseng TK, Holloway PH. Quantum dots and their multimodal applications: a review. Materials (Basel) 2010; 3(4): 2260-345.
[http://dx.doi.org/10.3390/ma3042260]

[74]   Abd Ellah NH, Tawfeek HM, John J, Hetta HF. Nanomedicine as a future therapeutic approach for Hepatitis C virus. Nanomedicine (Lond) 2019; 14(11): 1471-91.
[http://dx.doi.org/10.2217/nnm-2018-0348] [PMID: 31166139]

[75]   Barrientos K, Arango JP, Moncada MS, *et al.* Carbon dot-based biosensors for the detection of communicable and non -communicable diseases. Talanta 2023; 251: 123791.
[http://dx.doi.org/10.1016/j.talanta.2022.123791] [PMID: 35987023]

[76]   Venkateswarlu V, Manjunath K. Preparation, characterization and *in vitro* release kinetics of clozapine solid lipid nanoparticles. J Control Release 2004; 95(3): 627-38.
[http://dx.doi.org/10.1016/j.jconrel.2004.01.005] [PMID: 15023472]

[77]   Jaiswal JK, Mattoussi H, Mauro JM, Simon SM. Long-term multiple color imaging of live cells using quantum dot bioconjugates. Nat Biotechnol 2003; 21(1): 47-51.
[http://dx.doi.org/10.1038/nbt767] [PMID: 12459736]

[78]   Hu FQ, Jiang SP, Du YZ, Yuan H, Ye YQ, Zeng S. Preparation and characteristics of monostearin nanostructured lipid carriers. Int J Pharm 2006; 314(1): 83-9.
[http://dx.doi.org/10.1016/j.ijpharm.2006.01.040] [PMID: 16563671]

[79]   Venkateswarlu V, Manjunath K. Preparation, characterization and *in vitro* release kinetics of clozapine solid lipid nanoparticles. J Control Release 2004; 95(3): 627-38.
[http://dx.doi.org/10.1016/j.jconrel.2004.01.005] [PMID: 15023472]

[80]   Shidhaye SS, Saindane NS. Solid lipid nanoparticles for targeted drug delivery. Curr Drug Deliv 2018; 15(4): 494-503.

[81]   Agrawal Y, Patil K, Mahajan H, *et al. In vitro* and *in vivo* characterization of Entacapone-loaded nanostructured lipid carriers developed by quality-by-design approach. Drug Deliv 2022; 29(1): 1112-21.
[http://dx.doi.org/10.1080/10717544.2022.2058651] [PMID: 35380091]

[82]   Amiri M, Jafari S, Kurd M, *et al.* Engineered solid lipid nanoparticles and nanostructured lipid carriers as new generations of blood–brain barrier transmitters. ACS Chem Neurosci 2021; 12(24): 4475-90.
[http://dx.doi.org/10.1021/acschemneuro.1c00540] [PMID: 34841846]

[83]   Lu AH, Salabas EL, Schüth F. Magnetic nanoparticles: synthesis, protection, functionalization, and application. Angew Chem Int Ed 2007; 46(8): 1222-44.
[http://dx.doi.org/10.1002/anie.200602866] [PMID: 17278160]

[84]   Pankhurst QA, Connolly J, Jones SK, Dobson J. Applications of magnetic nanoparticles in biomedicine. J Phys D Appl Phys 2003; 36(13): R167-81.
[http://dx.doi.org/10.1088/0022-3727/36/13/201]

[85]   Shubayev VI, Pisanic TR II, Jin S. Magnetic nanoparticles for theragnostics. Adv Drug Deliv Rev 2009; 61(6): 467-77.
[http://dx.doi.org/10.1016/j.addr.2009.03.007] [PMID: 19389434]

[86]   Zhang Y, Kohler N, Zhang M. Surface modification of superparamagnetic magnetite nanoparticles and their intracellular uptake. Biomaterials 2002; 23(7): 1553-61.
[http://dx.doi.org/10.1016/S0142-9612(01)00267-8] [PMID: 11922461]

[87]   Gavilán H, Avugadda SK, Fernández-Cabada T, *et al.* Magnetic nanoparticles and clusters for magnetic hyperthermia: optimizing their heat performance and developing combinatorial therapies to tackle cancer. Chem Soc Rev 2021; 50(20): 11614-67.
[http://dx.doi.org/10.1039/D1CS00427A] [PMID: 34661212]

[88]   Javed R, Zia M, Naz S, Aisida SO, Ain N, Ao Q. Role of capping agents in the application of nanoparticles in biomedicine and environmental remediation: recent trends and future prospects. J Nanobiotechnology 2020; 18(1): 172.
[http://dx.doi.org/10.1186/s12951-020-00704-4] [PMID: 33225973]

[89]   Kaur P, Singh K. Magnetic nanoparticles: a boon to medical science. Curr Drug Metab 2018; 19(10): 821-33.
[http://dx.doi.org/10.2174/1389200219666181019110040]

[90]   Laurent S, Forge D, Port M, *et al.* Magnetic iron oxide nanoparticles: synthesis, stabilization, vectorization, physicochemical characterizations, and biological applications. Chem Rev 2008; 108(6): 2064-110.
       [http://dx.doi.org/10.1021/cr068445e] [PMID: 18543879]

[91]   Mornet S, Vasseur S, Grasset F, Duguet E. Magnetic nanoparticle design for medical diagnosis and therapy. J Mater Chem 2004; 14(14): 2161-75.
       [http://dx.doi.org/10.1039/b402025a]

[92]   Islam MA, Ahsan MZ. Plausible approach for rapid detection of SARS-CoV-2 virus by magnetic nanoparticle based biosensors. Am J Nanosci 2020; 6: 6.
       [http://dx.doi.org/10.2174/2405488X06666201019115734]

[93]   Koo KM, Soda N, Shiddiky MJA. Magnetic nanomaterial–based electrochemical biosensors for the detection of diverse circulating cancer biomarkers. Curr Opin Electrochem 2021; 25: 100645.
       [http://dx.doi.org/10.1016/j.coelec.2020.100645]

[94]   Torchilin VP. Recent advances with liposomes as pharmaceutical carriers. Nat Rev Drug Discov 2005; 4(2): 145-60.
       [http://dx.doi.org/10.1038/nrd1632] [PMID: 15688077]

[95]   Barenholz Y. Liposome application: problems and prospects. Curr Opin Colloid Interface Sci 2001; 6(1): 66-77.
       [http://dx.doi.org/10.1016/S1359-0294(00)00090-X]

[96]   Gabizon A, Shmeeda H, Barenholz Y. Pharmacokinetics of pegylated liposomal Doxorubicin: review of animal and human studies. Clin Pharmacokinet 2003; 42(5): 419-36.
       [http://dx.doi.org/10.2165/00003088-200342050-00002] [PMID: 12739982]

[97]   Wong TW. Liposome applications: cosmetic and therapeutic. J Microencapsul 2007; 24(2): 123-30.
       [http://dx.doi.org/10.1080/02652040601161255]

[98]   Herting MG, Kleinebudde P. Roll compaction/dry granulation: Effect of raw material particle size on granule and tablet properties. Int J Pharm 2007; 338(1-2): 110-8.
       [http://dx.doi.org/10.1016/j.ijpharm.2007.01.035] [PMID: 17324537]

[99]   Mohammadi R, Naderi-Manesh H, Farzin L, *et al.* Fluorescence sensing and imaging with carbon-based quantum dots for early diagnosis of cancer: A review. J Pharm Biomed Anal 2022; 212: 114628.
       [http://dx.doi.org/10.1016/j.jpba.2022.114628] [PMID: 35151068]

[100] Tuxen MK, Sölétormos G, Dombernowsky P. Tumor markers in the management of patients with ovarian cancer. Cancer Treat Rev 1995; 21(3): 215-45.
       [http://dx.doi.org/10.1016/0305-7372(95)90002-0] [PMID: 7656266]

[101] Hamanishi J, Mandai M, Ikeda T, *et al.* Safety and antitumor activity of anti–PD-1 antibody, nivolumab, in patients with platinum-resistant ovarian cancer. J Clin Oncol 2015; 33(34): 4015-22.
       [http://dx.doi.org/10.1200/JCO.2015.62.3397] [PMID: 26351349]

[102] Aydin S, Ugur K, Aydin S, Sahin İ, Yardim M. Biomarkers in acute myocardial infarction: current perspectives. Vasc Health Risk Manag 2019; 15: 1-10.
       [http://dx.doi.org/10.2147/VHRM.S166157] [PMID: 30697054]

[103] Choi S, Chae J. A microfluidic biosensor based on competitive protein adsorption for thyroglobulin detection. Biosens Bioelectron 2009; 25(1): 118-23.
       [http://dx.doi.org/10.1016/j.bios.2009.06.017] [PMID: 19577460]

[104] Rahmani H, Mansouri Majd S, Salimi A, Ghasemi F. Ultrasensitive immunosensor for monitoring of CA 19-9 pancreatic cancer marker using electrolyte-gated TiS$_3$ nanoribbons field-effect transistor. Talanta 2023; 257: 124336.
       [http://dx.doi.org/10.1016/j.talanta.2023.124336] [PMID: 36863296]

[105] Abdolrahim M, Rabiee M, Alhosseini SN, Tahriri M, Yazdanpanah S, Tayebi L. Development of

optical biosensor technologies for cardiac troponin recognition. Anal Biochem 2015; 485: 1-10.
[http://dx.doi.org/10.1016/j.ab.2015.06.003] [PMID: 26050627]

[106]   Saxena K, Kumar A, Chauhan N, Khanuja M, Malhotra BD, Jain U. Electrochemical immunosensor
for detection of h. Pylori secretory protein vaca on g-c3n4/zno nanocomposite-modified au electrode.
ACS Omega 2022; 7(36): 32292-301.
[http://dx.doi.org/10.1021/acsomega.2c03627] [PMID: 36120075]

[107]   Sun L, Zhong Y, Gui J, Wang X, Zhuang X, Weng J. A hydrogel biosensor for high selective and
sensitive detection of amyloid-beta oligomers. Int J Nanomedicine 2018; 13: 843-56.
[http://dx.doi.org/10.2147/IJN.S152163] [PMID: 29467574]

[108]   Sheen HJ, Panigrahi B, Kuo TR, *et al.* Electrochemical biosensor with electrokinetics-assisted
molecular trapping for enhancing C-reactive protein detection. Biosens Bioelectron 2022; 210:
114338.
[http://dx.doi.org/10.1016/j.bios.2022.114338] [PMID: 35550939]

[109]   Ribeiro JA, Sales MGF, Pereira CM. Electrochemistry-assisted surface plasmon resonance biosensor
for detection of CA 15–3. Anal Chem 2021; 93(22): 7815-24.
[http://dx.doi.org/10.1021/acs.analchem.0c05367] [PMID: 34038085]

[110]   Jeon HJ, Kim HS, Chung E, Lee DY. Nanozyme-based colorimetric biosensor with a systemic
quantification algorithm for noninvasive glucose monitoring. Theranostics 2022; 12(14): 6308-38.
[http://dx.doi.org/10.7150/thno.72152] [PMID: 36168630]

[111]   Hamad EM, Hawamdeh G, Jarrad NA, Yasin O, Al-Gharabli SI, Shadfan R. Detection of human
chorionic gonadotropin (HCG) hormone using digital lateral flow immunoassay. In2018 40th annual
international conference of the IEEE engineering in medicine and biology society (EMBC) 2018 Jul
18, 3845-3848.

[112]   Chow S, Berek JS, Dorigo O. Development of therapeutic vaccines for ovarian cancer. Vaccines
(Basel) 2020; 8(4): 657.
[http://dx.doi.org/10.3390/vaccines8040657] [PMID: 33167428]

[113]   Mas S, Badran AA, Juárez MJ, Fernández de Rojas DH, Morais S, Maquieira Á. Highly sensitive
optoelectrical biosensor for multiplex allergy diagnosis. Biosens Bioelectron 2020; 166: 112438.
[http://dx.doi.org/10.1016/j.bios.2020.112438] [PMID: 32755808]

[114]   Chianca M, Panichella G, Fabiani I, *et al.* Bidirectional Relationship Between Cancer and Heart
Failure: Insights on Circulating Biomarkers. Front Cardiovasc Med 2022; 9: 936654.
[http://dx.doi.org/10.3389/fcvm.2022.936654] [PMID: 35872912]

[115]   Li J, Liu D, Zhou D, Shao L, Chen X, Song H. Label-free photoelectrochemical biosensor for alpha-
fetoprotein detection based on Au/Cs$_x$WO$_3$ heterogeneous films. Talanta 2021; 225: 122074.
[http://dx.doi.org/10.1016/j.talanta.2020.122074] [PMID: 33592792]

[116]   Wang B, Wang C, Li Y, Liu X, Wu D, Wei Q. Electrochemiluminescence biosensor for cardiac
troponin I with signal amplification based on a MoS$_2$@Cu$_2$O–Ag-modified electrode and Ce:ZnO-
NGQDs. Analyst (Lond) 2022; 147(21): 4768-76.
[http://dx.doi.org/10.1039/D2AN01341J] [PMID: 36149312]

[117]   Dunn MW. Prostate cancer screening. InSeminars in oncology nursing 2017 May 1 (Vol. 33, No. 2,
pp. 156-164). WB Saunders.
[http://dx.doi.org/10.1016/j.soncn.2017.02.003]

[118]   Dasgupta P, Kumar V, Krishnaswamy PR, Bhat N. Serum creatinine electrochemical biosensor on
printed electrodes using monoenzymatic pathway to 1-methylhydantoin detection. ACS Omega 2020;
5(35): 22459-64.
[http://dx.doi.org/10.1021/acsomega.0c02997] [PMID: 32923804]

[119]   Pareek S, Jain U, Bharadwaj M, Saxena K, Roy S, Chauhan N. An ultrasensitive electrochemical DNA
biosensor for monitoring Human papillomavirus-16 (HPV-16) using graphene oxide/Ag/Au nano-

biohybrids. Anal Biochem 2023; 663: 115015.
[http://dx.doi.org/10.1016/j.ab.2022.115015] [PMID: 36496002]

[120] Zhu G, Lee HJ. Electrochemical sandwich-type biosensors for α−1 antitrypsin with carbon nanotubes and alkaline phosphatase labeled antibody-silver nanoparticles. Biosens Bioelectron 2017; 89(Pt 2): 959-63.
[http://dx.doi.org/10.1016/j.bios.2016.09.080] [PMID: 27816594]

[121] Tasić N, Paixão TRLC, Gonçalves LM. Biosensing of D-dimer, making the transition from the central hospital laboratory to bedside determination. Talanta 2020; 207: 120270.
[http://dx.doi.org/10.1016/j.talanta.2019.120270] [PMID: 31594601]

[122] Zhou B, Shi B, Jin D, Liu X. Controlling upconversion nanocrystals for emerging applications. Nat Nanotechnol 2015; 10(11): 924-36.
[http://dx.doi.org/10.1038/nnano.2015.251] [PMID: 26530022]

[123] Fu A, Gu W, Larabell C, Alivisatos AP. Semiconductor nanocrystals for biological imaging. Curr Opin Neurobiol 2005; 15(5): 568-75.
[http://dx.doi.org/10.1016/j.conb.2005.08.004] [PMID: 16150591]

[124] Alivisatos AP. Semiconductor clusters, nanocrystals, and quantum dots. Science 1996; 271(5251): 933-7.
[http://dx.doi.org/10.1126/science.271.5251.933]

[125] Scher EC, Manna L, Alivisatos AP. Shape control and applications of nanocrystals. Philosophical Transactions of the Royal Society of London. Series A: Mathematical, Physical and Engineering Sciences. 2003 Feb 15; 361(1803): 241-57.

[126] Wu H, Wang S, Li SFY, Bao Q, Xu Q. A label-free lead(II) ion sensor based on surface plasmon resonance and DNAzyme-gold nanoparticle conjugates. Anal Bioanal Chem 2020; 412(27): 7525-33.
[http://dx.doi.org/10.1007/s00216-020-02887-z] [PMID: 32829439]

[127] Alivisatos AP. Semiconductor clusters, nanocrystals, and quantum dots. Science. 1996 Feb 16; 271(5251): 933-7.
[http://dx.doi.org/10.1126/science.271.5251.933]

[128] Lu XY, Hu S, Jin Y, Qiu LY. Application of liposome encapsulation technique to improve anti-carcinoma effect of resveratrol. Drug Dev Ind Pharm 2012; 38(3): 314-22.
[http://dx.doi.org/10.3109/03639045.2011.602410] [PMID: 21851312]

[129] Yang Z, Zong S, Yang K, *et al.* Wavelength Tunable Aqueous $CsPbBr_3$-Based Nanoprobes with Ultrahigh Photostability for Targeted Super-Resolution Bioimaging. ACS Appl Mater Interfaces 2022; 14(15): 17109-18.
[http://dx.doi.org/10.1021/acsami.2c01638] [PMID: 35380800]

[130] He H, Zhang X, Du L, *et al.* Molecular imaging nanoprobes for theranostic applications. Adv Drug Deliv Rev 2022; 186: 114320.
[http://dx.doi.org/10.1016/j.addr.2022.114320] [PMID: 35526664]

CHAPTER 5

Current Applications of Nanoparticles in Tuberculosis Therapeutics

Bhabani Shankar Das[1], Gargi Balabantaray[2], Ashirbad Sarangi[1], Pradeepta Sekhar Patro[2] and Debapriya Bhattacharya[1,3,*]

[1] *Centre for Biotechnology, Siksha 'O' Anusandhan (Deemed to be University), Bhubaneswar, Odisha 751030, India*

[2] *Department of Immunology and Rheumatology, Institute of Medical Science, Sum Hospital, Siksha 'O' Anusandhan (Deemed to be University), Bhubaneswar, Odisha 751030, India*

[3] *Department of Biological Sciences, Indian Institute of Science Education and Research (IISER), Bhopal 462030, Madhya Pradesh, India*

Abstract: The global community is deeply concerned with the rapid spread of tuberculosis (TB), a highly contagious and potentially fatal disease. Current treatment regimens are often inadequate, leading to a poor quality of life. Moreover, the emergence of new antibiotics has necessitated the need for more effective therapeutic options. As such, research is being conducted around the world to develop novel strategies to combat TB, with nanotechnology playing a major role in these initiatives. Nanotechnology is an improved tool for existing treatments because of its unique properties and the capacity to enhance therapeutic efficacy. It is being used to target, deliver, and release drugs to infected tissue and cells to increase their absorption and efficacy. Nanoparticles (NPs) have also been shown to deliver anti-TB drugs to infected lungs, which may make the drugs more bioavailable and less harmful to the body as a whole. This book chapter provides a promising outlook on the potential uses of NPs for TB therapeutic development and serves as a guide for future research on infectious diseases.

Keywords: Antibiotics, Nanotechnology, Nanoparticles, Infectious diseases, Tuberculosis, Therapeutics.

INTRODUCTION

Tuberculosis (TB) is one of the oldest and most persistent infectious diseases in human history. Despite decades of efforts to control and eliminate TB, the disease

---
* **Corresponding author Debapriya Bhattacharya:** Centre for Biotechnology, Siksha 'O' Anusandhan (Deemed to be University), Bhubaneswar, Odisha 751030, India and Department of Biological Sciences, Indian Institute of Science Education and Research (IISER), Bhopal 462030, Madhya Pradesh, India;
E-mail: debapriyabhattacharyab@gmail.com

remains a major public health threat, especially in low- and middle-income countries. According to the World Health Organisation (WHO), TB is one of the top 10 causes of death worldwide, causing an estimated 1.5 million deaths in 2020 [1]. The emergence of drug-resistant strains of TB has further complicated the fight against the disease. Standard treatments for TB involve a combination of antibiotics taken over a period of several months. However, drug-resistant TB requires longer treatment regimens and often more expensive and less effective drugs. In addition, many people with TB do not take any treatment or are unable to complete the full course of antibiotics, leading to the spread of this infection [2]. The major challenges in TB treatment allow the bacilli of M. tuberculosis to survive and persist within host macrophages that engulf and destroy foreign particles. *M. tuberculosis* also evades the host immune system by forming clusters called granulomas, which prevent the antibiotics from reaching the bacteria [3]. Therefore, there is an urgent need to develop new and effective treatments for TB.

Nanotechnology has emerged as a promising approach for overcoming some of the challenges associated with TB. Nanoparticles are small particles with dimensions typically between 1 and 100 nanometers, and they have unique physical and chemical properties that make them attractive for use in medicine [4]. Researchers are exploring the use of nanoparticles to deliver drugs directly to the site of infection, enhance the immune response to TB, and develop new diagnostic tools for detecting the disease. One of the key advantages of using nanoparticles for TB treatment is their ability to target the bacteria selectively [5]. Antibiotics delivered using nanoparticles can accumulate in the lungs and release the drug gradually, increasing its efficacy while minimizing side effects. Nanoparticles can also be engineered to specifically target the immune cells that are involved in fighting TB, further enhancing the body's response to the infection [6].

Nanoparticles also improve the pharmacokinetics and pharmacodynamics of TB drugs [7]. For example, nanoparticles can increase the solubility, stability, and bioavailability of drugs, leading to enhanced drug efficacy and reduced toxicity. Furthermore, nanoparticles can prolong drug circulation in the body, allowing for sustained drug release and reducing drug administration frequency. By addressing the above potential of nanoparticles, this book chapter will provide an overview of the current state of TB treatment and the challenges and difficulties associated with controlling this disease [8]. It will also discuss the potential of nanoparticles to overcome these challenges and improve TB treatment outcomes.

Current Scenario and Challenges in Tuberculosis Disease

Tuberculosis is a major public health problem worldwide, with an estimated one-quarter of the world's population infected with *M. tuberculosis*. By 2030, the WHO End TB Strategy aims to reduce the incidence of TB by 80% and the number of TB deaths by 90%. However, the COVID-19 pandemic impacts TB control efforts, disrupting TB diagnosis, treatment, and prevention services [9]. In 2020, there were an estimated 1.4 million fewer TB cases reported than in 2019, and the number of people who started treatment for TB fell by 21% globally. The pandemic has highlighted the fragility of TB control efforts, as well as the need for a more robust and resilient system to address the TB epidemic.

TB diagnosis also presents a challenge, particularly in low-resource settings where diagnostic tools such as chest X-rays and sputum culture are not readily available [5]. The most common diagnostic method is sputum smear microscopy, which has a low sensitivity, particularly in people living with HIV. Newer diagnostic tools, such as the GeneXpert MTB/RIF test, which detects *M. tuberculosis* DNA in sputum samples, have improved sensitivity and specificity compared to smear microscopy. However, these tests are expensive and not widely available in low-resource settings [7]. The lack of access to accurate and timely TB diagnosis results in delayed treatment initiation, which can lead to poor treatment outcomes, including treatment failure and the development of drug-resistant TB.

Addressing Current Challenges in TB Using Nanoparticles

Nanoparticles have emerged as a promising tool for the management of TB disease. Nanoparticles are tiny particles, typically less than 100 nanometers in size, that have unique physicochemical properties that make them highly attractive for biomedical applications. One of the key advantages of nanoparticles is their ability to penetrate into cells and tissues, which is critical for the effective delivery of drugs to the site of infection [10]. Several different types of nanocarriers have been explored for the management of TB, including liposomes, polymeric nanoparticles, and metallic nanoparticles. Liposomes are lipid-based nanoparticles that can encapsulate drugs and release them at the site of infection.

Polymeric nanoparticles are made from biodegradable polymers and can be designed to release drugs over a prolonged period of time. Metallic nanoparticles, such as silver and gold nanoparticles, have antimicrobial properties and can kill bacteria directly [5]. In addition to drug delivery, nanoparticles can also be used for the diagnosis of TB. Nanoparticle-based biosensors can be designed to detect specific biomarkers of TB infection, such as mycobacterial DNA or proteins, in clinical samples. These biosensors can provide rapid and accurate diagnoses of

TB, which is critical for effective treatment and disease management [11]. Overall, the potential benefits of nanoparticles hold great promise for TB disease management. Nanoparticles provide a new and effective approach to TB disease management with continued innovation and research. The use of nanoparticles in TB disease management has shown a promising approach to addressing some of the current challenges in TB treatment and diagnosis [12].

CURRENT APPLICATIONS OF NANOCARRIERS IN TB MANAGEMENT

Use of Different Nanoparticles as Nanocarriers for Targeting Drug Delivery

Targeted drug delivery is an approach that aims to deliver drugs specifically to the site of infection while minimizing their distribution to other parts of the body. This approach has several advantages over traditional drug delivery methods, including increased drug efficacy, reduced toxicity, and a decreased likelihood of drug resistance [13]. Nanoparticles have emerged as a promising tool for targeted drug delivery in TB management, with several types of nanoparticles being developed and tested, as shown in Table **1.**

Table 1. Different nanoparticles used as carriers for drug delivery in TB infection.

Nanoparticle	Targeted Site	Targeted Drugs	References
Liposomes	Macrophages	Rifampicin, Isoniazid, Pyrazinamide, Ethambutol, Streptomycin	[14 - 18].
Dendrimers	Macrophages	Rifampicin, Isoniazid	[19, 20]
Chitosan nanoparticles	Lungs	Rifampicin, Isoniazid, Bedaquiline, Levofloxacin	[21 - 24]
Nanomicelles	Macrophages	Isoniazid, Rifampicin, Rifampicin-loaded chitosan nanomicelles	[25, 26]
Solid lipid nanoparticles	Lungs, Macrophages	Rifampicin, Isoniazid, Rifampicin-loaded mannosylated SLN, Bedaquiline	[27 - 30]
Carbon nanotubes	Macrophages	Isoniazid, Pyrazinamide	[31]

Some of the nanoparticles that have been used for targeted drug delivery in TB management are liposomes, dendrimers, and metallic nanoparticles. Liposomes are spherical vesicles composed of a lipid bilayer that can encapsulate both hydrophilic and hydrophobic drugs. They are biocompatible and biodegradable, making them ideal for drug delivery applications [32]. For example, liposomal isoniazid and rifampicin have been shown to improve drug delivery to the lungs and enhance therapeutic efficacy in TB [14, 15]. It has also been shown that liposomal pyrazinamide, ethambutol, and streptomycin reduce the number of bacteria more effectively than the usual drug regimen [16 - 18]. These results demonstrate the potential of liposomes as a drug delivery system for TB treatment.

Dendrimers are highly branched, monodisperse macromolecules that can encapsulate drugs and target specific cells or tissues. They have been explored for TB drug delivery and have shown promising results in both *in vitro* and *in vivo* studies [19]. Dendrimers can increase drug solubility, reduce drug toxicity, and enhance drug bioavailability. They can also target the lungs and macrophages, potentially overcoming drug resistance [20]. Another type of nanoparticle that has been explored for targeted drug delivery in TB management is chitosan nanoparticles, which are biocompatible and biodegradable particles that can enhance drug absorption and bioavailability. They have been investigated for TB treatment because of their ability to target macrophages, which are MTB's primary host cells.

Chitosan nanoparticles have been shown to improve the delivery of rifampicin and isoniazid to macrophages infected with MTB [21, 22]. Additionally, a study showed that chitosan nanoparticles containing bedaquiline and levofloxacin also show therapeutic efficacy against TB compared to standard drugs [23, 24].

Nanomicelles are one type of nanocarrier that has been explored for targeted drug delivery in TB management. Nanomicelles are composed of amphiphilic molecules that self-assemble in aqueous solutions to form micelles. The hydrophobic core of the micelle can be loaded with drugs, while the hydrophilic outer layer enhances its solubility and stability. Nanomicelles have been shown to improve drug bioavailability, enhance therapeutic efficacy, and reduce systemic toxicity in TB treatment [25]. For example, rifampicin-loaded nanomicelles have been shown to improve the pharmacokinetics and reduce the toxicity of the drug in animal models of TB infection [26].

Apart from that, solid lipid nanoparticles (SLNs) are also used as nanocarriers for targeted drug delivery in TB management. SLNs are composed of a solid lipid matrix and can be loaded with hydrophobic drugs. For example, rifampicin, isoniazid-loaded mannosylated SLN, and bedaquiline SLNs have been shown to improve the pharmacokinetics and therapeutic efficacy of the drug in TB infection [27 - 30].

Furthermore, carbon nanotubes (CNTs) are also used in nanocarriers for targeted drug delivery in TB management. CNTs are hollow cylinders made of carbon atoms with a large surface area that can be functionalized with drugs or target molecules. Isoniazid and pyrazinamide-loaded CNTs have been shown to improve the efficacy of drugs against TB [31]. Overall, the current use of different nanoparticles for targeted drug delivery in TB holds great promise for improving the treatment of this deadly disease (Fig. **1**). However, further research is needed to optimize the performance of nanoparticle-based drug delivery systems, improve their stability and shelf-life, and evaluate their safety and effectiveness in clinical settings.

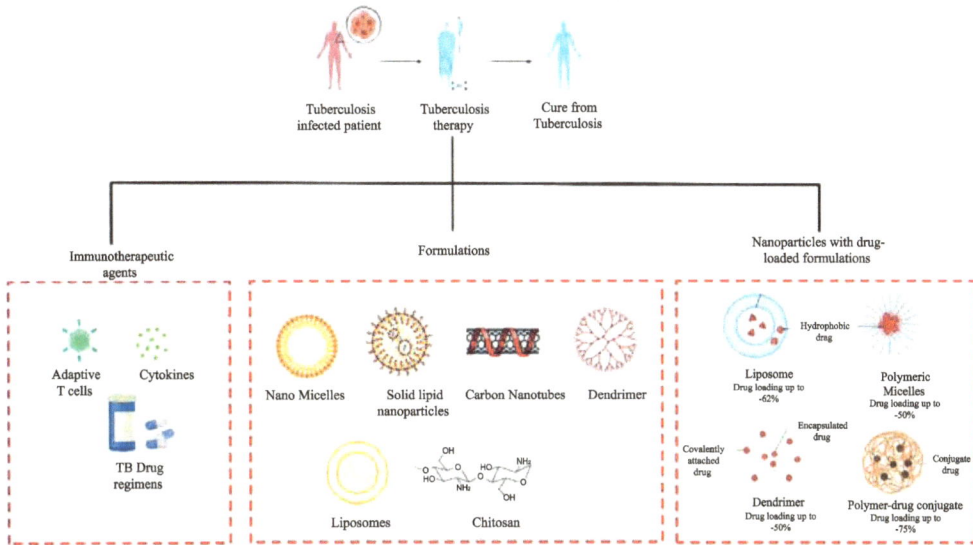

Fig. (1). Different nanoparticles used for drug delivery in TB.

Antimicrobial Activity of Metallic Nanoparticles Against *M. tuberculosis*

Different types of nanoparticles have been shown to exhibit antimicrobial activity against *M. tuberculosis*. Metallic nanoparticles, such as silver, gold, zinc oxide, iron oxide, titanium dioxide, selenium, chitosan, and graphene oxide, have shown extensive antimycobacterial activity against *M. tuberculosis*, as depicted in Table **2**.

Table 2. Different metallic nanoparticles used for TB administration.

Nanoparticle	Size	Antimicrobial Activity Against *M. tuberculosis*	References
Silver	50 nm, 5.4 nm	Inhibitory effect on *M. tuberculosis* growth *in vitro* with 50 mM, XDR strain of *M. tuberculosis* inhibits growth at 1 µg/mL	[33, 34]
Gold	16 nm	Inhibition of *M. tuberculosis* growth *in vitro* with 10 µg/mL	[35]

(Table 2) cont.....

Nanoparticle	Size	Antimicrobial Activity Against *M. tuberculosis*	References
Zinc oxide	9.3 nm	Antibacterial effect against XDR strain of *M. tuberculosis in vitro* with 1 μg/mL	[34]
Iron oxide	9 nm 9 nm	Enhances the antituberculosis drug rifampicin and inhibits *M. smegmatis* growth with a combination of 8 μg/mL rifampicin with 32 μg/mL Polyacrylic acid coated iron oxide nanoparticle biocompatibility enhancer of antituberculosis drugs isoniazid and rifampicin Inhibits *M. smegmatis* growth with a combination of 8 μg/mL rifampicin with 8 μg/mL citric acid coated iron oxide nanoparticle A combination of 16 μg/mL isoniazid with 8 μg/mL citric acid coated iron oxide nanoparticle enhances the activity of antitubercular drug against *M. smegmatis*	[36, 37]
Titanium dioxide	10–22 nm	Inhibitory biofilm effect shows against *M. tuberculosis* at 100 μg/mL TiO_2 nanoparticle	[38]
Selenium	105.7 nm	Inhibiting mycobacterial growth at MIC values of 0.400 μg/mL against *M. smegmatis* and for *M. tuberculosis* 0.195 μg/mL	[39]
Chitosan	500 nm	It shows anti-tubercular activity with MIC concentration of 1200 μg/mL and MBC concentration of 2400 μg/mL against *M. tuberculosis*	[40]
Graphene oxide	59 nm	Enhancing antitubercular drug ethambutol with graphene oxide shows an inhibitory mycobacterial effect at an MIC value of 1.5 μg/mL μg/mL against *M. smegmatis*	[41]

In brief, silver nanoparticles have shown potent antimycobacterial activity against *M. tuberculosis*, with minimum inhibitory concentrations (MICs) of 50 mM (50 μL), as shown in the XDR strain of *M. tuberculosis,* which inhibits growth at 1 μg/mL [33, 34]. Furthermore, it has been demonstrated that 10 μg/mL biogenic gold nanoparticles can stop the growth of *M. tuberculosis* by messing up the way cells work [35]. Aside from silver and gold nanoparticles, other types of nanoparticles like zinc oxide, iron oxide, titanium dioxide, selenium, chitosan, and graphene oxide are very good at killing M. smegmatis and *M. tuberculosis*, including their MDR and XDR strains, which can be seen in Table **2**. Overall, the above reports on nanoparticles suggest that they hold great promise for the development of new and effective TB therapies.

Development of Different Nanoparticles with medicinal Herbs for TB Management

For centuries, medicinal herbs have been used to treat a variety of diseases. Recently, there has been a growing interest in using medicinal herbs in combination with nanoparticle drug delivery systems to improve drug therapeutic efficacy and reduce toxicity. Several studies have looked at how to treat tuberculosis (TB) using different nanoparticles that contain medicinal herbs to deliver drugs to the site of infection and keep them there for a long time [43]. Some of those potentially delivered nanoparticles, as briefly described in the context below, are biodegradable polymers. This is one approach to developing nanoparticle-based drug delivery systems using medicinal herbs. Biodegradable polymers are shown to be attractive for drug delivery because they are biocompatible and can be tailored to release drugs over a specific time period [43]. Curcumin, a medicinal herb, has been used with rifampicin-encapsulated biodegradable polymer nanoparticles to clear TB pathogens in macrophages. The study's findings state that polymer nanoparticles can be a new avenue for the implementation of host-protective roles in TB treatment, with improved efficacy of existing TB drugs [44]. In addition to biodegradable polymers, other types of nanoparticles, such as silver nanoparticles, have shown potential effects against the TB pathogen. Garlic extract contains several bioactive compounds with antimicrobial properties that may be useful for TB treatments [1].

Silver nanoparticles have also been shown to have antimicrobial activity against *M. tuberculosis* [45]. In the development of garlic extract-based silver nanoparticles, the extract is used as a reducing agent to synthesize the nanoparticles. The result shows that garlic with isoniazid-loaded silver nanoparticles is used as an effective drug against Mycobacterium tuberculosis [46]. Other than silver nanoparticles, zinc oxide and gold nanoparticles also show effective activity in TB management. As an example, 25 µg/mL of *Canthium dicoccum* leaf extract mixed with zinc oxide synthesis nanoparticles is enough to stop M. tuberculosis from growing [47]. However, gold nanoparticles synthesized from the ethanolic extract of *Ocimum gratissimum* exhibited profound efficiency in inhibiting TB pathogens with an effective concentration of 50 µg/ml [48]. As an overview, the use of medicinal herbs in the development of nanoparticles for sustained drug release in TB management holds great promise for improving the efficacy and reducing the side effects of TB drugs. However, further research is needed to optimize the performance of these nanoparticles, evaluate their safety and effectiveness in clinical settings, and develop scalable production methods for clinical use.

Nanoparticle-based Biosensors for TB Diagnosis

Nanoparticles have emerged as a promising tool for the diagnosis of tuberculosis (TB), a major global health challenge caused by the bacterium *Mycobacterium tuberculosis*. Conventional TB diagnostic methods, such as sputum smear microscopy and culture, are slow and often fail to detect TB in a timely manner. On the other hand, nanoparticle-based diagnostic methods offer several advantages over conventional methods, including increased sensitivity, specificity, penetration, retention, insoluble drug transportation, and controlled drug delivery releases [49]. The use of biosensors is one of the most promising nanoparticle-based diagnostic methods for TB. Biosensors are analytical devices that can detect specific biomolecules, such as TB-specific antigens or nucleic acids, in clinical samples. Nanoparticles can be used as signal amplifiers or transducers in biosensors, increasing the sensitivity and accuracy of the assay [50]. Moreover, nanoparticle-based biosensors have several advantages over traditional TB diagnostic methods. They can detect TB-specific biomarkers at much lower concentrations than conventional methods, enabling early detection of TB infection. They also require less sample volume, making them more convenient and cost-effective. Additionally, nanoparticle-based biosensors can be designed to be highly specific, reducing the likelihood of false positives or false negatives [51].

Similarly, another example of a nanoparticle-based biosensor for TB diagnosis is the gold nanoparticle-based colorimetric biosensor. This biosensor uses gold nanoparticles that are functionalized with TB-specific antibodies to capture and detect TB biomarkers in clinical samples. The detection of TB biomarkers is achieved through a colorimetric reaction, which changes the color of the gold nanoparticles in the presence of the TB biomarkers. The color change is then measured using a simple spectrophotometer, providing a quick and easy-t--interpret diagnostic result. This method has been shown to be highly sensitive and specific for the detection of TB antigens in sputum and serum samples [52].

Apart from two methods of nanoparticle-based biosensors for TB diagnosis, magnetic nanoparticles (MNPs) are used for the isolation and detection of TB-specific nucleic acids. MNPs can be functionalized with probes that specifically bind to TB DNA or RNA sequences. The MNPs are then used to capture and isolate TB-specific nucleic acids from clinical samples, which can be detected using PCR or other nucleic acid amplification methods. This method has been shown to be highly sensitive and specific for the detection of TB DNA and RNA in sputum and blood samples [53].

In addition to biosensors, nanoparticles have also been explored for the development of imaging agents for TB diagnosis. Nanoparticle-based imaging agents have several advantages over traditional imaging agents, including increased sensitivity and specificity and the ability to target specific tissues and cells. Fluorescent silica nanoparticles are one example of a nanoparticle-based imaging agent for TB diagnosis. These nanoparticles can be functionalized with TB-specific antibodies or peptides and used to target TB-infected cells or tissues.

The fluorescent properties of the nanoparticles allow for real-time imaging of TB infection, providing a rapid and non-invasive diagnostic tool [54].

Overall, the current use of nanoparticles as TB diagnostic tools in management holds great promise for improving the diagnosis and treatment of this deadly disease. These methods offer several advantages over conventional diagnostic methods, including increased sensitivity, specificity, and speed. Nanoparticle-based diagnostic methods are likely to play an increasingly important role in TB infection management as further research and development continue in this field.

Future Aspects of Nanoparticle Applications in TB Management

The future of nanoparticle applications for TB disease holds great promise. As research in this field continues to advance, there are potential avenues for the development of novel nanoparticle-based therapies for TB. One area of potential development is the use of nanoparticles as targeted drug delivery systems for TB drugs. Encapsulating TB drugs within nanoparticles may be possible to increase drug efficacy and reduce toxicity by delivering the drugs directly to the site of infection [55]. Additionally, by targeting specific immune cells within the lung, nanoparticle-based drug delivery systems could potentially enhance the immune response to TB and improve treatment outcomes.

Another potential application for nanoparticles in TB treatment is the development of vaccines. Nanoparticle-based vaccines offer several advantages over traditional drugs, including the ability to enhance immune responses and deliver antigens directly to immune cells. By encapsulating TB antigens within nanoparticles, it may be possible to develop more effective TB vaccines that offer long-lasting protection against infection [56]. In addition to drug delivery and vaccine development, nanoparticles also hold potential for use in TB diagnostics. By using nanoparticles as biosensors, it may be possible to develop highly sensitive and specific diagnostic tests for TB that can detect the presence of *M. tuberculosis* in patient samples [54]. This could help to improve TB diagnosis and treatment outcomes by enabling earlier detection and treatment initiation. Furthermore, nanoparticles can be used to develop new therapies for drug-resistant TB [57]. Drug-resistant TB is a growing problem worldwide, and there is a need for new treatments that can effectively target drug-resistant strains of *M. tuberculosis*. Nanoparticles offer a potential solution by providing a versatile platform for developing new TB therapies [58]. For example, nanoparticles can be used to deliver combination therapies that target multiple pathways in drug-resistant TB, potentially enhancing treatment outcomes.

As with any new technology, there are challenges that need to be addressed in the development of nanoparticle-based interventions for TB disease. For example, the potential toxicity of nanoparticles and their impact on the host immune system need to be carefully evaluated to ensure their safety and effectiveness in clinical settings. Additionally, the scalability and cost-effectiveness of nanoparticle-based interventions are considered to be widespread adoption [59, 60]. In conclusion, potential applications of nanoparticles for TB disease are vast, and ongoing research is providing exciting opportunities for improving TB treatment and management. While there are still challenges that need to be addressed, the future of nanoparticle-based interventions for TB disease looks promising, and it is likely that we will see many new developments in this area in the coming years.

CONCLUSION

The use of nanoparticles has emerged as a promising approach for TB management due to their unique physicochemical properties and potential to enhance drug efficacy, reduce toxicity, and overcome drug resistance. Nanoparticles offer several advantages, such as targeted drug delivery, improved bioavailability, and reduced toxicity, which can help overcome the limitations of conventional TB treatments. However, more research is needed to optimize the design and formulation of nanoparticles for TB treatment, as well as ensure their safety and efficacy in clinical settings. Despite the challenges, the use of nanoparticles holds great promise for improving TB treatment outcomes and reducing the global burden of this deadly disease. As a result, the book chapter highlights the potential uses of nanoparticles in TB management and provides a better way to improve patient outcomes.

ACKNOWLEDGEMENTS

We are thankful to the esteemed editors for the invitation to contribute a book chapter in the book titled "*Nanomaterials in Biological Milieu: Biomedical Applications and Environmental Sustainability*". We acknowledge the Center for Biotechnology, Siksha O Anusandhan University, Bhubaneswar, Odisha, for overwhelming support.

REFERENCES

[1]   Sarangi A, Das BS, Patnaik G, *et al.* Potent anti-mycobacterial and immunomodulatory activity of some bioactive molecules of Indian ethnomedicinal plants that have the potential to enter in TB management. J Appl Microbiol 2021; 131(4): 1578-99.
[http://dx.doi.org/10.1111/jam.15088] [PMID: 33772980]

[2]   Das BS, Sarangi A, Bhattacharya D. Potential of Curcumin Nanoparticles in Tuberculosis Management. In: Arakha M, Pradhan AK, Jha S, Eds. Bio-Nano Interface. Singapore: Springer 2021; pp. 225-49.

[3]   Sharma R, Jhorar R, Kumar R. Nanomedicine in therapeutic intervention of tuberculosis meningitis. Curr Nanosci 2015; 11(1): 15-22.
[http://dx.doi.org/10.2174/1573413710666141016000110]

[4]   Griffiths G, Nyström B, Sable SB, Khuller GK. Nanobead-based interventions for the treatment and prevention of tuberculosis. Nat Rev Microbiol 2010; 8(11): 827-34.
[http://dx.doi.org/10.1038/nrmicro2437] [PMID: 20938454]

[5]   Chopra H, Mohanta YK, Rauta PR, *et al.* An Insight into Advances in Developing Nanotechnology Based Therapeutics, Drug Delivery, Diagnostics and Vaccines: Multidimensional Applications in Tuberculosis Disease Management. Pharmaceuticals (Basel) 2023; 16(4): 581.
[http://dx.doi.org/10.3390/ph16040581] [PMID: 37111338]

[6]   Mazlan MKN, Mohd Tazizi MHD, Ahmad R, *et al.* Antituberculosis targeted drug delivery as a potential future treatment approach. Antibiotics (Basel) 2021; 10(8): 908.
[http://dx.doi.org/10.3390/antibiotics10080908] [PMID: 34438958]

[7]   Nasiruddin M, Neyaz MK, Das S. Nanotechnology-based approach in tuberculosis treatment. Tuberc Res Treat 2017; 2017: 1-12.
[http://dx.doi.org/10.1155/2017/4920209] [PMID: 28210505]

[8]   Sarkar K, Kumar M, Jha A, Bharti K, Das M, Mishra B. Nanocarriers for tuberculosis therapy: design of safe and effective drug delivery strategies to overcome the therapeutic challenges. J Drug Deliv Sci Technol. 2022; 67(2022): 102850.
[http://dx.doi.org/10.1016/j.jddst.2021.102850]

[9]   Esmail H, Barry CE III, Wilkinson RJ. Understanding latent tuberculosis: the key to improved diagnostic and novel treatment strategies. Drug Discov Today 2012; 17(9-10): 514-21.
[http://dx.doi.org/10.1016/j.drudis.2011.12.013] [PMID: 22198298]

[10]   Debjit B, Chandira RM, Jayakar B, Kumar KS. Recent trends of drug used treatment of tuberculosis. J Chem Pharm Res 2009; 1(1): 113-33.

[11]   Vedha Hari BN, Chitra K, Bhimavarapu R, Karunakaran P, Muthukrishnan N, Rani BS. Novel technologies: A weapon against tuberculosis. Indian J Pharmacol 2010; 42(6): 338-44.
[http://dx.doi.org/10.4103/0253-7613.71887] [PMID: 21189901]

[12]   Sosnik A, Carcaboso ÁM, Glisoni RJ, Moretton MA, Chiappetta DA. New old challenges in tuberculosis: Potentially effective nanotechnologies in drug delivery. Adv Drug Deliv Rev 2010; 62(4-5): 547-59.
[http://dx.doi.org/10.1016/j.addr.2009.11.023] [PMID: 19914315]

[13]   Tewabe A, Abate A, Tamrie M, Seyfu A, Abdela Siraj E. Targeted drug delivery—from magic bullet to nanomedicine: principles, challenges, and future perspectives. J Multidiscip Healthc 2021; 14: 1711-24.
[http://dx.doi.org/10.2147/JMDH.S313968] [PMID: 34267523]

[14]   Patil J. Significance of Particulate Drug Delivery System in Anti-microbial Therapy. Adv Pharmacoepidemiol Drug Saf 2016; 5(1): 139.
[http://dx.doi.org/10.4172/2167-1052.1000e139]

[15]   Nkanga CI, Krause RW, Noundou XS, Walker RB. Preparation and characterization of isoniazid-loaded crude soybean lecithin liposomes. Int J Pharm 2017; 526(1-2): 466-73.
[http://dx.doi.org/10.1016/j.ijpharm.2017.04.074] [PMID: 28461265]

[16]   Rojanarat W, Nakpheng T, Thawithong E, Yanyium N, Srichana T. Inhaled pyrazinamide proliposome for targeting alveolar macrophages. Drug Deliv 2012; 19(7): 334-45.
[http://dx.doi.org/10.3109/10717544.2012.721144] [PMID: 22985352]

[17]   Wiens T, Redelmeier T, Av-Gay Y. Development of a liposome formulation of ethambutol. Antimicrob Agents Chemother 2004; 48(5): 1887-8.
[http://dx.doi.org/10.1128/AAC.48.5.1887-1888.2004] [PMID: 15105152]

[18]   Su FY, Chen J, Son HN, *et al.* Polymer-augmented liposomes enhancing antibiotic delivery against intracellular infections. Biomater Sci 2018; 6(7): 1976-85.
[http://dx.doi.org/10.1039/C8BM00282G] [PMID: 29850694]

[19]   Bellini RG, Guimarães AP, Pacheco MAC, *et al.* Association of the anti-tuberculosis drug rifampicin with a PAMAM dendrimer. J Mol Graph Model 2015; 60(60): 34-42.
[http://dx.doi.org/10.1016/j.jmgm.2015.05.012] [PMID: 26093506]

[20]   Rodrigues B, Shende P. Monodispersed Metal-Based Dendrimeric Nanoclusters for Potentiation of Anti-tuberculosis Action. J Mol Liquids. 2020; 304(2020): 112731.
[http://dx.doi.org/10.1016/j.molliq.2020.112731]

[21]   Rawal T, Parmar R, Tyagi RK, Butani S. Rifampicin loaded chitosan nanoparticle dry powder presents an improved therapeutic approach for alveolar tuberculosis. Colloids Surf B Biointerfaces 2017; 154(154): 321-30.
[http://dx.doi.org/10.1016/j.colsurfb.2017.03.044] [PMID: 28363192]

[22]   Garg T, Rath G, Goyal AK. Inhalable chitosan nanoparticles as antitubercular drug carriers for an effective treatment of tuberculosis. Artif Cells Nanomed Biotechnol 2016; 44(3): 997-1001.
[PMID: 25682840]

[23]   Rawal T, Patel S, Butani S. Chitosan nanoparticles as a promising approach for pulmonary delivery of bedaquiline. Eur J Pharm Sci 2018; 124(124): 273-87.
[http://dx.doi.org/10.1016/j.ejps.2018.08.038] [PMID: 30176365]

[24]   Shah S, Ghetiya R, Soniwala M, Chavda J. Development and Optimization of Inhalable Levofloxacin Nanoparticles for The Treatment of Tuberculosis. Curr Drug Deliv 2021; 18(6): 779-93.
[http://dx.doi.org/10.2174/1567201817999201103194626] [PMID: 33155907]

[25]   Praphakar RA, Shakila H, Azger Dusthackeer VN, Munusamy MA, Kumar S, Rajan M. A mannose-conjugated multi-layered polymeric nanocarrier system for controlled and targeted release on alveolar macrophages. Polym Chem 2018; 9(5): 656-67.
[http://dx.doi.org/10.1039/C7PY02000G]

[26]   Praphakar RA, Munusamy MA, Rajan M. Development of extended-voyaging anti-oxidant Linked Amphiphilic Polymeric Nanomicelles for Anti-Tuberculosis Drug Delivery. Int J Pharm 2017; 524(1-2): 168-77.
[http://dx.doi.org/10.1016/j.ijpharm.2017.03.089] [PMID: 28377319]

[27]   Vieira ACC, Chaves LL, Pinheiro M, *et al.* Mannosylated Solid Lipid Nanoparticles for the Selective Delivery of Rifampicin to Macrophages. Artif Cell Nanomed Biotechnol. 2018; 46(sup1): 653–663.
[http://dx.doi.org/10.1080/21691401.2018.1434186]

[28]   Ma C, Wu M, Ye W, *et al.* Correction: Inhalable solid lipid nanoparticles for intracellular tuberculosis infection therapy: macrophage-targeting and pH-sensitive properties. Drug Deliv Transl Res 2022; 12(11): 2893.
[http://dx.doi.org/10.1007/s13346-022-01160-3] [PMID: 35441322]

[29]   Truzzi E, Nascimento TL, Iannuccelli V, *et al.* *In vivo* biodistribution of Respirable Solid Lipid Nanoparticles Surface-Decorated with a Mannose-Based Surfactant: a Promising Tool for Pulmonary Tuberculosis Treatment?. Nanomater. 2020; 10(3): 568.
[http://dx.doi.org/10.3390/nano10030568]

[30]   De Matteis L, Jary D, Lucía A, *et al.* New Active Formulations against M. tuberculosis: Bedaquiline Encapsulation in Lipid Nanoparticles and Chitosan Nanocapsules. Chem Eng J 2018; 2018(340): 181-91.
[http://dx.doi.org/10.1016/j.cej.2017.12.110]

[31]   Zomorodbakhsh S, Abbasian Y, Naghinejad M, Sheikhpour M. The Effects Study of Isoniazid Conjugated Multi-Wall Carbon Nanotubes Nanofluid on *Mycobacterium tuberculosis*. Int J Nanomedicine 2020; 15(15): 5901-9.
[http://dx.doi.org/10.2147/IJN.S251524] [PMID: 32884258]

[32]   Guo A. Metal-Organic Frameworks as Bacteria Mimicking Delivery Systems for Tuberculosis. South Dakota State University. ProQuest Dissertations Publishing, 2021: 28867181.

[33]   Praba VL, Kathirvel M, Vallayyachari K, *et al.* Bactericidal effect of silver nanoparticles against *Mycobacterium tuberculosis.* Journal of Bionanoscience 2013; 7(3): 282-7.
[http://dx.doi.org/10.1166/jbns.2013.1138]

[34]   Heidary M, Zaker Bostanabad S, Amini SM, *et al.* The anti-mycobacterial activity of Ag, ZnO, and Ag-ZnO nanoparticles against MDR-and XDR-*Mycobacterium tuberculosis.* Infect Drug Resist 2019; 12(12): 3425-35.
[http://dx.doi.org/10.2147/IDR.S221408] [PMID: 31807033]

[35]   Govindaraju K, Vasantharaja R, Suganya KU, *et al.* Unveiling the anticancer and antimycobacterial potentials of bioengineered gold nanoparticles. Process Biochem. 2020, 2020; 96: 213-219.
[http://dx.doi.org/10.1016/j.procbio.2020.06.016]

[36]   Padwal P, Bandyopadhyaya R, Mehra S. Polyacrylic acid-coated iron oxide nanoparticles for targeting drug resistance in mycobacteria. Langmuir 2014; 30(50): 15266-76.
[http://dx.doi.org/10.1021/la503808d] [PMID: 25375643]

[37]   Padwal P, Bandyopadhyaya R, Mehra S. Biocompatible citric acid-coated iron oxide nanoparticles to enhance the activity of first-line anti- TB drugs in *Mycobacterium smegmatis.* J Chem Technol Biotechnol 2015; 90(10): 1773-81.
[http://dx.doi.org/10.1002/jctb.4766]

[38]   Ramalingam V, Sundaramahalingam S, Rajaram R. Size-dependent antimycobacterial activity of titanium oxide nanoparticles against *Mycobacterium tuberculosis.* J Mater Chem B Mater Biol Med 2019; 7(27): 4338-46.
[http://dx.doi.org/10.1039/C9TB00784A]

[39]   Estevez H, Palacios A, Gil D, *et al.* Antimycobacterial effect of selenium nanoparticles on *Mycobacterium tuberculosis.* Front Microbiol 2020; 11: 800.
[http://dx.doi.org/10.3389/fmicb.2020.00800] [PMID: 32425916]

[40]   Wardani G, M M, Sudjarwo SA. *In vitro* antibacterial activity of chitosan nanoparticles against *Mycobacterium tuberculosis.* Pharmacogn J 2017; 10(1): 162-6.
[http://dx.doi.org/10.5530/pj.2018.1.27]

[41]   Saifullah B, Chrzastek A, Maitra A, *et al.* Novel anti-tuberculosis nanodelivery formulation of ethambutol with graphene oxide. Molecules 2017; 22(10): 1560.
[http://dx.doi.org/10.3390/molecules22101560] [PMID: 29023399]

[42]   Patra JK, Das G, Fraceto LF, *et al.* Nano based drug delivery systems: recent developments and future prospects. J Nanobiotechnology 2018; 16(1): 71.
[http://dx.doi.org/10.1186/s12951-018-0392-8] [PMID: 30231877]

[43]   Sung YK, Kim SW. Recent advances in polymeric drug delivery systems. Biomater Res 2020; 24(1): 12.
[http://dx.doi.org/10.1186/s40824-020-00190-7] [PMID: 32537239]

[44]   Jahagirdar PS, Gupta PK, Kulkarni SP, Devarajan PV. Intramacrophage delivery of dual drug loaded nanoparticles for effective clearance of *Mycobacterium tuberculosis.* J Pharm Sci 2020; 109(7): 2262-70.
[http://dx.doi.org/10.1016/j.xphs.2020.03.018] [PMID: 32240695]

[45]   Tăbăran AF, Matea CT, Mocan T, *et al.* Silver Nanoparticles for the Therapy of Tuberculosis. Int J Nanomedicine 2020; 15(31): 2231-58.
[http://dx.doi.org/10.2147/IJN.S241183] [PMID: 32280217]

[46]   Mohamed JMM, Alqahtani A, Kumar TVA, *et al.* Superfast Synthesis of Stabilized Silver Nanoparticles Using Aqueous *Allium sativum* (Garlic) Extract and Isoniazid Hydrazide Conjugates: Molecular Docking and *In-Vitro* Characterizations. Molecules 2021; 27(1): 110.
[http://dx.doi.org/10.3390/molecules27010110] [PMID: 35011342]

[47]   Mahendra C, Chandra MN, Murali M, *et al.* Phyto-fabricated ZnO nanoparticles from Canthium

dicoccum (L.) for antimicrobial, anti-tuberculosis and antioxidant activity. Process Biochem 2020; 89(89): 220-6.
[http://dx.doi.org/10.1016/j.procbio.2019.10.020]

[48]   Gupta A, Pandey S, Variya B, Shah S, Yadav JS. Green synthesis of gold nanoparticles using different leaf extracts of Ocimum gratissimum Linn for anti-tubercular activity. Curr Nanomed 2019; 9(2): 146-57.
[http://dx.doi.org/10.2174/2468187308666180807125058]

[49]   Carissimi G, Montalbán GM, Fuster GM, Víllora G. Nanoparticles as Drug Delivery Systems. In: Pham PV, Ed. Century Nanostructured Materials-Physics, Chemistry, Classification, and Emerging Applications in Industry, Biomedicine, and Agriculture. IntechOpen 2021; pp. 227-49.

[50]   Gupta AK, Singh A, Singh S. Diagnosis of Tuberculosis: Nanodiagnostics Approaches. NanoBiomed. Springer Publisher. 2019; 25: 261–83.

[51]   Rabti A, Raouafi A, Raouafi N. DNA markers and nano-biosensing approaches for tuberculosis diagnosis. In: Kesharwani P, Ed. Nanotechnology Based Approaches for Tuberculosis Treatment. Elsevier 2020; pp. 207-30.
[http://dx.doi.org/10.1016/B978-0-12-819811-7.00013-8]

[52]   Teengam P, Siangproh W, Tuantranont A, Vilaivan T, Chailapakul O, Henry CS. Multiplex paper-based colorimetric DNA sensor using pyrrolidinyl peptide nucleic acid-induced AgNPs aggregation for detecting MERS-CoV, MTB, and HPV oligonucleotides. Anal Chem 2017; 89(10): 5428-35.
[http://dx.doi.org/10.1021/acs.analchem.7b00255] [PMID: 28394582]

[53]   Gupta S, Kakkar V. Recent technological advancements in tuberculosis diagnostics – A review. Biosens Bioelectron 2018; 115(115): 14-29.
[http://dx.doi.org/10.1016/j.bios.2018.05.017] [PMID: 29783081]

[54]   El-Samadony H, Althani A, Tageldin MA, Azzazy HME. Nanodiagnostics for tuberculosis detection. Expert Rev Mol Diagn 2017; 17(5): 427-43.
[http://dx.doi.org/10.1080/14737159.2017.1308825] [PMID: 28317400]

[55]   Khizar S, Elaissari A, Al-Dossary AA, Zine N, Jaffrezic-Renault N, Errachid A. Advancement in Nanoparticle-based Biosensors for Point-of-care *In vitro* Diagnostics. Curr Top Med Chem 2022; 22(10): 807-33.
[http://dx.doi.org/10.2174/1568026622666220401160121] [PMID: 35366774]

[56]   Luo X, Zeng X, Gong L, *et al.* Nanomaterials in tuberculosis DNA vaccine delivery: historical perspective and current landscape. Drug Deliv 2022; 29(1): 2912-24.
[http://dx.doi.org/10.1080/10717544.2022.2120565] [PMID: 36081335]

[57]   Dahiya B, K Mehta P. Utility of nanoparticle-based assays in the diagnosis of tuberculosis. Nanomedicine (Lond) 2021; 16(15): 1263-8.
[http://dx.doi.org/10.2217/nnm-2021-0077] [PMID: 33988032]

[58]   Bhusal N, Shrestha S, Pote N, Alocilja EC. Nanoparticle-based biosensing of tuberculosis, an affordable and practical alternative to current methods. Biosensors (Basel) 2018; 9(1): 1.
[http://dx.doi.org/10.3390/bios9010001] [PMID: 30586842]

[59]   Moreira AC, Batiha GE, Abdellah NH, Cruz-Martins N. Nanoparticle-Based Therapy in Chronic Obstructive and Infectious Lung Diseases: Past, Present, and Future Perspectives. In: Thangadurai D, Islam S, Sangeetha J, Cruz-Martins N, Eds. Biogenic Nanomaterials. Apple Academic Press 2023; pp. 145-62.

[60]   Bakishzade A, Nasibova A. Future prospects of biomaterials in nanomedicine. Adv Bio Earth Sci 2023; 2023(8): 5-10.

CHAPTER 6

Nanoparticles and Amalgamation of Stem Cells: A Novel Approach of Treatment in Current Medicine

Seema Tripathy[1,*] and **Pratyush Kumar Behera**[2]

[1] *School of Biological Sciences, National Institute of Science Education and Research, Bhubaneshwar-752050, India*

[2] *Department of Zoology, Maharaja Sriram Chandra Bhanja Deo University, Takatpur, Baripada-757003, Mayurbhanj, Odisha, India*

Abstract: Stem cells (SCs) are fundamental entities in multicellular organisms, possessing remarkable potential for self-renewal and differentiating into various cell types, including neurocytes, osteoblasts, chondrocytes, and hepatocytes. The therapeutic applications of differentiated cells derived from both embryonic and adult SCs hold promise for addressing a spectrum of disorders, ranging from Alzheimer's disease to liver cirrhosis. Despite significant strides in triggering cellular differentiation, the complexities of directing stem cells toward desired cell types remain a formidable challenge. In this context, integrating nanotechnology offers a promising avenue and reproducible therapeutic approach. Recent advancements in the development of inorganic nanoparticles (NPs), metal NPs, and carbon-based NPs, characterized by unique physical and chemical properties, present an opportunity to enhance the differentiation, expansion, and proliferation of stem cells. By leveraging the distinct characteristics of these NPs, researchers can exert control over cellular behaviors and differentiation, thereby augmenting the efficacy of stem cell-based therapeutic approaches. This chapter provides a comprehensive overview of the latest developments in leveraging amalgamated stem cells and NP-based systems for clinical translation. By harnessing the synergistic potential of stem cells and NPs, this innovative approach holds immense promise for revolutionizing treatment modalities in contemporary medicine.

Keywords: Alzheimer's disease, Cellular behaviors, Liver cirrhosis, Nanotechnology, Osteoporosis, Osteoarthritis, Stem cells.

INTRODUCTION

Stem cells are the quiescent primary cells mainly accountable for the development, repair, and replenishment of deteriorated cells and tissues in multi-

---

[*] **Corresponding author Seema Tripathy:** School of Biological Sciences, National Institute of Science Education and Research, Bhubaneshwar-752050, India; E-mail: stripathy.seema@gmail.com

Manoranjan Arakha & Arun Kumar Pradhan (Eds.)

cellular organisms. They have a substantial capacity for self-renewing and competence for differentiation. Upon the availability of external cues, stem cells get differentiated into various types of cells.

Based on the stages of development in animals, stem cells can be classified into embryonic stem cells (ES cells), somatic stem cells, or adult stem cells. A total of three kinds of stem cells can be characterized based on their developmental potential: totipotent stem cells, pluripotent stem cells, multipotent stem cells, and unipotent stem cells [1]. Currently, stem cells such as mesenchymal stem cells or MSCs are extensively being used in therapeutic trials [2]. Even though MSCs can be derived from various tissues like bone marrow, endometrial polyps, umbilical cord (UC), menstrual blood (MB), and adipose tissue (AT), these cells share some common stem cell characteristics that significantly vary in population numbers inside host tissues and their aptitude to proliferate and differentiate *in vivo* counterpart. Owing to the presence of a smaller number of MSCs present in tissues, it is foremost important to expand them *in vitro* before therapeutic intervention [3]. This can be achieved by manipulation of *in vitro* conditions or by using nanoparticles (NPs) to achieve an adequate number of MSCs that facilitate them as excellent candidates for forthcoming implementations in regenerative medicine, clinical sciences, and scientific experiments. MSCs can be differentiated into a range of cell types like neuronal cells, cardiomyocytes, hepatocytes, osteocytes, chondrocytes, adipocytes, and so on, which are representatives of three primary germ layers under *in vivo* and *in vitro* conditions (Fig. **1**).

Stem cell differentiation to corresponding cells and tissues, when engrafted into the body, can repair and replace deteriorated cells or tissues that occur due to disease or injuries. This is why stem cells show greater potential to treat major diseases in humans.

One of the most complicated aspects of deploying stem cell therapies is figuring out how to foster stem cell differentiation *in vivo* whilst witnessing transplanted stem cells or comprehending stem cell differentiation. In this scenario, different nanoparticles (NPs), such as inorganic, organic, and metal NPs, can be used with stem cells for efficient expansion and reliable differentiation to desired cell types. It is documented that various NPs are used as remedial drug therapy for several diseases in the field of biomedicine due to their specific properties, such as controllable size, optical features, magnetic properties, catalytic activities, electrical conductivity, and excellent biocompatible nature [4]. These characteristics of NPs, along with characteristic features of stem cells, create an amalgamated system that is efficiently used to ameliorate specific conditions associated with neuronal, bone-related, and liver diseases.

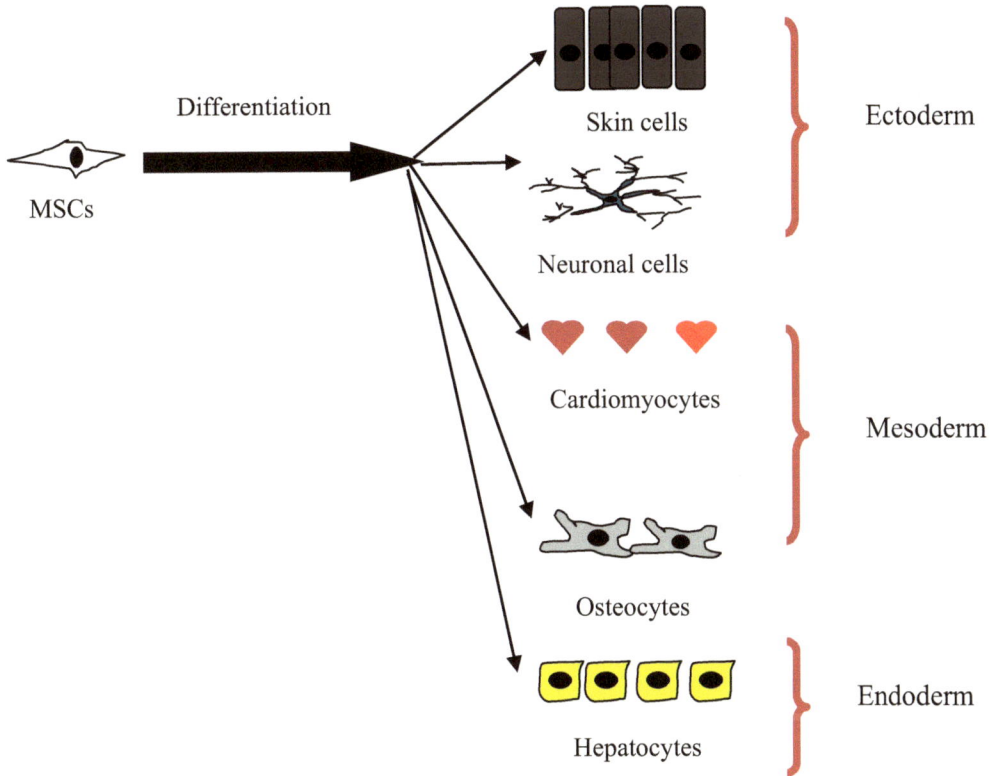

Fig. (1). Differentiation of MSCs into various cell types.

NPs

NPs act as easy transport vehicles that deliver several bioactive and targeted molecules to their desired location. It facilitates the production of most drugs that can be used in multiple regenerative medicines. These are the most supportive molecules that participate in the trapping of small particles that act as active reagents used in tissue engineering [5]. NPs have a significant surface-to-volume ratio since they are regarded as nanomaterials with at least one exterior dimension ranging from 1 nm to 100 nm [6]. Depending on the method of preparation, there are three types of nano-structures available: NPs, nanospheres (NSs), and nanocapsules (NCs). NSs are organized milieu in which the drug is uniformly and physically distributed, whereas NCs are special polymer membrane systems that enclose the drug within a cavity. The NPs exhibit distinct biological, physical, and chemical properties at the nanoscale level, contrary to their size at greater scales. This impact is brought about by a surface area to volume ratio that is significantly higher than average, as well as greater chemical reactivity or stability, improved mechanical strength, *etc.* A wide range of attributes, such as being stronger, lighter, more durable, reactive, sieve-like, or improved electrical conductors, can be productively customized in materials. A well-controlled particle size, silent surface features, and the liberation of pharmacologically dynamic molecules are the major objectives while designing NPs as a delivery vehicle to accomplish drug delivery at target-specific locations with ideal rate and optimal dosage. Biodegradable polymeric NPs coated with hydrophilic polymers, notably polyethylene glycol (PEG), are utilized as efficient machines for drug delivery. They can move for longer periods and have the capacity for organ-specific targeting and the potential to act as DNA carriers in gene therapy for the effective delivery of genes, peptides, and proteins. There are varieties of NPs, such as magnetic NPs and quantum dots, *etc.*, which are utilized for tracking and visualizing stem cells within tissues [7].

Types of NPs

NPs can be categorized into four main types: (1) Inorganic NPs, (2) Carbon-based NPs, (3) Organic NPs, and (4) Composite NPs (Fig. **2**).

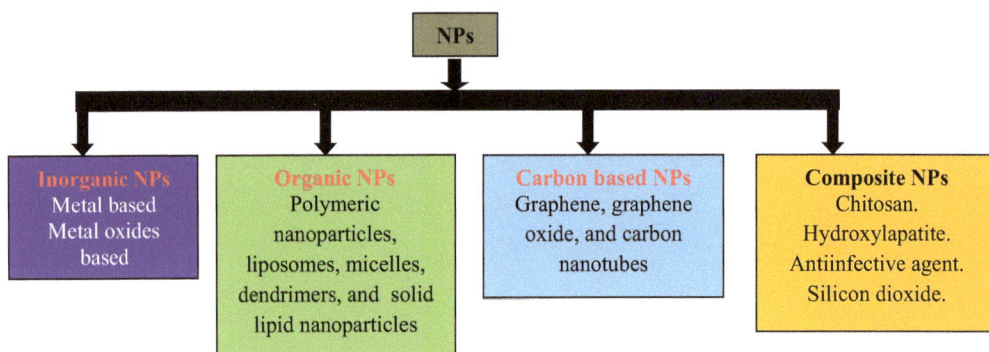

Fig. (2). Types of NPs.

Inorganic NPs

NPs without carbon content are known as inorganic NPs. Inorganic NPs are typically defined as those that are made of metal or metal oxide. Metal oxides, ceramics, and semiconductors, such as silicon and metal NPs that include Au or Ag, can all be manufactured from these NPs.

Metal-based

Metals are synthesized to nanometric scales using constructive or destructive methods to generate metal-based nanoparticles. Mostly, the NPs of the metals are synthetically manufactured. The most commonly utilized metals for the synthesis of NP are zinc (Zn), aluminium (Al), cobalt (Co), gold (Au), lead (Pb), silver (Ag), cadmium (Cd), iron (Fe), and copper (Cu).

Metal oxides based

The synthesis of the metal oxide-based NPs modifies the characteristics of the accompanying metal-based NPs. The most frequently produced materials are iron oxide (Fe_2O_3), magnetite (Fe_3O_4), titanium oxide (TiO_2), aluminium oxide (Al_2O_3), silicon dioxide (SiO_2), zinc Oxide (ZnO), and cerium Oxide (CeO_2). Superparamagnetic iron oxide nanoparticles, also abbreviated as SPIONs, are the versatile section of MRI-based contrast agents.

Organic NPs

The biomedical industry has increasingly used organic nanoparticles (NPs) for targeted systems for drug delivery. The term "organic NPs" or "polymers" refers to substances like ferritin, dendrimers, micelles, and liposomes. Some of the particles, including liposomes and micelles, include hollow cores that resemble nanocapsules, which are susceptible to heat and light along with electromagnetic radiation. These NPs are toxin-free and biodegradable.

Carbon-based NPs

NPs comprised of carbon are the carbon-based NPs. They can be obtained as spheres, ellipsoids, or hollow tubes. Fullerenes (C60), graphene (Gr), carbon nanotubes (CNTs), carbon black (CB), and carbon nanofibers (CNFs) can be allocated under them.

Composite NPs (Nanocomposites)

NPs can be coupled with other NPs, with larger or denser materials, or with more complex structures, like a metalorganic framework, to create composite NPs. The bulk materials for the composites may be made from any type of metal, a semiconducting material, ceramic, polymer, carbon-based, metal, or organic NPs.

Sources of NPs

Classification of NPs sources can be done based on their origin. There are three major sections:

• Incidental NPs are by-products of industrial operations that are created accidentally, such as NPs from forest fire natural processes, combustion processes, and welding gases.

- NPs that are engineered are produced for certain required applications in the industries.
- NPs that are naturally manufactured can be discovered in plants, insects, and human tissues.

Synthesis of NPs

Recent developments in nanotechnology have fabricated different physical, chemical, and biological processes that synthesize different types of NPs (Fig. **3**).

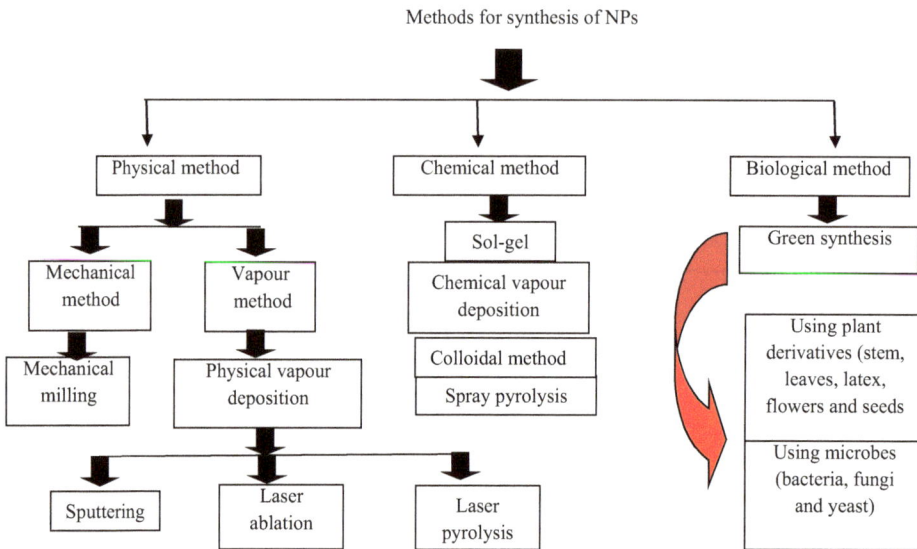

Fig. (**3**). Synthesis methods for NPs.

NPs used in Stem cell-mediated Therapy

NPs and their structural equivalents have been used by humans since the fourth century AD by the Romans, who demonstrated nanotechnology as an advanced technology to be used in medicine [8]. It is demonstrated that NPs assist in overcoming the limitations associated with stem cell therapy. Various NPs, such as nanofiber scaffolds, polymeric NPs, CNTs, nano-engineered scaffolds, etc, are used to induce differentiation and growth, along with *in vivo* and *in vitro* processes of stem cell generation. Nonetheless, these progressive nanotechnological advances can recreate a suitable microenvironment that is comparable to stem cell niches in accordance to the environment and concentration of the cells that support the rearrangement of cells in *in vivo* conditions. Now, the use of nanotechnological approaches can rebuild a suitable microenvironment comparable to niches of stem cells surrounding the separated cells according to the environment as well as the concentration of the cells that support the rearrangement of cells in *in vivo* conditions [9]. The nanofibers and CNT are used as potential nanocarriers present within the stem cells, which imitate the formation of ECM. The different types of stem cells respond differently to ECM. In the course of events, neural stem cell differentiation is facilitated, and MSC proliferation is elevated. *In vitro* isolation, along with *in vivo* proliferation, growth, differentiation, and migration of stem cells, is governed by nanotechnology. This leads to elevated efficiency and the potential for stem cell therapy [10].

Stem Cell and Nanoparticle Amalgamation Facilitates Stem Cell-mediated Therapy

The breakthrough in nanotechnology initiated the establishment of several nanobiomaterials that may facilitate stem cell-associated drug delivery systems, gene therapies, diagnosis, and regenerative purposes (Fig. **4**).

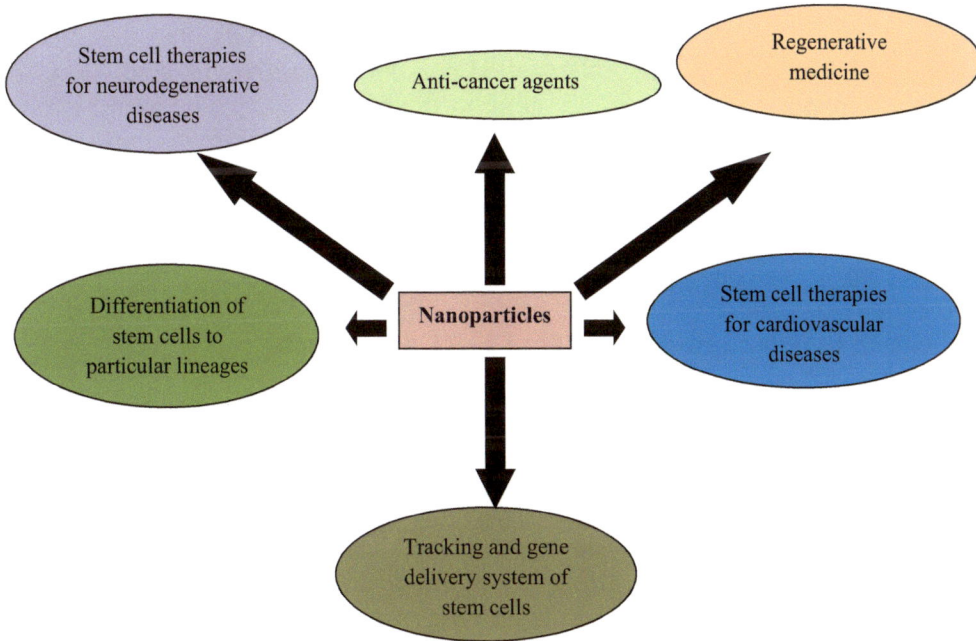

Fig. (4). Nanoarticles in different application of stem cells.

In vivo Tracking of Stem Cells and Gene Delivery and the Role of NPs

The capability of self-renewal, differentiation, and proliferation of stem cells makes them an important therapeutic and diagnostic tool for different untreatable disorders or diseases. Isolation of required stem cells for *in vitro* culture is the primary measure while accounting for stem cell-mediated therapy. The successful implication of gold NPs, magnetic NPs, and quantum dots are used in the segregation of stem cells and *in vivo* tracing of stem cells and their descendants. In an experiment conducted on rat brains, SPION-labelled stem cells are used to examine the migration and *in vivo* localization of neural cells that have been grafted in it [11]. *In vivo* detection of labeled stem cells is done by magnetic resonance imaging (MRI), which is an incredibly precise method. PFC or perfluorocarbon NP-labeled UCEndoSC or human UC-derived endothelial stem cells are precisely detected through an MRI scan. Correspondingly, quantum dots and gold NPs are used for competent labeling and *in vivo* trailing of stem cells. By applying labeling to reveal the embedded cells at the site of action with minimal damage, NPs are widely employed to trace stem cells [12]. In stem cell research, a novel approach that affects *ex vivo* cell differentiation sometimes alters stem cell proliferation and differentiation in the integration of a specific gene into a particular cell type. *In vivo* and *in vitro* gene transfer to stem cells procedures are collaborated by NPs, and NPs have a vital role in the process.

In the procedure of gene transfer to the MSCs, magnetic NPs play a significant role [13, 14]. Moreover, PGLA, also called poly lactic-co-glycolic acid NP, is also used as a transport carrier of SOX9 gene to the hMSCs [15]. The administration of NPs assists in the reprogramming of particular genes, which are attributed to the intended characteristics of the culminating cell, in adult stem cells to generate iPSCs. It is seen that the predation efficiency of iPSCs from human fibroblast-derived stem cells by transfection Oct4, LIN28, Nanog, and Sox2 genes is increased by using polyamidoamine dendrimer-based magnetic NPs [16].

Application of NPs to Regulate Stem Cells Commitment

The primary goal of NPs is to imitate and effectively surpass the effect of biological molecules. Growth factors like fibroblast growth factors (bFGF) can be used during the preparation of scaffold by electrospinning techniques, resulting in faster healing of injury by skin stem cell induction [17]. From certain experimental observations, the use of nanofibers can increase the possibilities of human UC blood stem cell proliferation without inducing differentiation. Fibrin-based nano-scaffolds, when encapsulated with growth factors and delivered at the target site, accelerate the differentiation of stem cells and proliferation where the injury occurs [18]. It is also associated with NGF or nerve growth factor, VEGF or vascular endothelial growth factor, and BMP or bone morphogenetic protein expression vector, causing the stimulation of nerve repair, formation of blood vessels, and bone regeneration, respectively [19]. Nano-scaffolds also mimic the ECM that most often interacts with cellular receptors, which can transform the shape with altered gene expression patterns and, ultimately stem cell fate.

Neuronal marker expression, alignment, and elongation along the grafting axis are reported when iPSCs are incorporated onto silicon substrates made of nanostructures [20]. NPs can imitate the ECM and govern the stem cell fate in a nano environment by controlling stem cells' alignment, survival, motility, proliferation, and differentiation [21]. hMSCs differentiate from the estrogenic lineage in the presence of TiO2 nanotubes [22].

Nanoslits made like square depressions 50 to 100 nm deep are called nanopits. They have a depth of 100 nm and normal width in the range of 100-500 nm [23]. The collection of square and hexagonal nanopits hampered the process of spreading of cells as well as mature focal adhesion development in primary human osteoblastic cells. hMSCs maintain their renewal capacity in a square array of nanopits and regulate their differentiation into osteogenic cells. A common glucocorticoid, namely dexamethasone, is being employed to stimulate the differentiation of MSCs into osteoblasts [24].

Differentiation of primary MSCs into bone occurs when they are cultured on nano pits and raised on nano mounds [25]. Neurons selectively develop from neural progenitors when implanted in a nanofiber matrix with the laminin-specific cell-binding domain (IKVAV) [26]. The differentiation of MSCs to osteoblasts was greatly improved by a scaffold of nanofiber comprising RGD, which can be seen in several ECM proteins. RGD acts as a coupling location for cellular integrins [27].

APPLICATIONS OF DIFFERENT TYPES OF NPS IN REGENERATIVE MEDICINE

Nanofibers

Peptides are used to construct noncovalent nanofiber scaffolds. Hippocampal neurons grow on scaffolds consisting of the RADA16 peptide, which is made of repetitive sequences of the amino acids alanine, lysine, and glutamate. These neurons later generate neurites and synapses.

In addition to being utilized as a framework for RADA16-I nanofibers, brain and spinal cord injuries are also regenerated substantially [28]. The combination of stem cells and NPs yields successful outcomes in regenerative medicine. The cells that are transplanted proceed with the course of differentiation as they develop into oligodendrocytes, astrocytes, and neurons once the combination of Schwann cells and neural progenitor cells are introduced in the RADA16-I scaffold engrafted to the location of the injured spinal cord [29]. Using osteogenic media, embryoid bodies derived from ES cells, which are implanted into RADA16-I, get differentiated into osteocytes [30]. Induction of stem cell homing, cell differentiation, and adhesion are observed when some functional motifs get associated with RADA16-I [31]. The matrix environment of neural stem cells enables them to attach well, survive, and develop into both neuronal and glial phenotypes. Using natural or synthetic biopolymers, long continuous strands of nanofibers can be created by electrospinning technique. A mesh comprised of nanofibers promotes cell migration and growth. MSC-based bone and cartilage tissue engineering is an extensively used scaffold that consists of different types of electrospun nanofibers [32]. Using airflow bubble-spinning, composite nanofibers of PLGA, also called polylactic-co-glycolic acid and MWCNT or multi-walled carbon nanotubes, have been successfully manufactured [33]. MWCNTs are incorporated into PLA to strengthen electrospun nanofibers.

The objective of the use of composite electrospun fibers is to create scaffolds embedded with drugs that can aid the formation of osteoblasts. Furthermore, neural stem cells that have been electrospun into gelatin or poly-L-lactic acid fibers are differentiated into motor neurons using these scaffolds [34].

Nanotubes

Strong electrical conductivity and mechanical attributes are well-known features that distinguish carbon nanotubes, and they are also subject to modification. Carbon nanotubes exhibit significant levels of electrical conductivity and mechanical properties, which make it relatively easier for stem cells to adhere to the nanotubes [35]. When single-walled carbon nanotubes (SWCNTs) or MWCNTs are either used alone or combined with poly-(L-lactic acid) nanofibers, it improves chondrogenic differentiation of MSCs [36]. Studies have demonstrated that the use of hydroxyapatite or chitosan along with MWCNTs predominantly simulates bone tissue niche with admissible pore size, which is responsible for periosteal stem cell proliferation [37, 38].

SWCNTs, in association with polylactic-co-glycolic acid, result in a biocompatible substrate that contributes to the differentiation and proliferation of pre-osteoblasts and consequently adds to the successful expression of mature osteoblast markers [39]. Furthermore, MWCNTs are being used extensively as scaffolds for the differentiation and maturation of brain stem cells from MSCs [40]. According to certain experiments, 2D thin film scaffolds with biocompatible polymer-labelled carbon nanotubes or CNTs facilitate hES cell development into neuronal cells while sustaining higher cell survival. It was found that, in contrast to polyacrylic acid thin films and the traditional poly-L-ornithine surfaces, they are widely used for neural development. Neuron differentiation came about on polyacrylic acid grafted CNT thin films at a substantially higher proportion [41].

SWCNTs and MWCNTs scaffolds are used prevalently for the optimal differentiation and growth of neural stem cells.

The integrated system of thin films SWCNTs and laminin-induced neural stem cells differentiate into an operational excitatory neural network with synaptic connections [42]. CNTs, when conjugated with subventricular zone neural progenitor cells (SVNPCs), proactively repair injured neural tissues in case of brain stroke [43]. CNTs amalgamated polyurethane or poly-L-lactic acid leads to improved generation of myotube from myoblasts.

It is also documented that the differentiation process of MSCs into cardiomyocytes was enhanced when they were associated with modified CNTs in the presence of electrical impulse stimulation [36]. Biocompatible CNTs usage was found to play a vital role in the repair of injured nerve and muscle tissues [44] (Boni. *et al.*, 2018).

Nanoscaffolds

The generation of cardiomyocytes from ES cells may be induced in combination or alone by nanofibers consisting of polycaprolactone (PCL), biodegradable polyurethane, and polyethylene glycol [25, 45]. When ES cells are cultured in the presence of retinoic acid on electrospun PCL, remarkable differentiation is observed in neuronal lineages [46]. It was observed that fibrous scaffolds of electrospun polyurethane were converted to neurons involving dopamine as a neurotransmitter when ES cells were cultured without growth factors or hormones [47]. ES cells seeded on PCL scaffolds cultured in the presence of retinoic acid, insulin, and triiodothyronine (T3) were significantly differentiated into adipocytes [48].

METALLIC NPS USED IN STEM CELL THERAPY

Diagnosis

It has been found that the sensitivity, precision, and rate of diagnosis of diseases are improved by using nanomaterials. NPs are regarded as efficient labeling materials of cells [49]. Metals like gold (Au), silver (Ag), and platinum (Pt) are noble elements that have been popularly used as nanosensors. AgNPs are most preferred over AuNPs due to their affordable cost and employment in the development of nanocomposite-based biosensors [50]. Metallic NPs may be considered useful instruments for immunoassay-based fast protein analysis [51]. The NPs prepared from Au, Ag, Pt, palladium (Pd), copper (Cu), iron (Fe), and cobalt (Co) contain metal oxide (ZnO, SnO_2, TiO_2, and MnO_2) and demonstrate excellent electronic, optical, chemical, magnetic, mechanical, and catalytic properties. The performance of biosensing metallic NPs is greatly increased by coating their surfaces with various matrices, such as metal oxides, silica networks, polymers, graphene, fibers, and dendrimers [52]. Cancer like lymphocytic leukemia and breast cancer can be easily detected by using metallic NPs [53]. The photochemical stability of metallic NPs makes them more versatile for modifying their emission spectrum.

Devices based upon NPs imaging techniques are often utilized for *in vivo* imaging and the diagnosis of complex diseases. Due to their substantial scattering and absorbance capacity, metallic NPs are employed as contrast materials for several imaging machineries such as MRI, surface-enhanced Raman scattering (SERS), photoacoustic imaging (PAI), and computed tomography (CT). Pancreatic cancer cells are labeled using superparamagnetic iron oxide (SPIO). Paramagnetic AuNPs based on gadolinium coated with sugar are utilized to track the cancer cells [54].

Drug Delivery

NPs are considered efficient drug delivery systems for efficient and continuous drug release, negligible toxicity, and successful cross-host immunological responses [55]. One of the most important methods for changing the carrier interface occurs when PEG is associated with the NP's surface. It is well documented that by creating an obstacle between the NPs and plasma proteins, this conjugation reduced the phagocytosis and opsonization of NPs [56]. AuNP serves as a literal depiction of a metallic NP, demonstrating excellent effectiveness in the delivery of medicinal molecules, recombinant proteins, and nucleic acids to a target location [57].

Stem Cells and NPs Amalgamated Systems as Therapeutic Agents

The nanoengineered MSCs carrying the anticancer drug paclitaxel (PTX) are reserved inside the lung tumors after they migrate to the tumor site. The nanoengineered MSCs are created by incorporating small molecules of PTX in MSCs [58]. Polymeric nanoparticles formulated with a biodegradable and biocompatible polymer, poly(DL-lactide-co-glycolide) (PLGA), were used to load PTX in MSCs. This led to predominantly improved anti-cancer activity at a lower dose of PTX drug, ultimately inhibiting tumor growth and enhancing the survivability of patients [59]. Glioblastoma (GBM) is a malignant brain tumor with a poor diagnosis record. Evidence suggests that polymeric nanoparticles (PNPs) are regarded as the most suitable delivery vehicle of chemotherapeutic agents to the brain. PNPs consist of poly(lactide) (PLA), poly(lactide-c--glycolide) (PLGA) copolymers, poly (ε-caprolactone) (PCL), and poly(amino acids), and also some natural polymers like alginate, chitosan, gelatin, and albumin [60 - 62]. The therapeutic agents constructed either by chemical conjugation or encapsulation within the PNPs provide an improved drug release outline [63, 64].

Potential Therapeutic Outcomes of Stem Cells and NPs Amalgamated Systems

Several studies have found that stem cells and NPs amalgamated systems are exceptionally effective against certain lethal diseases. A few of the investigations are briefly discussed below.

SCs and NPs Amalgamated System Therapy for Neurodegenerative Diseases

It is demonstrated that the injection of MSCs loaded with paclitaxel (Ptx)-encapsulated with poly(d,l-lactide-co-glycolide) (PLGA) NPs is considered an effective treatment for glioma [65]. The cascade of neurogenesis follows a complex mechanism that leads to the formation of neurons, which configure the central nervous system (CNS). When stem cells combine with NPs, they become a good candidate for neurogenesis [66]. A recent study reported that artificial polymer NPs combined with stem cells enhance neural stem cell differentiation [67]. Yet another study reported loading dopamine onto PLGA NPs and then administering it into a rat model of Parkinson's disease, improving the trafficking of drugs crossing the blood-brain barrier (BBB) with sustained release of dopamine into brain lesions; improvements in neurobehavioral function were observed [68]. Alzheimer's disease (AD) is a complex neurodegenerative disease. The development of iron oxide NPs integrated with Wharton's jelly-derived MSCs shows elevated therapeutic efficacy in treating AD [69].

SCs and NPs Amalgamated System Therapy for Chondrocyte Disorders

It was observed that when PLGA NPs coated with SOX-9 plasmid DNA and antiCbfa1 siRNA were supplemented to hMSCs before transplantation into nude mice, the expression of genes is effectively augmented that support differentiation and promote chondrogenesis, acting as an effective tool for chondrocyte disorders [70, 71].

SCs and NPs Amalgamated System Therapy for Cardiovascular Diseases

Recent evidence suggests that Au, iron oxide, and Ag NPs are used to treat cardiovascular diseases. The iron oxide NPs facilitate recognizing and identifying atherosclerotic plaques. In addition, these NPs enhance stem cell delivery at the damage sites, thereby promoting the regenerative ability of cells. Further, the antihypertrophic and antioxidative properties of AuNPs are helpful in the detection of plaques and the identification of inflammatory markers [72]. It was confirmed that the bone marrow-derived MSCs in the presence of Ag-NPs show an increase in cardiomyogenic differentiation, which was detected by cardiac markers [73]. So, an amalgamated system of stem cells and NPs ameliorates many cardiovascular diseases. This may be due to NPs' enhanced retention of stem cells for longer by providing a precise extracellular matrix (ECM) niche for cardiac differentiation signal pathways [74]. NPs facilitate the proangiogenic effect of stem cells.

SCs and NPs Amalgamated System Therapy for Hepatic Diseases

Liver fibrosis is a chief cause of impairment and death globally due to persistent worsening of the liver, resulting in liver cancer, cirrhosis, and liver failure. *In vivo,* the recognition and tracking of engrafted stem cells by using NPs in the target tissues of a treated patient will become a vital tool for the efficient delivery of cells and evaluation of optimal doses of cells. SPION can endure, proliferate, and differentiate MSCs into hepatocytes, facilitating liver repair [75].

CONCLUDING REMARKS

Taking stem cells and NPs amalgamated systems, incredible progress has been made in current medicine in comparison to conventional therapeutic approaches. NPs influence stem cell functional activity like self-renewal and differentiation by regulating or modifying the signaling pathways. NPs and stem cell, in combination, might have an important role in human treatments, thereby meeting the challenges that occur during the development of personalized medicine. Future research may be in the direction of developing cost-effective, biocompatible NPs that show the sustainability and efficacy of stem cells to improve the condition of patients suffering from different chronic diseases.

ACKNOWLEDGEMENTS

The authors are grateful to Dr. Cuckoo Mahapatra. Assistant Professor of Zoology, Maharaja Sriram Chandra Bhanja Deo University, Baripada, Odisha, India, for her support and encouragement during the preparation of the manuscript.

REFERENCES

[1]   Zakrzewski W, Dobrzyński M, Szymonowicz M, Rybak Z. Stem cells: past, present, and future. Stem Cell Res Ther 2019; 10(1): 68.
[http://dx.doi.org/10.1186/s13287-019-1165-5] [PMID: 30808416]

[2]   Pittenger MF, Mackay AM, Beck SC, *et al.* Multilineage potential of adult human mesenchymal stem cells. Science. 1999 Apr 2; 284(5411): 143-7.
[http://dx.doi.org/10.1126/science.284.5411.143]

[3]   Thirumala S, Goebel WS, Woods EJ. Clinical grade adult stem cell banking. Organogenesis 2009;
        5(3): 143-54.
        [http://dx.doi.org/10.4161/org.5.3.9811] [PMID: 20046678]

[4]   Anselmo AC, Mitragotri S. A review of clinical translation of inorganic nanoparticles. AAPS J 2015;
        17(5): 1041-54.
        [http://dx.doi.org/10.1208/s12248-015-9780-2] [PMID: 25956384]

[5]   Arora P, Sindhu A, Dilbaghi N, Chaudhury A, Rajakumar G, Rahuman AA. Nano-regenerative
        medicine towards clinical outcome of stem cell and tissue engineering in humans. J Cell Mol Med
        2012; 16(9): 1991-2000.
        [http://dx.doi.org/10.1111/j.1582-4934.2012.01534.x] [PMID: 22260258]

[6]   European Commission. Commission Recommendation of 18 October 2011 on the definition of
        nanomaterial 2011/696/EU. Off. J Eur Union 2011; 275: 38-40.

[7]   Wang Z, Ruan J, Cui D. Advances and prospect of nanotechnology in stem cells. Nanoscale Res Lett
        2009; 4(7): 593-605.
        [http://dx.doi.org/10.1007/s11671-009-9292-z] [PMID: 20596412]

[8]   Jeevanandam J, Barhoum A, Chan YS, Dufresne A, Danquah MK. Review on nanoparticles and
        nanostructured materials: history, sources, toxicity and regulations. Beilstein J Nanotechnol 2018;
        9(1): 1050-74.
        [http://dx.doi.org/10.3762/bjnano.9.98] [PMID: 29719757]

[9]   Kaur S, Singhal B. When nano meets stem: The impact of nanotechnology in stem cell biology. J
        Biosci Bioeng 2012; 113(1): 1-4.
        [http://dx.doi.org/10.1016/j.jbiosc.2011.08.024] [PMID: 21956156]

[10]   Hashi CK, Zhu Y, Yang GY, *et al.* Antithrombogenic property of bone marrow mesenchymal stem
        cells in nanofibrous vascular grafts. Proc Natl Acad Sci USA 2007; 104(29): 11915-20.
        [http://dx.doi.org/10.1073/pnas.0704581104] [PMID: 17615237]

[11]   Alvarim LT, Nucci LP, Mamani JB, *et al.* Therapeutics with SPION-labeled stem cells for the main
        diseases related to brain aging: a systematic review. Int J Nanomedicine 2014; 9: 3749-70.
        [PMID: 25143726]

[12]   Narayanan K, Mishra S, Singh S, Pei M, Gulyas B, Padmanabhan P. Engineering concepts in stem cell
        research. Biotechnol J 2017; 12(12): 1700066.
        [http://dx.doi.org/10.1002/biot.201700066] [PMID: 28901712]

[13]   Pickard MR, Barraud P, Chari DM. The transfection of multipotent neural precursor/stem cell
        transplant populations with magnetic nanoparticles. Biomaterials 2011; 32(9): 2274-84.
        [http://dx.doi.org/10.1016/j.biomaterials.2010.12.007] [PMID: 21193228]

[14]   Chen D, Tang Q, Xue W, Wang X. The feasibility of using magnetic nanoparticles modified as gene
        vector. West Indian Med J 2010; 59(3): 300-5.
        [PMID: 21291111]

[15]   Kim JH, Park JS, Yang HN, *et al.* The use of biodegradable PLGA nanoparticles to mediate SOX9
        gene delivery in human mesenchymal stem cells (hMSCs) and induce chondrogenesis. Biomaterials
        2011; 32(1): 268-78.
        [http://dx.doi.org/10.1016/j.biomaterials.2010.08.086] [PMID: 20875683]

[16]   Ruan J, Shen J, Wang Z, *et al.* Efficient preparation and labeling of human induced pluripotent stem
        cells by nanotechnology. Int J Nanomedicine 2011; 6: 425-35.
        [http://dx.doi.org/10.2147/IJN.S16498] [PMID: 21499432]

[17]   Nosrati H, Aramideh Khouy R, Nosrati A, *et al.* Nanocomposite scaffolds for accelerating chronic
        wound healing by enhancing angiogenesis. J Nanobiotechnology 2021; 19(1): 1-21.
        [http://dx.doi.org/10.1186/s12951-020-00755-7] [PMID: 33397416]

[18]   Rajangam T, An SS. Fibrinogen and fibrin based micro and nano scaffolds incorporated with drugs, proteins, cells and genes for therapeutic biomedical applications. Int J Nanomedicine 2013; 8: 3641-62.
[PMID: 24106425]

[19]   Aswin SK, Jothishwar S, Nayagam PVC, Priya G. Scaffolds for biomolecule delivery and controlled release-A Review. Research Journal of Pharmacy and Technology 2018; 11(10): 4719-30.
[http://dx.doi.org/10.5958/0974-360X.2018.00861.2]

[20]   Pan F, Zhang M, Wu G, *et al.* Topographic effect on human induced pluripotent stem cells differentiation towards neuronal lineage. Biomaterials 2013; 34(33): 8131-9.
[http://dx.doi.org/10.1016/j.biomaterials.2013.07.025] [PMID: 23891397]

[21]   Griffin MF, Butler PE, Seifalian AM, Kalaskar DM. Control of stem cell fate by engineering their micro and nanoenvironment. World J Stem Cells 2015; 7(1): 37-50.
[http://dx.doi.org/10.4252/wjsc.v7.i1.37] [PMID: 25621104]

[22]   Kong K, Chang Y, Hu Y, *et al.* TiO2 nanotubes promote osteogenic differentiation through regulation of Yap and Piezo1. Front Bioeng Biotechnol 2022; 10: 872088.
[http://dx.doi.org/10.3389/fbioe.2022.872088] [PMID: 35464728]

[23]   Reisner W, Larsen NB, Flyvbjerg H, Tegenfeldt JO, Kristensen A. Directed self-organization of single DNA molecules in a nanoslit *via* embedded nanopit arrays. Proc Natl Acad Sci USA 2009; 106(1): 79-84.
[http://dx.doi.org/10.1073/pnas.0811468106] [PMID: 19122138]

[24]   Ito S, Suzuki N, Kato S, Takahashi T, Takagi M. Glucocorticoids induce the differentiation of a mesenchymal progenitor cell line, ROB-C26 into adipocytes and osteoblasts, but fail to induce terminal osteoblast differentiation. Bone 2007; 40(1): 84-92.
[http://dx.doi.org/10.1016/j.bone.2006.07.012] [PMID: 16949358]

[25]   Hofmann MC. Stem cells and nanomaterials. Nanomaterial: Impacts on Cell Biology and Medicine. 2014:255-75.
[http://dx.doi.org/10.1007/978-94-017-8739-0_13]

[26]   Silva GA, Czeisler C, Niece KL, *et al.* Selective differentiation of neural progenitor cells by high-epitope density nanofibers. Science 2004; 303(5662): 1352-5.
[http://dx.doi.org/10.1126/science.1093783] [PMID: 14739465]

[27]   Hosseinkhani H, Hosseinkhani M, Tian F, Kobayashi H, Tabata Y. Osteogenic differentiation of mesenchymal stem cells in self-assembled peptide-amphiphile nanofibers. Biomaterials 2006; 27(22): 4079-86.
[http://dx.doi.org/10.1016/j.biomaterials.2006.03.030] [PMID: 16600365]

[28]   Zhang S. Discovery and design of self-assembling peptides. Interface Focus 2017; 7(6): 20170028.
[http://dx.doi.org/10.1098/rsfs.2017.0028] [PMID: 29147558]

[29]   Guo J, Su H, Zeng Y, *et al.* Reknitting the injured spinal cord by self-assembling peptide nanofiber scaffold. Nanomedicine 2007; 3(4): 311-21.
[http://dx.doi.org/10.1016/j.nano.2007.09.003] [PMID: 17964861]

[30]   Sutha K, Schwartz Z, Wang Y, Hyzy S, Boyan BD, McDevitt TC. Osteogenic embryoid body-derived material induces bone formation in vivo. Sci Rep 2015; 5(1): 9960.
[http://dx.doi.org/10.1038/srep09960] [PMID: 25961152]

[31]   Horii A, Wang X, Gelain F, Zhang S. Biological designer self-assembling peptide nanofiber scaffolds significantly enhance osteoblast proliferation, differentiation and 3-D migration. PLoS One 2007; 2(2): e190.
[http://dx.doi.org/10.1371/journal.pone.0000190] [PMID: 17285144]

[32]   Owida HA, Al-Nabulsi JI, Alnaimat F, *et al.* Recent applications of electrospun nanofibrous scaffold in tissue engineering. Appl Bionics Biomech 2022; 2022: 1-15.

[http://dx.doi.org/10.1155/2022/1953861] [PMID: 35186119]

[33]   Fang Y, Liu F, Xu L, Wang P, He J. Preparation of PLGA/MWCNT composite nanofibers by airflow bubble-spinning and their characterization. Polymers (Basel) 2018; 10(5): 481.
[http://dx.doi.org/10.3390/polym10050481] [PMID: 30966515]

[34]   Binan L, Tendey C, De Crescenzo G, El Ayoubi R, Ajji A, Jolicoeur M. Differentiation of neuronal stem cells into motor neurons using electrospun poly-l-lactic acid/gelatin scaffold. Biomaterials 2014; 35(2): 664-74.
[http://dx.doi.org/10.1016/j.biomaterials.2013.09.097] [PMID: 24161168]

[35]   Stout DA, Webster TJ. Carbon nanotubes for stem cell control. Mater Today 2012; 15(7-8): 312-8.
[http://dx.doi.org/10.1016/S1369-7021(12)70136-0]

[36]   Mooney E, Dockery P, Greiser U, Murphy M, Barron V. Carbon nanotubes and mesenchymal stem cells: biocompatibility, proliferation and differentiation. Nano Lett 2008; 8(8): 2137-43.
[http://dx.doi.org/10.1021/nl073300o] [PMID: 18624387]

[37]   Zhang C, Hu YY, Cui FZ, Zhang SM, Ruan DK. A study on a tissue-engineered bone using rhBMP-2 induced periosteal cells with a porous nano-hydroxyapatite/collagen/poly(L-lactic acid) scaffold. Biomed Mater 2006; 1(2): 56-62.
[http://dx.doi.org/10.1088/1748-6041/1/2/002] [PMID: 18460757]

[38]   Turnbull G, Clarke J, Picard F, *et al.* 3D bioactive composite scaffolds for bone tissue engineering. Bioact Mater 2018; 3(3): 278-314.
[http://dx.doi.org/10.1016/j.bioactmat.2017.10.001] [PMID: 29744467]

[39]   Gupta A, Woods MD, Illingworth KD, *et al.* Single walled carbon nanotube composites for bone tissue engineering. J Orthop Res 2013; 31(9): 1374-81.
[http://dx.doi.org/10.1002/jor.22379] [PMID: 23629922]

[40]   Chen YS, Hsiue GH. Directing neural differentiation of mesenchymal stem cells by carboxylated multiwalled carbon nanotubes. Biomaterials 2013; 34(21): 4936-44.
[http://dx.doi.org/10.1016/j.biomaterials.2013.03.063] [PMID: 23578561]

[41]   Chao TI, Xiang S, Chen CS, *et al.* Carbon nanotubes promote neuron differentiation from human embryonic stem cells. Biochem Biophys Res Commun 2009; 384(4): 426-30.
[http://dx.doi.org/10.1016/j.bbrc.2009.04.157] [PMID: 19426708]

[42]   Kam NWS, Jan E, Kotov NA. Electrical stimulation of neural stem cells mediated by humanized carbon nanotube composite made with extracellular matrix protein. Nano Lett 2009; 9(1): 273-8.
[http://dx.doi.org/10.1021/nl802859a] [PMID: 19105649]

[43]   Moon SU, Kim J, Bokara KK, *et al.* Carbon nanotubes impregnated with subventricular zone neural progenitor cells promotes recovery from stroke. Int J Nanomedicine 2012; 7: 2751-65.
[PMID: 22701320]

[44]   Boni R, Ali A, Shavandi A, Clarkson AN. Current and novel polymeric biomaterials for neural tissue engineering. J Biomed Sci 2018; 25(1): 90.
[http://dx.doi.org/10.1186/s12929-018-0491-8] [PMID: 30572957]

[45]   Llorens E, Armelin E, Del Mar Pérez-Madrigal M, Del Valle L, Alemán C, Puiggalí J. Nanomembranes and nanofibers from biodegradable conducting polymers. Polymers (Basel) 2013; 5(3): 1115-57.
[http://dx.doi.org/10.3390/polym5031115]

[46]   Xie J, Willerth SM, Li X, *et al.* The differentiation of embryonic stem cells seeded on electrospun nanofibers into neural lineages. Biomaterials 2009; 30(3): 354-62.
[http://dx.doi.org/10.1016/j.biomaterials.2008.09.046] [PMID: 18930315]

[47]   Carlberg B, Axell MZ, Nannmark U, Liu J, Kuhn HG. Electrospun polyurethane scaffolds for proliferation and neuronal differentiation of human embryonic stem cells. Biomed Mater 2009; 4(4): 045004.

[http://dx.doi.org/10.1088/1748-6041/4/4/045004] [PMID: 19567936]

[48]   Kang X, Xie Y, Powell HM, *et al.* Adipogenesis of murine embryonic stem cells in a three-dimensional culture system using electrospun polymer scaffolds. Biomaterials 2007; 28(3): 450-8.
[http://dx.doi.org/10.1016/j.biomaterials.2006.08.052] [PMID: 16997371]

[49]   Bhirde A, Xie J, Swierczewska M, Chen X. Nanoparticles for cell labeling. Nanoscale 2011; 3(1): 142-53.
[http://dx.doi.org/10.1039/C0NR00493F] [PMID: 20938522]

[50]   Malekzad H, Sahandi Zangabad P, Mirshekari H, Karimi M, Hamblin MR. Noble metal nanoparticles in biosensors: recent studies and applications. Nanotechnol Rev 2017; 6(3): 301-29.
[http://dx.doi.org/10.1515/ntrev-2016-0014] [PMID: 29335674]

[51]   Liu L, Hao Y, Deng D, Xia N. Nanomaterials-based colorimetric immunoassays. Nanomaterials (Basel) 2019; 9(3): 316.
[http://dx.doi.org/10.3390/nano9030316] [PMID: 30818816]

[52]   Naresh V, Lee N. A Review on Biosensors and Recent Development of Nanostructured Materials-Enabled Biosensors. Sensors (Basel) 2021; 21(4): 1109.
[http://dx.doi.org/10.3390/s21041109] [PMID: 33562639]

[53]   Chinen AB, Guan CM, Ferrer JR, Barnaby SN, Merkel TJ, Mirkin CA. Nanoparticle probes for the detection of cancer biomarkers, cells, and tissues by fluorescence. Chem Rev 2015; 115(19): 10530-74.
[http://dx.doi.org/10.1021/acs.chemrev.5b00321] [PMID: 26313138]

[54]   Abdal Dayem A, Lee SB, Cho SG. The impact of metallic nanoparticles on stem cell proliferation and differentiation. Nanomaterials (Basel) 2018; 8(10): 761.
[http://dx.doi.org/10.3390/nano8100761] [PMID: 30261637]

[55]   Patra JK, Das G, Fraceto LF, *et al.* Nano based drug delivery systems: recent developments and future prospects. J Nanobiotechnology 2018; 16(1): 71.
[http://dx.doi.org/10.1186/s12951-018-0392-8] [PMID: 30231877]

[56]   Shi L, Zhang J, Zhao M, *et al.* Effects of polyethylene glycol on the surface of nanoparticles for targeted drug delivery. Nanoscale 2021; 13(24): 10748-64.
[http://dx.doi.org/10.1039/D1NR02065J] [PMID: 34132312]

[57]   Jeong EH, Jung G, Hong CA, Lee H. Gold nanoparticle (AuNP)-based drug delivery and molecular imaging for biomedical applications. Arch Pharm Res 2014; 37(1): 53-9.
[http://dx.doi.org/10.1007/s12272-013-0273-5] [PMID: 24214174]

[58]   Sadhukha T, O'Brien TD, Prabha S. Nano-engineered mesenchymal stem cells as targeted therapeutic carriers. J Control Release 2014; 196: 243-51.
[http://dx.doi.org/10.1016/j.jconrel.2014.10.015] [PMID: 25456830]

[59]   Layek B, Sadhukha T, Panyam J, Prabha S. Nano-engineered mesenchymal stem cells increase therapeutic efficacy of anticancer drug through true active tumor targeting. Mol Cancer Ther 2018; 17(6): 1196-206.
[http://dx.doi.org/10.1158/1535-7163.MCT-17-0682] [PMID: 29592881]

[60]   Olivier JC. Drug transport to brain with targeted nanoparticles. NeuroRx 2005; 2(1): 108-19.
[http://dx.doi.org/10.1602/neurorx.2.1.108] [PMID: 15717062]

[61]   Tosi G, Costantino L, Ruozi B, Forni F, Vandelli MA. Polymeric nanoparticles for the drug delivery to the central nervous system. Expert Opin Drug Deliv 2008; 5(2): 155-74.
[http://dx.doi.org/10.1517/17425247.5.2.155] [PMID: 18248316]

[62]   Fazil M, Md S, Haque S, *et al.* Development and evaluation of rivastigmine loaded chitosan nanoparticles for brain targeting. Eur J Pharm Sci 2012; 47(1): 6-15.
[http://dx.doi.org/10.1016/j.ejps.2012.04.013] [PMID: 22561106]

[63]   Fazil M, Md S, Haque S, *et al.* Development and evaluation of rivastigmine loaded chitosan nanoparticles for brain targeting. Eur J Pharm Sci 2012; 47(1): 6-15.
[http://dx.doi.org/10.1016/j.ejps.2012.04.013] [PMID: 22561106]

[64]   Huang HT, Seo HS, Zhang T, *et al.* MELK is not necessary for the proliferation of basal-like breast cancer cells. eLife 2017; 6: e26693.
[http://dx.doi.org/10.7554/eLife.26693] [PMID: 28926338]

[65]   Wang X, Gao JQ, Ouyang X, Wang J, Sun X, Lv Y. Mesenchymal stem cells loaded with paclitaxel–poly(lactic-*co*-glycolic acid) nanoparticles for glioma-targeting therapy. Int J Nanomedicine 2018; 13: 5231-48.
[http://dx.doi.org/10.2147/IJN.S167142] [PMID: 30237710]

[66]   In Choi J, Tae Cho H, Ki Jee M, Kyung Kang S. Core-shell nanoparticle controlled hATSCs neurogenesis for neuropathic pain therapy. Biomaterials 2013; 34(21): 4956-70.
[http://dx.doi.org/10.1016/j.biomaterials.2013.02.037] [PMID: 23582861]

[67]   Zhang B, Yan W, Zhu Y, *et al.* Nanomaterials in neural-stem-cell-mediated regenerative medicine: Imaging and treatment of neurological diseases. Adv Mater 2018; 30(17): 1705694.
[http://dx.doi.org/10.1002/adma.201705694] [PMID: 29543350]

[68]   Pahuja R, Seth K, Shukla A, *et al.* Trans-blood brain barrier delivery of dopamine-loaded nanoparticles reverses functional deficits in parkinsonian rats. ACS Nano 2015; 9(5): 4850-71.
[http://dx.doi.org/10.1021/nn506408v] [PMID: 25825926]

[69]   Jung M, Kim H, Hwang JW, *et al.* Iron oxide nanoparticle-incorporated mesenchymal stem cells for Alzheimer's disease treatment. Nano Lett 2023; 23(2): 476-90.
[http://dx.doi.org/10.1021/acs.nanolett.2c03682] [PMID: 36638236]

[70]   Jeon SY, Park JS, Yang HN, Woo DG, Park KH. Co-delivery of SOX9 genes and anti-Cbfa-1 siRNA coated onto PLGA nanoparticles for chondrogenesis of human MSCs. Biomaterials 2012; 33(17): 4413-23.
[http://dx.doi.org/10.1016/j.biomaterials.2012.02.051] [PMID: 22425025]

[71]   André EM, Passirani C, Seijo B, Sanchez A, Montero-Menei CN. Nano and microcarriers to improve stem cell behaviour for neuroregenerative medicine strategies: Application to Huntington's disease. Biomaterials 2016; 83: 347-62.
[http://dx.doi.org/10.1016/j.biomaterials.2015.12.008] [PMID: 26802487]

[72]   Zhu K, Li J, Wang Y, Lai H, Wang C. Nanoparticles-assisted stem cell therapy for ischemic heart disease. Stem Cells Int 2016; 2016(1): 1384658.
[http://dx.doi.org/10.1155/2016/1384658] [PMID: 26839552]

[73]   Adibkia K, Ehsani A, Jodaei A, Fathi E, Farahzadi R, Barzegar-Jalali M. Silver nanoparticles induce the cardiomyogenic differentiation of bone marrow derived mesenchymal stem cells *via* telomere length extension. Beilstein J Nanotechnol 2021; 12(1): 786-97.
[http://dx.doi.org/10.3762/bjnano.12.62] [PMID: 34395152]

[74]   Bejleri D, Streeter BW, Nachlas ALY, *et al.* A printed cardiac patch composed of cardiac-specific extracellular matrix and progenitor cells for heart repair. Adv Healthc Mater 2018; 7(23): 1800672.
[http://dx.doi.org/10.1002/adhm.201800672] [PMID: 30379414]

[75]   Fahmy HM, Abd El-Daim TM, Eid HA. El qassem Mahmoud EA, Abdallah EA, Hassan FE, Maihop DI, Amin AE, Mustafa AB, Hassan FM, Mohamed DM. Multifunctional nanoparticles in stem cell therapy for cellular treatment of kidney and liver diseases. Tissue Cell 2020; 65: 101371.
[http://dx.doi.org/10.1016/j.tice.2020.101371] [PMID: 32746989]

CHAPTER 7

The Therapeutic Potential of Nanoparticles in Healthcare

Twinkle Rout[1], Soumyashree Rout[2], Bhangyashree Nanda[3] and Arun Kumar Pradhan[4,*]

[1] *Department of Surgical Oncology, Siksha 'O' Anusandhan (Deemed to be University), Bhubaneswar, Odisha 751030, India*

[2] *Department of Neurology, Siksha 'O' Anusandhan (Deemed to be University), Bhubaneswar, Odisha 751030, India*

[3] *Department of Psychiatry, Siksha 'O' Anusandhan (Deemed to be University), Bhubaneswar, Odisha 751030, India*

[4] *Centre for Biotechnology, Siksha 'O' Anusandhan (Deemed to be University), Bhubaneswar, Odisha 751030, India*

Abstract: Healthcare has undergone an important change as a result of nanotechnology, which promises historically unseen possibilities for therapeutic, preventative, and diagnostic nanointerventions An overview of the therapeutic potential of nanoparticles (NPs) in healthcare is given in this abstract, with particular attention on how they may be used in targeted treatment, imaging, and drug delivery. Because of their special physicochemical characteristics, nanoparticles are a great option to get beyond conventional medication delivery restrictions. Their control over drug release kinetics is made possible by their customizable size, surface chemistry, and biocompatibility, which maximizes therapeutic efficacy and reduces adverse effects. Additionally, because NPs may cross biological barriers like the blood-brain barrier, therapeutic medicines can be delivered to certain tissues or cells with precision. By enhancing solubility, preventing drug degradation, and promoting prolonged release, these nanocarriers maximize therapeutic effects. Additionally essential to medical imaging, nanoparticles offer improved contrast for diagnostic imaging. This improves monitoring, early identification, and individualized treatment plans. Targeted therapy has also been transformed by the development of nanotherapeutics, which enable controlled release and site-specific drug delivery in diseased tissues. By using ligands or antibodies on NP surfaces, active targeting technologies allow for the selective identification and binding to certain cell receptors, which enhances medication absorption by sick cells while reducing off-target effects. Notwithstanding these encouraging developments, issues with scalability, toxicity, and biocompatibility still need to be resolved before nanotherapeutics may be widely and safely used in clinical settings. This abstract underscores the transformative potential of nanoparticles in revo-

* **Corresponding author Arun Kumar Pradhan:** Centre for Biotechnology, Siksha 'O' Anusandhan (Deemed to be University), Bhubaneswar, Odisha 751003, India; E-mail: arunpradhan@soa.ac.in

lutionizing healthcare, paving the way for personalized and more effective therapeutic interventions across a spectrum of diseases.

Keywords: Cancer therapy, Drug delivery, Liposome, Magnetic nanoparticles, Vaccine.

INTRODUCTION

Nanoparticles, materials with dimensions ranging from 1 to 100 nanometers, have emerged as promising entities in the realm of therapeutic applications due to their unique physical and chemical properties. The field of nanotechnology has provided a platform for the design and engineering of nanoparticles with tailored characteristics, enabling targeted and controlled delivery of therapeutic agents [1]. Nanoparticles in therapeutic strategies represent a paradigm shift in drug delivery and treatment modalities. These minuscule structures offer several advantages, including increased surface area, tunable reactivity, and the ability to navigate biological barriers that may impede traditional drug formulations. In the context of therapeutic applications, nanoparticles have demonstrated remarkable potential in drug delivery, diagnostics, imaging, and various other medical interventions [2]. One of the key areas where nanoparticles have shown considerable promise is in drug delivery systems. The ability to encapsulate drugs within nanoparticles facilitates their transport to specific target sites, minimizing systemic side effects and enhancing therapeutic efficacy. Additionally, the unique physicochemical properties of nanoparticles can be exploited to overcome biological barriers, such as the blood-brain barrier, enabling the delivery of therapeutic agents to previously inaccessible areas [3]. In diagnostic imaging, nanoparticles serve as versatile contrast agents, enhancing the visibility of tissues and improving the accuracy of medical imaging techniques. The precise control over the composition and surface characteristics of nanoparticles allows for their customization to meet the specific demands of various imaging modalities, including magnetic resonance imaging (MRI), computed tomography (CT), and ultrasound [3]. Despite the tremendous potential of nanoparticles in therapeutic applications, challenges such as potential toxicity, biocompatibility, and scalability must be addressed [4]. The multifaceted landscape of nanoparticles in therapeutic applications sets the stage for a comprehensive exploration of their role in drug delivery, imaging, and other innovative medical interventions. As research in nanotechnology advances, the potential for transformative breakthroughs in healthcare continues to grow, marking an exciting era in the development of novel therapeutic strategies.

Nanoparticles have gained significant attention in therapeutic applications due to their unique properties at the nanoscale. These particles, typically ranging from 1 to 100 nanometers in size, offer distinct advantages in drug delivery, imaging, and diagnostics. Here are some ways in which nanoparticles are used in therapeutic applications (Fig. **1**):

DRUG DELIVERY

The role of nanoparticles in targeted drug delivery is transformative, offering a paradigm shift in the way we approach the treatment of various diseases. The precision, enhanced pharmacokinetics, and ability to overcome biological barriers make nanoparticles a promising tool for delivering therapeutic agents with unprecedented accuracy and efficacy [5]. Nanoparticles can be designed to carry therapeutic agents, such as drugs or genes, directly to the target site in the body. This minimizes side effects and enhances the therapeutic effect. Targeted drug delivery represents a cornerstone in the application of nanoparticles within the realm of nanomedicine. This high specificity ensures that therapeutic agents are delivered precisely to the intended target, minimizing collateral damage to healthy cells [4, 5]. The size and surface properties of nanoparticles can be tailored to optimize their pharmacokinetics. This allows for prolonged circulation times in the bloodstream, avoiding rapid clearance and improving the overall bioavailability of the drug. As a result, the therapeutic payload can reach the target site in sufficient concentrations, enhancing treatment outcomes. Additionally, certain nanoparticles can traverse cellular barriers, such as the blood-brain barrier, enabling the delivery of drugs to the central nervous system [6]. It can be designed to facilitate controlled drug release, providing a sustained and prolonged therapeutic effect. This is particularly valuable in chronic conditions where maintaining a consistent drug concentration is crucial for optimal treatment. This opens avenues for personalized medicine, where treatment regimens can be tailored to the specific needs of each patient, considering factors such as genetic variations and disease heterogeneity. By delivering drugs directly to the target site, nanoparticles help minimize systemic exposure and, consequently, reduce side effects associated with off-target effects. This targeted approach not only enhances the safety profile of therapeutic agents but also improves the overall tolerability of treatments. As research on nanomedicine advances, the potential for developing innovative and patient-centric drug delivery systems continues to grow, ushering in a new era in healthcare [7].

CANCER THERAPY

Cancer stands as a leading cause of both mortality and morbidity, characterized by a complex pathophysiology. Conventional cancer treatments encompass chemotherapy, radiation therapy, targeted therapy, and immunotherapy; however, challenges like lack of specificity, cytotoxicity, and multi-drug resistance present formidable obstacles to effective cancer management. The emergence of nanotechnology has brought about a revolutionary shift in the landscape of cancer diagnosis and treatment. Nanoparticles, ranging in size from 1 to 100 nm, offer distinct advantages, including biocompatibility, reduced toxicity, enhanced stability, improved permeability and retention effects, and precise targeting capabilities, making them well-suited for cancer treatment. Various categories of nanoparticles exist, with the nanoparticle drug delivery system leveraging the specific characteristics of tumors and their microenvironments. Nanoparticles not only address the limitations of traditional cancer treatments but also surmount challenges related to multidrug resistance. Furthermore, as novel mechanisms of multidrug resistance are unveiled and studied, nanoparticles are becoming the focus of intensified research efforts. The diverse therapeutic applications of nano-formulations are opening up new perspectives for the treatment of cancer [8]. Nanoparticles have revolutionized the landscape of cancer therapy by offering innovative solutions to longstanding challenges in the treatment of this complex and heterogeneous disease. Their unique properties, such as size, surface characteristics, and customizable features, empower nanoparticles to play a crucial role in various aspects of cancer therapy. The utilization of nanoparticles in targeted chemotherapy represents a significant advancement in cancer therapeutics. Their ability to enhance drug delivery specificity, reduce systemic toxicity, and enable combination therapies contributes to more effective and patient-tailored treatments, fostering the evolution of cancer care toward precision medicine. As research in nano-medicine progresses, the potential for further innovations in targeted chemotherapy continues to expand [8, 9]. Nanoparticles can be engineered to selectively accumulate in tumor tissues, delivering chemotherapy drugs directly to cancer cells while minimizing damage to healthy tissues [9, 10]. Some nanoparticles can convert light into heat, allowing for localized hyperthermia to destroy cancer cells [10].

DIAGNOSTIC IMAGING

The continuous demand for early disease detection and diagnosis has been a driving force for advancements in imaging modalities and contrast agents. Despite current challenges in achieving rapid and detailed imaging of tissue microstructures and lesion characterization, the development of non-toxic contrast agents with extended circulation time holds great potential. Nanoparticle technology emerges as a promising solution to meet these needs. This article provides a comprehensive review of nanoparticle-based contrast agents utilized in various biomedical imaging modalities, such as fluorescence imaging, MRI, CT, US, PET, and SPECT [11]. The discussion encompasses their structural features, advantages, and limitations. Additionally, the article explores the applications of these contrast agents in each imaging modality, citing commonly studied examples. Future research directions are anticipated to focus on multifunctional nanoplatforms to address safety, efficacy, and theranostic capabilities. The use of nanoparticles as imaging contrast agents holds the promise of significantly enhancing clinical practices. Nanoparticles can serve as contrast agents in various imaging techniques (*e.g.*, MRI, CT, and ultrasound), enhancing the visibility of tissues and improving diagnostic accuracy. Nanoparticles can be functionalized to detect specific biomarkers, enabling early diagnosis of diseases [11].

VACCINES

Vaccination remains the most cost-effective strategy for combating infectious diseases. Traditional vaccinations often exhibit low immunogenicity and typically offer only partial protection in many cases. A novel category of vaccinations based on nanoparticles displays significant potential in overcoming the limitations associated with conventional and subunit vaccines [12]. Recent advancements in chemical and biological engineering have enabled precise control over nanoparticle size, shape, functionality, and surface properties, leading to enhanced antigen presentation and robust immunogenicity. The characterization of nanovaccines can be accurately achieved through a combination of physicochemical, immunological, and toxicological experiments. Nanoparticles can enhance the immunogenicity of vaccines by improving antigen stability, facilitating antigen uptake by immune cells, and promoting a more robust immune response [12].

GENE THERAPY

Gene therapy is one type of medical approach that includes recombinant DNA technology and gene cloning technology. Gene therapy can modify and displace the disease-causing genes at the molecular level as well as restore or repair various imperfect or non-functioning proteins to cure genetic diseases. According to the Food and Drug Administration (FDA), there are different gene therapy drugs like small interfering RNA (siRNA), microRNA(miRNA), plasmids DNA, an antisense oligonucleotide (ASO), CRISPR/Cas9 system, and short hairpin RNA (shRNA) . For delivery, these drugs require a specific intracellular location in the body.

Viral vectors are generally used to deliver therapeutic genes, but despite their high efficiency, they showed different disadvantages like immunological problems, insertional mutagenesis, replication competence, non-integration, *etc.* Nonviral carriers commonly include cationic liposomes, micelles, polymers, and inorganic nanoparticles [13].

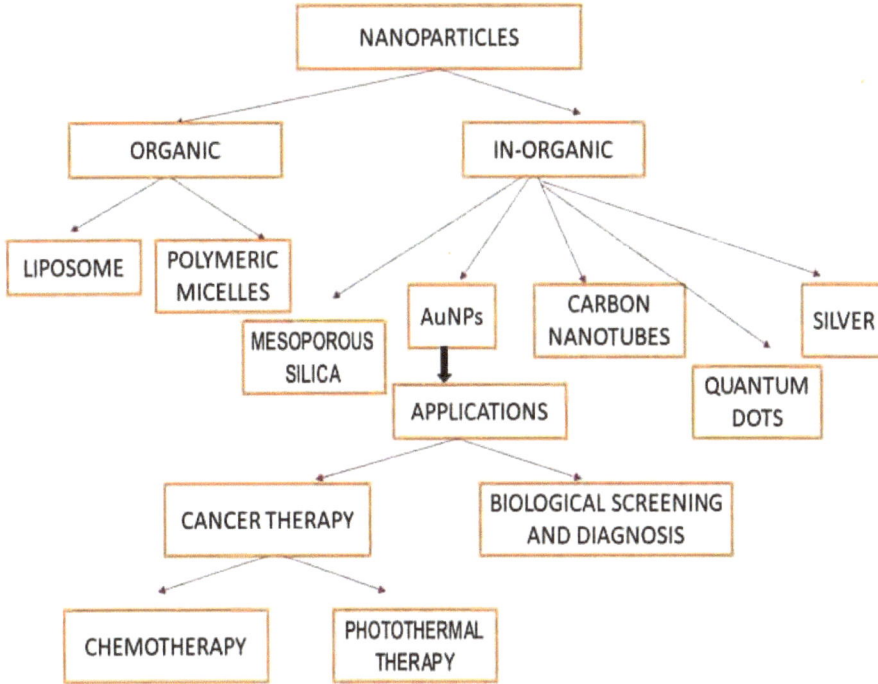

Fig. (1). Types of nanoparticles with their different applications.

LIPOSOME

Liposomal systems have been regarded as the most promising genetic delivery tools with the advantages of marvelous pharmacological targets, biocompatibility efficiency, decreased medication toxicity, and excellent capacity to overcome drug resistance and promote endosomal escape. Liposome nanoparticles are formed by the alignment of phospholipid biomolecules that have a high-quality encapsulation to fat and water-soluble medication, which transports various genetic drugs to the body through cell membrane fusion [13]. Liposomes possess following remarkable characteristics in DNA delivery: (a) poor immunogenic reaction, (b) a great amount of DNA can be loaded easily, (c) developed DNA stability within the body, and (d) lack of clearance that improves the blood circulation [14, 15]. Different gene therapy drugs like Lipofectamine, TurboFect, and StemFect have been widely used for the delivery of gene therapy drugs. Lipolexes (cationic liposomes), developed by the interaction between its positively charged surface area negatively charged DNA enhance circulation time and increase transfection efficacy, resulting in better cellular internalization. Not only a positive charge and neutral lipid of cationic liposomes but also different patterns of other liposomal non-viral vectors like exosome-liposome hybrid nanoparticles, DOTAP liposome (N-(1-(2,3-dioleoyloxy) propyl)-N, N, Ntrimethylammonium), *etc.*, are supportive and increase the supply of various gene therapy drugs and vaccines into cells [13, 16].

According to recent research, breast cancer was inhibited through the delivery of miRNA-34a to cells by using liposomal carriers interacting with cyclic RGD (Arginyl-glycyl-aspartic acid) peptide for cell adhesion [17]. In another research, delivering cationic liposomes-loaded miR-7 is also helpful in inhibiting ovarian cancer cell growth, invasion, and migration [18].

Gold Nanoparticles (AuNPs)

Gold nanoparticles are inorganic nanomaterials having different shapes like nano-sphere, nano-shells, nanostars, and nano-rods and possess various unique photothermal properties, which are helpful for a wide range of medical applications [19]. AuNPs play an important role in drug delivery due to their (a) good compatibility and nontoxicity, (b) conjugation ability with different drug molecules due to a negative charge and bonding mechanism, and (c) modification and processing of surface comfortably [19, 20]. AuNPs are efficient in defending nucleic acid from deterioration by nucleases and are useful for gene transfection [19, 20]. Oligonucleotide-modified functionalized AuNPs not only possess themselves as intracellular gene regulatory candidates but also motivate immune genes [20].

AuNPs are widely used for DNA and RNA delivery for different disease management. AuNPs are easily conjugated with DNA and siRNA because of their covalent bonding with nucleic acid strands by thiol moieties. The binding and interaction capacity of nanoparticles with different cell surface receptors can happen effortlessly because these nanoparticle surfaces can be coated by alternating layers of polycations and anionic nucleic acid [21]. For example miR155 was comfortably delivered into the human body to improvise cardiac ability [22]. According to Zhao, AuNPs with PSS (poly-sodium 4 styrenesulfonate) and PAH (poly-allylamine hydrochloride) were used to supply siRNA to the body [23]. For better antimicrobial therapy, AuNPs deliver genes to mesenchymal stem cells by interacting with antimicrobial peptides [24].

Magnetic Nanoparticle (MNP)

MNPs possess super magnetic characteristics, higher surface energy, and a large surface-volume ratio and are used in different biological applications. The size and shape of MNPs control their properties. MNPs have the capacity to communicate and change the outside and nearby magnetic fields. Usually, MNPs are developed by breaking down materials, nucleating atoms, and using the growth method. These nanoparticles use two ways for drug delivery: (a) using magnetic field and (b) permeability and retention [19, 25]

Natural polymers include proteins and carbohydrates, and synthetic organic polymers include polyethylene glycol. Poly-lactic acid and gold can be coated with MNPs. Even MNPs can be coated with polyethyleneimine (PEI) that uses charge interactions to affix DNA to the particle surface [25].

Mesoporous Silica Nanoparticles (MSNs)

According to the U.S. Food and Drug Administration (FDA), silica has been identified as a secure and unharmed nanoparticle over 50 years because of its large pore volume (1 cm^3/g) and surface area (700 m^2/g), size of particles (50 to 300 nm), two functional surfaces (a. cylindrical pore surfaces, b. exterior particle surface), nontoxicity, and good compatibility characteristics [19, 26]. MSNs are helpful to improve (a) drug solubility and intestinal absorption rate due to its (i) medication in noncrystalline conditions captured by mesopores and (ii) wettability and high dispersibility of MSNs large hydrophilic surface area, (b) selective targeting for localized therapy, and (c) stimuli-responsive drug delivery (internal and external) to control drug release [19, 27].

MSNs' large pores can enclose the genes and show protection from nucleases during gene delivery. Through cationic modification, these nanoparticles can be complexed with the genes and transfected to various cells. Dr. Meka confirms that fabricated MSN possesses higher drug loading ability and is much more efficient in delivering to cancer cells by using ethanol as a solvent and fluorocarbon-hydrocarbon as a template [27].

Nanoparticle offers different properties for different medical applications, including chemical stability, scalability in synthesis, facile functionalization, normal thermal conditions, stability, *etc.* These characteristics are important for sterilization, as well as the presence of a lower level of toxicity. However, nanoparticles may also cause some health problems. It remains a challenge to improve the current delivery system to avoid toxicity. To get a better delivery process, new nanoparticles should be researched that will be suitable for gene delivery.

Anti-Inflammatory Applications

Nanomaterials play a crucial role in disease treatment, drug synthesis, and delivery, with nanomedicine being a key branch that enhances diagnosis accuracy and efficacy in various fields like cancer treatment, AIDS detection, and diabetes treatment [28, 29]. Numerous areas, including chemistry, biology, engineering, and medicine, may see major shifts thanks to the rapidly growing field called nanotechnology [30]. This makes it possible to develop new nanomaterials with rich structures that may be utilized to treat diseases while providing drugs just where they are required [31]. Nanoparticles offer tremendous potential in therapeutic applications, but there are challenges such as potential toxicity, immunogenicity, and the need to understand their long-term effects further. Nanoparticles offer tremendous potential in therapeutic applications, but there are challenges such as potential toxicity, immunogenicity, and the need to understand their long-term effects further.

NPs have numerous advantages in giving medications, and they continue to have problems requiring to be solved before being used in medical facilities. PLGA, a widely used vehicle, may result in *in vivo* inflammation and minor harm to the gut and liver. Exosomes, released by cells in the body, have strong compatibility and the ability to focus but need improvement in uniformity.

Liposomes reduce drug toxicities but are not ideal for drugs that melt in water. Metal carriers like ZnO NPs and GNP are unstable, while CNTs have promising biocompatibility but need improvement in targeting ability.

Silver nanoparticles (AgNPs) have an influence on brain immune cells and microglia and their effects on neurotoxicity. AgNPs dissolve and integrate into microglia, increasing cystathionine-γ-lyase production and reducing inflammatory effects caused by microbiota [32]. The existence of silver nanoparticles (AgNPs) in everyday products like water filters and food containers caused concerns about probable negative health impacts on persons. AgNPs can penetrate the barrier between blood and brain, cause cytotoxicity in human cells, and cause dementia both *in vivo* and *in vitro*. It will take more study to comprehend the effects on various types of CNS cells [33]. Ag+ ions damage DNA, thiol protein groups, and cell walls, which results in AgNP toxicity. The study examines how Ag+ ion release triggers H2S-synthesizing enzymes, hence reducing toxicity, using citrate-capped AgNPs and the murine microglial N9 cell line [34]. This research additionally explores the effects of AgNP regulation on microglia-mediated neurodegeneration and the effect of AgNPs on inflammatory activation [35].

As silver nanoparticles (AgNPs) may be found in apoptosis in human cells and cross the blood-brain barrier, their inclusion in everyday items, including food packaging and water purification devices, has caused concerns about potential negative health impacts [36]. When given orally or intravenously, AgNPs may result in neurotoxicity *in vivo* and apoptosis in axons *in vitro* [37].

Nanotechnology, a rapidly growing field, has the potential to create novel nanomaterials in the domains of medical care, engineering, life sciences, and chemicals. These inorganic and organic nanomaterials possess rich structural properties, offering the potential for targeted drug delivery and disease treatment.

A Drug Transport Method Built Upon Nanoparticles in Anti-inflammatory Treatment

Heterogeneous nanoparticles, also known as NPs, are currently being researched as an anti-inflammatory treatment as they can better target and biocompatibly synthesize elements into nanostructures. These nanoparticles (NPs) are filled with anti-inflammatory drugs and used as transporters.

Nanomaterials are crucial in disease treatment, drug synthesis, and delivery, particularly in nanomedicine. They improve diagnosis accuracy and efficacy in various fields, including cancer treatment, AIDS detection, diabetes treatment, and anti-inflammatory treatments [38]. Recent studies show that AuNPs, an emerging type of nanoparticle, can penetrate the blood-brain barrier and reduce inflammatory responses, making them a promising option for Alzheimer's treatment [39].

Organic nanoparticles (NPs) are biocompatible and often made of biomaterials or polymers. A potential polymer for drug delivery, poly (lactic-glycolic acid)(PLGA) has uses in target and treatment [40]. Studies have shown that PLGA can reduce inflammation, inhibit the growth of bacteria, and release drugs in a controlled manner. PLGA is being utilized for the treatment of bone illnesses like osteosarcomas, osteoarthritis, and metastases of bone cancer. A biodegradable thermosensitive implant made of m-PEG and PLGA has shown advantages in osteomyelitis treatment [41].

Crohn's disease (CD) and ulcerative colitis (UC) are types of long-term digestive disorders (IBD). An oral medication delivery system was developed *via* the combination of nuclear factor-kB (NF-kB) deception oligonucleotide (ODN) with chitosan (CS)-modified PLGA nanospheres. ODN is a good choice for UC mouse models. When CS-PLGA NS was loaded with Decoy ODN, it reduced bloody stools and diarrhea and increased ODN stability [42]. Another common chronic inflammatory disease that can be treated with small amounts of methylene (MTX) in addition to a condition-modifying anti-rheumatic medicine (DMARD) is rheumatoid arthritis, also known as RA [43]. Packing lutetium-177 (177Lu) with MTX and PLGA may lower inflammation of the synovial tissue. As 177Lu-DOTA-HA-PLGA (MTX) showed hope as a drug delivery method for anti-rheumatic treatment, additional *in vivo* models and research are required to confirm the findings [44]. Exosomes occur naturally as 30–100 nm tiny particles are released by various kinds of cells. Their native concentrating abilities have been explored. Curcumin is a naturally occurring polyphenol that has anti-inflammatory, antineoplastic, and antioxidant properties. It is derived from the roots of the Curcuma plant longa [45]. Murine IBD caused by dextran sulfate sodium (DSS) may be effectively controlled by exosomes generated from TGF-β1 gene-modified BMDC (TGF-β1-EXO). Phospholipidserine (PSL) and apoptotic body liposomes are both examples of liposomal nanoparticles that have shown interest as osseointegrative, antibacterial, and anti-inflammatory drugs [46].

INORGANIC

Inorganic nanoparticles (NPs) are commonly used to treat inflammation, with metal oxide NPs showing potential in nanomedicine as a drug delivery vehicle. Zinc oxide nanoparticles (ZnO NPs) have excellent biomedical properties due to their role in transmitting genetic messages and strong antibacterial properties [47].

Studies have shown that ZnO NPs can penetrate bacterial cells, forming ROS reactions and leading to cell death. Gold nanoparticles (GNCs) have been used to stimulate immune responses and induce monocyte recruitment, demonstrating anti-inflammatory effects in the case of osteoarthritis in animals produced by MRSA. One of the unique characteristics of gold nanoparticles (GNPs) for improved treatment is their ability to lessen the effects of toll-like receptor (TLR) activity and inhibit inflammation *in vivo* [48]. Quantum dot (QD) is a fluorescent semiconductor nanoparticle with potential for drug delivery, with studies showing that QD-Celecoxib conjugates can target inflammation and improve bioimaging effects.

N-doped graphene semiconductor quantum dots (N-GQDs) combined with ammonium 10-amino-2-methoxyundecanoate (SAM) have been shown to have effective anti-inflammatory potential [49].

Numerous studies have explored the use of carbon nanotubes (CNTs) as a promising nanocarrier for anti-inflammatory drugs, such as NSAIDs, in a low-voltage, non-irritating manner. CNTs have been used to modulate and enhance the release of these drugs, enhancing their anti-inflammatory effect. A transdermal drug delivery system with MWCNTs, water-soluble polyethylene oxide, and water-insoluble pentaerythritol triacrylate was developed, demonstrating the potential of CNTs in drug delivery [50].

Essential in wound healing and maintaining homeostasis, inflammation is the immune systems' response to tissue harm or germs. A pro-inflammatory response includes a rapid induction phase, the initial stage of a series of actions [51]. Multiple inflammatory diseases, like cancer, obesity, sepsis, and cardiovascular, neurological, and immunological barrier disorders, can be set on by irregular inflammation [52]. Multiple phases of dysregulated chronic inflammation occur, with early inflammation activating immune cells to generate pro-inflammatory responses. If insufficient or excessive, it can lead to late-stage inflammation, causing tissue damage and excessive danger signal production [53]. The understanding of inflammation's connection to homeostasis offers opportunities for designing effective therapeutic strategies. Biomaterials can be used as drug carriers for controlled delivery, scavenging pro-inflammatory factors, inhibiting inflammation, offering versatility, targeting different pathways, and providing high spatiotemporal control of anti-inflammatory activities [54].

Carbon monoxide (CO) gas therapy can treat inflammatory diseases by selectively inhibiting pro-inflammatory cytokines and increasing anti-inflammatory cytokine expression. However, its clinical applications are limited due to targeting and potential toxicity. Photoactive CO-releasing nanoplatforms, ester formation, and CO products that reacted to pH might overcome this limitation, enabling prolonged CO release and treating inflammation in mice [55].

The gas H_2 may reduce gene transcription factors and anti-inflammatory cytokines, helping to repair damaged tissue. Nanoplatforms like liposomes and polymer nanoparticles can deliver H_2 gas, while high concentrations can be generated in inflammatory tissue [56].

A comparative study of organic and inorganic nanoparticles, or NPs (Quantum dots (QDs), carbon nanotubes (CNTs), poly (lactic-co-glycolic acid), and nanoparticles (NPs) all have related concepts) (Table **1**).

Table 1. Comparative analysis of organic and inorganic nanoparticles: types, advantages, and disadvantages.

Types	Carrier	Advantages	Disadvantages
ORGANIC NPS	PLGA	Multiple kinds of stable structures	Improving biocompatibility is crucial.
-	EXOSOMES	Effect targeting an organic level, a suitable source	Hard to get high purity, low unity, and the exosome
-	LIPOSOMES	Simple to get, highly stable, and non-toxic	Volatile for drugs that dissolve in water, limiting lifespan
INORGANIC NPS	-	-	-
-	Au/Zn	Bacteriostatic therapy or antibacterial qualities	Volatile
-	CNT	Strong biological compatibility	Unsatisfactory to focus
-	QD	Biomedical imaging ability and selectivity	Toxicity related to biological substances

Anti-bacterial Applications

Antibacterial nanoparticles kill bacteria through various mechanisms like APT depletion, membrane disruption, and inhibiting DNA synthesis. Nanoparticles kill bacteria by attaching to the bacterial surface. This changes the membrane potential and increases permeability. Nanoparticle adsorption to the cell surface increases permeability through depolarization by changing the normally negative charge. Nanoparticles cause the dissipation of the cell membrane potential, which causes the outer membrane to become unstable. Anti-bacterial nanoparticles, particularly metallic nanocomposites, destroy bacteria through ROS generation. ROS such as hydrogen peroxide, superoxide anion, hydroxide, *etc.*, are harmful to bacteria due to their relatively high reactivity and oxidizing potential. ROS directly affects the proteins, lipids, and DNA present in the cell [57 - 59].

It has been established that the dimensions (shape and size) of the nanoparticles have an impact on their interaction with the bacteria. Smaller nanoparticles exhibit stronger antibacterial properties. The rod and cube-shaped nanoparticles exhibit more successful interaction with bacterial surfaces. Nanoparticles with rod and wire morphology have very strong bactericidal effects as they easily penetrate bacteria walls [57, 58].

Nowadays, long-term and high dosages of antibiotics are preferred for the treatment of bacterial infections. Different nanoparticles are used for the delivery of antibiotics to the human body.

Polymeric Nanoparticles

Polymeric nanoparticles are organic nanomaterials that may be shaped like nanocapsules or nanosphere-like structures. These nanoparticles are biodegradable, biocompatible, and stable. Polymeric nanoparticles are developed from different varieties of natural or synthetic polymers like collagen, chitosan, albumin, *etc.* Synthetic polymers include poly(esters), polylactic acid (PLA), poly(alk-L-cyanoacrylates), *etc.* Polymeric nanosystems are used as carriers for drug delivery due to (a) their high drug solubility capacity, (b) storage ability, (c) long-term antibacterial performance, and (d) drug release at the exact location [60].

Cpolymers are widely used for drug delivery. Chitosan is a polycationic alkaline polysaccharide obtained from chitin, which is a natural polysaccharide. Hydrogel is made from chitosan, which is later used as a carrier for drug delivery. As per recent research, chitosan with phosphatidylcholine captures gentamicin to kill Listeria monocytogenes and Pseudomonas aeruginosa by destroying biofilms [61].

When researchers researched PLA (Polylactic acid) nanoparticles with rifampicin containing functionalized Poly-L-Lysine, theyfound a significant prevention growth of planktonic bacteria by changing the negative nanoparticle surface charge to a positive charge [62].

Solid Nanoparticles

Solid nanoparticles (SLN) are submicron (1000nm) drug carriers and are composed of biocompatible lipids and surfactants. SLNs are solid at body temperature and possess better stability. SLNs have excellent characteristics that comprise (a) bioavailability, (b)biodegradability, (c) prominent encapsulation efficiency, (d) low toxicity, (d) significant protection to different medications (lipophilic and hydrophilic) against harsh environments, (e) high biocompatibility, (f) mass production by high-pressure homogenization method, and (g) can easily be used [60, 63].

SLNs serve as better antibiotic carriers and provide better protection against intracellular bacterial infection. This ability of SLNs to serve as better antibiotic carriers has been effectively demonstrated by many studies. The antibiotic drugs, when carried by SLNs, provide better inhibitory effects in comparison with SLN-free drugs. In a comparative study by Xie *et al.*, two cohorts are given the same antibiotic enrofloxacin; one with an SLN carrier and the other with a free drug (without SLN). It is observed that the concentration of the antibiotic enrofloxacin is higher by 27.06-37.71 times in salmonella-infected mouse macrophages through the usage of behenic acid SLNs. Also, after treatment of the infected cell for 48 hours, it is observed that the intracellular bacterial colonies are higher in the free drug cohort (enrofloxacin concentration 0.6 g mL $^{-1}$)as compared to the SLN carrier-based drug (3.80 CFU Ml^{-1}) [64].

Inorganic Non-metallic Nanomaterials

Inorganic non-metallic nanomaterials have different dimensions (size and shape) ranging from 1 to 100 nm. Non-metallic nanomaterials are made up of inorganic materials by using various physical and chemical techniques. For conjugation with a variety of drugs, inorganic nanomaterials use different interaction forces, such as hydrophobic, electrostatic, *etc.* These nanoparticles show various physical and chemical characteristics: (a) good biocompatibility, (b) high storage ability, (c) drug loading capability, (d) low toxicity, and (e) facile preparation for drug delivery and release [60].

Dr. Pi *et al.* prepared a mannosylated and PEGlyated modified GO (graphene oxide)(GO-PEG-MAN) with loaded rifampicin drug to kill intracellular mycobacterium tuberculosis and found 50% inhibition of bacterial extension with reduced volume of RIF (10 ng ML^{-1}) than free RIF capacity (40%) of inhibition [65]. Subramanium was investigated upon mesoporous SiO_2 nanoparticles with rifampicin, which exhibited a prominent intracellular bactericidal efficiency (28.6%) compared to free RIF [66]. Additionally, Hussain developed a loaded vancomycin and CARG peptide with mesoporous silica nanoparticles to treat staphylococcus aureus-infected tissues. He observed that this nanoparticle is highly efficient for antibiotic loading and also suppresses infection growth, which shows its higher antibacterial efficacy. This nanoparticle also prevents frequent infection rates [67].

Neurological Disorders

Worldwide, hundreds of millions of people are affected by various neurological disorders like Alzheimer's, Parkinson's, epilepsy, multiple sclerosis, Huntington's, *etc.* Till now, drug delivery to the central nervous system is the most significant challenge despite advancements in different drugs and nanoparticles. The blood-brain barrier (BBB) is one of the barriers to drug delivery as therapeutic drugs do not cross this barrier due to their impermeability and reach the distinct location. BBB not only controls the cognitive method but also regulates the accumulation and excretion of various proteins, nutrients, and oxygen for a healthy brain. It also prevents the entering of various neurotoxic agents present in the bloodstream into the CNS, resulting in normal brain functioning. BBB regulates sodium (Na), Potassium (K), and calcium (Ca) levels for synapse processes as well as controls neurotransmitter movement, ionic homeostasis, and transportation systems of different substances from within and outside the brain [68 - 70].

The brain capillaries can be as small as 7-10 µm in diameter. Only tiny particles that have a molecular weight of < 400 Da can cross the BBB. Nowadays, researchers are investigating different nano-based drug delivery pathways to CNS through BBB. Nanoparticles can encapsulate drugs and deliver them through BBB. But sometimes encapsulated drugs cannot reach distinct locations. Therefore, an intermediator is needed to bind with nanoparticles so that they can be identified at specific sites where medication is required [68] (Fig. **2**).

Alzheimer Disorder

Alzheimer's disorder is one type of abnormal brain condition where the patient experiences declining symptoms in thinking, learning, memory, concentrating, and understanding in different situations, *etc.* Abnormal accumulation of amyloid beta plaques and neurofibrillary tangles, which are made up of tau protein, became the main reason of Alzheimer's disease. A variety of nanoparticles are used for drug delivery to treat this disorder [71].

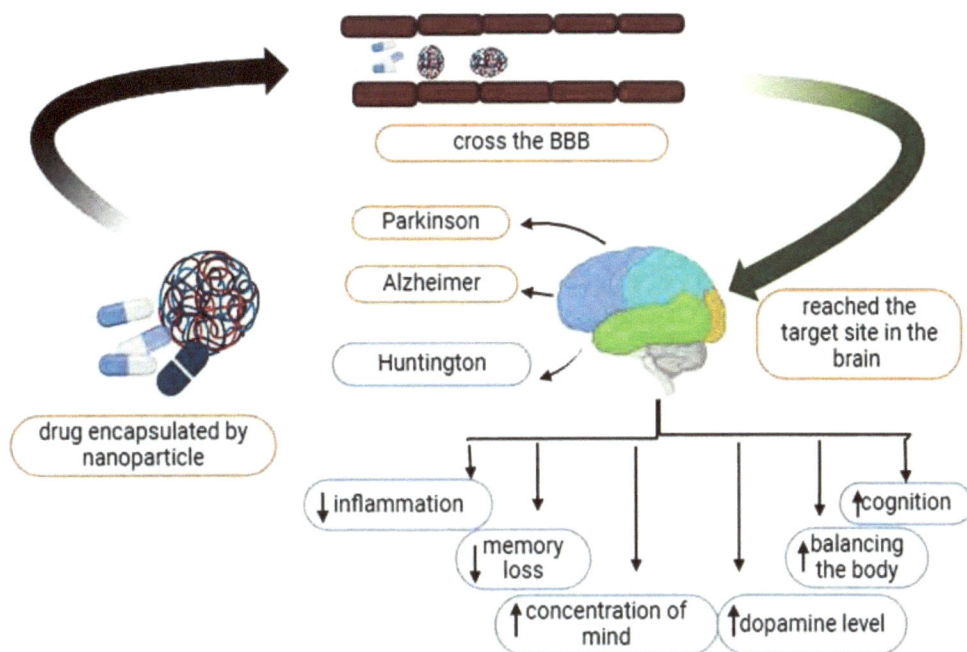

Fig. (2). Use of nanoparticles in the treatment of neurological disorder.

Lipid nanoparticles are used to deliver Huperzine A, a drug that is a cholinesterase inhibitor. Huperzine A effectively enhances memory and cognitive levels as well as protects nerve cells by raising neurotransmitter levels and functioning in Alzheimer's patient's brain. HUPA faces some difficulties in the solubility and clearance process, which is solved by lipid nanoparticles that possess good biocompatibility and stability [72]. Another drug, galantamine, helps to treat memory-based issues in different stages (mild to moderate) of Alzheimer's patients. By using lipid film hydration or solvent evaporation process, galantamine is enclosed into lipid bilayers, resulting in a better drug delivery to the specific site of the brain because of its unique properties like small size, solubility capacity, good storage, etc [72].

Another therapeutic agent, memantine, is used to decrease the inflammation rate and memory loss of AD patients but is delivered to the brain through polymer-based nanoparticles [73]. Researchers found that Alzheimer's patients who took sitagliptin with zinc nanoparticles exhibited an effective reduction in different AD symptoms (inflammation and cognitive) [74]. Scientists found another bioactive compound named thymoquinone (TQ), which is produced from Nigella sativa seeds and is widely used in different therapeutic processes. TQ is supplied to the brain by encapsulated poly (lactic-co-glycolic acid) NPs in the presence of polysorbate-80 (P-80], which helps NPs to cross the BBB. PLGA-NPs are protected from phagocytosis by the body due to being coated by P-80 surfactant [71].

Parkinson Disease

Parkinson's disease shows neurodegenerative brain conditions where patients possess slow movements, unstable walking, tremors in muscles, rigidity, trouble in swallowing, *etc.* Loss of dopamine level, disposition of α-synuclein, and death of most dopaminergic neurons causes Parkinson's disease. Therefore, the treatment's central point is to increase dopamine levels as well as to protect dopaminergic neurons by various therapeutic strategies [75].

Levopoda (L-DOPA) prefers organic nanoparticles (lipids, micelles, polymeric) to bedelivered into the brain due to their nature and composition (low toxicity, biocompatibility, modifying capacity, *etc.*). Organic nanoparticles compress and encapsulate L-DOPA, raise their circulation capacity, and obstruct degradation pathways successfully, resulting in more uptake of medication by Parkinson's patients' brains through BBB [76]. According to researchers, ropinirole, which is used for improving patients' motor disturbances, is encapsulated by polymeric nanoparticles (PLGA) and successfully transported to the brain (animal model) [77]. According to recent research, other PD medicines like pramipexole and hesperidin are supplied to the brain by chitosan nanoparticles prepared by Somasundaram. He conducted this research by taking nanoparticles that had a size (188nm) below 200 nm, for which NP can cross BBB easily [78].

CONCLUSION

In summary, the therapeutic potential of nanoparticles holds immense promise, marking an exciting era in healthcare innovation. As research in nanotechnology advances, the continued exploration of nanoparticles in drug delivery, imaging, and other therapeutic interventions opens up new possibilities for more effective and targeted healthcare solutions, ushering in a future where nanoparticles play a pivotal role in improving patient outcomes and advancing medical treatment modalities.

REFERENCES

[1]   Murthy SK. Nanoparticles in modern medicine: state of the art and future challenges. Int J Nanomedicine 2007; 2(2): 129-41.
       [PMID: 17722542]

[2]   Yetisgin AA, Cetinel S, Zuvin M, Kosar A, Kutlu O. Therapeutic Nanoparticles and Their Targeted Delivery Applications. Molecules 2020; 25(9): 2193.
       [http://dx.doi.org/10.3390/molecules25092193] [PMID: 32397080]

[3]   Yusuf A, Almotairy ARZ, Henidi H, Alshehri OY, Aldughaim MS. Nanoparticles as Drug Delivery Systems: A Review of the Implication of Nanoparticles' Physicochemical Properties on Responses in Biological Systems. Polymers (Basel) 2023; 15(7): 1596.
        [http://dx.doi.org/10.3390/polym15071596] [PMID: 37050210]

[4]   Malik S, Muhammad K, Waheed Y. Emerging Applications of Nanotechnology in Healthcare and Medicine. Molecules 2023; 28(18): 6624.
        [http://dx.doi.org/10.3390/molecules28186624] [PMID: 37764400]

[5]   Afzal O, Altamimi ASA, Nadeem MS, *et al.* Nanoparticles in Drug Delivery: From History to Therapeutic Applications. Nanomaterials (Basel) 2022; 12(24): 4494.
        [http://dx.doi.org/10.3390/nano12244494] [PMID: 36558344]

[6]   Hoshyar N, Gray S, Han H, Bao G. The effect of nanoparticle size on *in vivo* pharmacokinetics and cellular interaction. Nanomedicine (Lond) 2016; 11(6): 673-92.
        [http://dx.doi.org/10.2217/nnm.16.5] [PMID: 27003448]

[7]   Mitchell MJ, Billingsley MM, Haley RM, *et al.* Engineering precision nanoparticles for drug delivery. Nat Rev Drug Discov, 2021 20, 101–124.
        [http://dx.doi.org/10.1038/s41573-020-0090-8]

[8]   Gavas S, Quazi S, Karpiński TM. Nanoparticles for Cancer Therapy: Current Progress and Challenges. Nanoscale Res Lett 2021; 16(1): 173.
        [http://dx.doi.org/10.1186/s11671-021-03628-6] [PMID: 34866166]

[9]   Yao Y, Zhou Y, Liu L, *et al.* Nanoparticle-Based Drug Delivery in Cancer Therapy and Its Role in Overcoming Drug Resistance. Front Mol Biosci 2020; 7: 193.
        [http://dx.doi.org/10.3389/fmolb.2020.00193] [PMID: 32974385]

[10]   Doughty ACV, Hoover AR, Layton E, Murray CK, Howard EW, Chen WR. Nanomaterial Applications in Photothermal Therapy for Cancer. Materials (Basel) 2019; 12(5): 779.
        [http://dx.doi.org/10.3390/ma12050779] [PMID: 30866416]

[11]   Han X, Xu K, Taratula O, Farsad K. Applications of nanoparticles in biomedical imaging. Nanoscale 2019; 11(3): 799-819.
        [http://dx.doi.org/10.1039/C8NR07769J] [PMID: 30603750]

[12]   Bezbaruah R, Chavda VP, Nongrang L, *et al.* Nanoparticle-Based Delivery Systems for Vaccines. Vaccines (Basel) 2022; 10(11): 1946.
        [http://dx.doi.org/10.3390/vaccines10111946] [PMID: 36423041]

[13]   Pan X, Veroniaina H, Su N, *et al.* Applications and developments of gene therapy drug delivery systems for genetic diseases. Asian Journal of Pharmaceutical Sciences 2021; 16(6): 687-703.
        [http://dx.doi.org/10.1016/j.ajps.2021.05.003] [PMID: 35027949]

[14]   D. Ibraheem *et al.* Gene therapy and DNA delivery systems, International Journal of Pharmaceutics, 2014; 459:70-83.
        [http://dx.doi.org/10.1016/j.ijpharm.2013.11.027]

[15]   Mali S. Delivery systems for gene therapy. Indian J Hum Genet 2013; 19(1): 3-8.
        [http://dx.doi.org/10.4103/0971-6866.112870] [PMID: 23901186]

[16]   Nsairat H, Alshaer W, Odeh F, Esawi E, *et al.* Recent advances in using liposomes for delivery of nucleic acid based therapeutics. Open Nano 2023; Vol. 11.
        [http://dx.doi.org/10.1016/j.onano.2023.100132]

[17]   Vakhshiteh F, Khabazian E, Atyabi F, Ostad SN, *et al.* Peptide-conjugated liposomes for targeted miR-34a delivery to suppress breast cancer and cancer stem-like population. J Drug Delivery Sci Tech 2020; 57: 101687.
        [http://dx.doi.org/10.1016/j.jddst.2020.101687]

[18]   Cui X, Song K, Lu X, Feng W, Di W. Liposomal Delivery of MicroRNA-7 targeting EGFR to inhibit

the growth, invasion, and migration of ovarian cancer. ACS Omega 2021; 6(17): 11669-11678.
[http://dx.doi.org/10.1021/acsomega.1c00565]

[19]   Miron-Barroso. S, Domenech. E.B, Trigueros. S. Nanotechnology-Based Strategies to Overcome
        Current Barriers in Gene Delivery. Int J Mol Sci. 2021; 22(16): 8537.
        [http://dx.doi.org/10.3390/ijms22168537]

[20]   Kong F.Y, Zhang J.W, Li R.F, Wang Z.X, Wang W.J. Unique Roles of Gold Nanoparticles in Drug
        Delivery, Targeting, and Imaging Applications. Molecules 2017; 22(9): 1445.
        [http://dx.doi.org/10.3390/molecules22091445]

[21]   Mendes BB, Conniot J, Avital A, *et al.* Nanodelivery of nucleic acids. Nature Reviews Methods
        Primers 2022; 2(1): 24.
        [http://dx.doi.org/10.1038/s43586-022-00104-y] [PMID: 35480987]

[22]   Zhao X, Huang Q, Jin Y. Gold nanorod delivery of LSD1 siRNA induces human mesenchymal stem
        cell differentiation. Mater Sci Eng C 2015; 54: 142-9.
        [http://dx.doi.org/10.1016/j.msec.2015.05.013] [PMID: 26046277]

[23]   Hamimed S, Jabberi M, Chatti A. Nanotechnology in drug and gene delivery. Naunyn Schmiedebergs
        Arch Pharmacol 2022; 395(7): 769-87.
        [http://dx.doi.org/10.1007/s00210-022-02245-z] [PMID: 35505234]

[24]   Peng LH, Huang YF, Zhang CZ, *et al.* Integration of antimicrobial peptides with gold nanoparticles as
        unique non-viral vectors for gene delivery to mesenchymal stem cells with antibacterial activity.
        Biomaterials 2016; 103: 137-49.
        [http://dx.doi.org/10.1016/j.biomaterials.2016.06.057] [PMID: 27376562]

[25]   Majidi S, Zeinali Sehrig F, Samiei M, *et al.* Magnetic nanoparticles: Applications in gene delivery and
        gene therapy. Artif Cells Nanomed Biotechnol 2016; 44(4): 1186-93.
        [PMID: 25727710]

[26]   Carvalho AM, Cordeiro RA, Faneca H. Silica-Based Gene Delivery Systems: From Design to
        Therapeutic Applications. Pharmaceutics 2020; 12(7): 649.
        [http://dx.doi.org/10.3390/pharmaceutics12070649] [PMID: 32660110]

[27]   Zhou Y, Quan G, Wu Q, *et al.* Mesoporous silica nanoparticles for drug and gene delivery. Acta
        Pharm Sin B 2018; 8(2): 165-77.
        [http://dx.doi.org/10.1016/j.apsb.2018.01.007] [PMID: 29719777]

[28]   Cook AB, Decuzzi P. Harnessing endogenous stimuli for responsive materials in theranostics. ACS
        Nano 2021; 15(2): 2068-98.
        [http://dx.doi.org/10.1021/acsnano.0c09115] [PMID: 33555171]

[29]   Wang Y, Wang C, Li K, *et al.* Recent advances of nanomedicine-based strategies in diabetes and
        complications management: Diagnostics, monitoring, and therapeutics. J Control Release 2021; 330:
        618-40.
        [http://dx.doi.org/10.1016/j.jconrel.2021.01.002] [PMID: 33417985]

[30]   Pearce AK, Wilks TR, Arno MC, O'Reilly RK. Synthesis and applications of anisotropic nanoparticles
        with precisely defined dimensions. Nat Rev Chem 2020; 5(1): 21-45.
        [http://dx.doi.org/10.1038/s41570-020-00232-7] [PMID: 37118104]

[31]   Snider C, Grant D, Grant SA. Investigation of an injectable gold nanoparticle extracellular matrix. J
        Biomater Appl 2022; 36(7): 1289-300.
        [http://dx.doi.org/10.1177/08853282211051586] [PMID: 34672227]

[32]   Liu J, Sonshine DA, Shervani S, Hurt RH. Controlled release of biologically active silver from
        nanosilver surfaces. ACS Nano 2010; 4(11): 6903-13.
        [http://dx.doi.org/10.1021/nn102272n] [PMID: 20968290]

[33]   Bagheri-Abassi F, Alavi H, Mohammadipour A, Motejaded F, Ebrahimzadeh-Bideskan A. The effect
        of silver nanoparticles on apoptosis and dark neuron production in rat hippocampus. Iran J Basic Med

Sci 2015; 18(7): 644-8.
[PMID: 26351553]

[34]   De Matteis V, Malvindi MA, Galeone A, *et al.* Negligible particle-specific toxicity mechanism of silver nanoparticles: The role of Ag+ ion release in the cytosol. Nanomedicine 2015; 11(3): 731-9.
[http://dx.doi.org/10.1016/j.nano.2014.11.002] [PMID: 25546848]

[35]   Sivarajah A, Collino M, Yasin M, *et al.* Anti-apoptotic and anti-inflammatory effects of hydrogen sulfide in a rat model of regional myocardial I/R. Shock 2009; 31(3): 267-74.
[http://dx.doi.org/10.1097/SHK.0b013e318180ff89] [PMID: 18636044]

[36]   Xu L, Dan M, Shao A, *et al.* Silver nanoparticles induce tight junction disruption and astrocyte neurotoxicity in a rat blood-brain barrier primary triple coculture model. Int J Nanomedicine 2015; 10: 6105-18.
[PMID: 26491287]

[37]   Yin N, Yao X, Zhou Q, Faiola F, Jiang G. Vitamin E attenuates silver nanoparticle-induced effects on body weight and neurotoxicity in rats. Biochem Biophys Res Commun 2015; 458(2): 405-10.
[http://dx.doi.org/10.1016/j.bbrc.2015.01.130] [PMID: 25661000]

[38]   Wu N, Lu C, Wang Y, *et al.* Semiconducting Polymer Nanoparticles-Manganese Based Chemiluminescent Platform for Determining Total Antioxidant Capacity in Diabetic Mice. Anal Chem 2023; 95(16): 6603-11.
[http://dx.doi.org/10.1021/acs.analchem.2c05624] [PMID: 37043629]

[39]   Ling L, Jiang Y, Liu Y, *et al.* Role of gold nanoparticle from Cinnamomum verum against 1-methyl-4-phenyl-1, 2, 3, 6-tetrahydropyridine (MPTP) induced mice model. J Photochem Photobiol B 2019; 201: 111657.
[http://dx.doi.org/10.1016/j.jphotobiol.2019.111657] [PMID: 31706085]

[40]   Mir M, Ahmed N, Rehman A. Recent applications of PLGA based nanostructures in drug delivery. Colloids Surf B Biointerfaces 2017; 159: 217-31.
[http://dx.doi.org/10.1016/j.colsurfb.2017.07.038] [PMID: 28797972]

[41]   Peng KT, Chen CF, Chu IM, *et al.* Treatment of osteomyelitis with teicoplanin-encapsulated biodegradable thermosensitive hydrogel nanoparticles. Biomaterials 2010; 31(19): 5227-36.
[http://dx.doi.org/10.1016/j.biomaterials.2010.03.027] [PMID: 20381140]

[42]   Tahara K, Samura S, Tsuji K, *et al.* Oral nuclear factor-κB decoy oligonucleotides delivery system with chitosan modified poly(d,l-lactide-co-glycolide) nanospheres for inflammatory bowel disease. Biomaterials 2011; 32(3): 870-8.
[http://dx.doi.org/10.1016/j.biomaterials.2010.09.034] [PMID: 20934748]

[43]   Trujillo-Nolasco RM, Morales-Avila E, Ocampo-García BE, *et al.* Preparation and *in vitro* evaluation of radiolabeled HA-PLGA nanoparticles as novel MTX delivery system for local treatment of rheumatoid arthritis. Mater Sci Eng C 2019; 103: 109766.
[http://dx.doi.org/10.1016/j.msec.2019.109766] [PMID: 31349410]

[44]   Wu G, Zhang J, Zhao Q, *et al.* Molecularly engineered macrophage-derived exosomes with inflammation tropism and intrinsic heme biosynthesis for atherosclerosis treatment. Angew Chem Int Ed 2020; 59(10): 4068-74.
[http://dx.doi.org/10.1002/anie.201913700] [PMID: 31854064]

[45]   Yan F, Zhong Z, Wang Y, *et al.* Exosome-based biomimetic nanoparticles targeted to inflamed joints for enhanced treatment of rheumatoid arthritis. J Nanobiotechnology 2020; 18(1): 115.
[http://dx.doi.org/10.1186/s12951-020-00675-6] [PMID: 32819405]

[46]   Xu X, Li Y, Wang L, *et al.* Triple-functional polyetheretherketone surface with enhanced bacteriostasis and anti-inflammatory and osseointegrative properties for implant application. Biomaterials 2019; 212: 98-114.
[http://dx.doi.org/10.1016/j.biomaterials.2019.05.014] [PMID: 31112825]

[47]   Agarwal H, Shanmugam V. A review on anti-inflammatory activity of green synthesized zinc oxide nanoparticle: Mechanism-based approach. Bioorg Chem 2020; 94: 103423.
        [http://dx.doi.org/10.1016/j.bioorg.2019.103423] [PMID: 31776035]

[48]   Gao W, Wang Y, Xiong Y, *et al.* Size-dependent anti-inflammatory activity of a peptide-gold nanoparticle hybrid *in vitro* and in a mouse model of acute lung injury. Acta Biomater 2019; 85: 203-17.
        [http://dx.doi.org/10.1016/j.actbio.2018.12.046] [PMID: 30597258]

[49]   Sameer Kumar R, Shakambari G, Ashokkumar B, Nelson DJ, John SA, Varalakshmi P. Nitrogen-doped graphene quantum dot-combined sodium 10-amino-2-methoxyundecanoate: Studies of proinflammatory gene expression and live cell imaging. ACS Omega 2018; 3(9): 11982-92.
        [http://dx.doi.org/10.1021/acsomega.8b02085] [PMID: 30320283]

[50]   Spizzirri U G, Hampel S, Cirillo G, Nicoletta F P, Hassan A, Vittorio O, *et al.* Spherical gelatin/CNTs hybrid microgels as electro-responsive drug delivery systems. Int. J. Pharm. 2013; 448, 115–122.
        [http://dx.doi.org/10.1016/j.ijpharm.2013.03.013]

[51]   Schett G, Neurath MF. Resolution of chronic inflammatory disease: universal and tissue-specific concepts. Nat Commun 2018; 9(1): 3261.
        [http://dx.doi.org/10.1038/s41467-018-05800-6] [PMID: 30111884]

[52]   Furman D, Campisi J, Verdin E, *et al.* Chronic inflammation in the etiology of disease across the life span. Nat Med 2019; 25(12): 1822-32.
        [http://dx.doi.org/10.1038/s41591-019-0675-0] [PMID: 31806905]

[53]   Rajendran P, Chen YF, Chen YF, *et al.* The multifaceted link between inflammation and human diseases. J Cell Physiol 2018; 233(9): 6458-71.
        [http://dx.doi.org/10.1002/jcp.26479] [PMID: 29323719]

[54]   Zindel J, Kubes P. DAMPs, PAMPs, and LAMPs in Immunity and Sterile Inflammation. Annu Rev Pathol 2020; 15(1): 493-518.
        [http://dx.doi.org/10.1146/annurev-pathmechdis-012419-032847]

[55]   Fujita K, Tanaka Y, Abe S, Ueno T. A Photoactive Carbon-Monoxide-Releasing Protein Cage for Dose-Regulated Delivery in Living Cells. Angew Chem Int Ed 2016; 55(3): 1056-60.
        [http://dx.doi.org/10.1002/anie.201506738] [PMID: 26332099]

[56]   Meneksedag-Erol D, Kizhakkedathu JN, Tang T, Uludağ H. Molecular Dynamics Simulations on Nucleic Acid Binding Polymers Designed To Arrest Thrombosis. ACS Appl Mater Interfaces 2018; 10(34): 28399-411.
        [http://dx.doi.org/10.1021/acsami.8b09914] [PMID: 30085650]

[57]   Aflakian. F, Mirzavi. F, Aiyelabegan, Soleimani. A, Navashenaq. J.G, *et al.* Nanoparticles-based therapeutics for the management of bacterial infections: A special emphasis on FDA-approved products and clinical trials. Eur J Pharm Sci 2023; 188: 106515.
        [http://dx.doi.org/10.1016/j.ejps.2023.106515]

[58]   Wang L, Hu C, Shao L. The antimicrobial activity of nanoparticles: present situation and prospects for the future. Int J Nanomedicine 2017; 12: 1227-49.
        [http://dx.doi.org/10.2147/IJN.S121956] [PMID: 28243086]

[59]   Shaikh S, Nazam N, Rizvi SMD, *et al.* Mechanistic Insights into the Antimicrobial Actions of Metallic Nanoparticles and Their Implications for Multidrug Resistance. Int J Mol Sci 2019; 20(10): 2468.
        [http://dx.doi.org/10.3390/ijms20102468] [PMID: 31109079]

[60]   Wang C, Yang Y, Cao Y, *et al.* Nanocarriers for the delivery of antibiotics into cells against intracellular bacterial infection. Biomater Sci 2023; 11(2): 432-44.
        [http://dx.doi.org/10.1039/D2BM01489K] [PMID: 36503914]

[61]   Qiu Y, Xu D, Sui G, *et al.* Gentamicin decorated phosphatidylcholine-chitosan nanoparticles against biofilms and intracellular bacteria. Int J Biol Macromol 2020; 156: 640-7.

[http://dx.doi.org/10.1016/j.ijbiomac.2020.04.090] [PMID: 32304789]

[62]   Spirescu VA, Chircov C, Grumezescu AM, Andronescu E. Polymeric Nanoparticles for Antimicrobial Therapies: An up-to-date Overview. Polymers (Basel) 2021; 13(5): 724.
[http://dx.doi.org/10.3390/polym13050724] [PMID: 33673451]

[63]   Mohammadi-Samani S, Ghasemiyeh P. Solid lipid nanoparticles and nanostructured lipid carriers as novel drug delivery systems: applications, advantages and disadvantages. Res Pharm Sci 2018; 13(4): 288-303.
[http://dx.doi.org/10.4103/1735-5362.235156] [PMID: 30065762]

[64]   Xie S, Yang F, Tao Y, *et al.* Enhanced intracellular delivery and antibacterial efficacy of enrofloxacin-loaded docosanoic acid solid lipid nanoparticles against intracellular Salmonella. Sci Rep 2017; 7(1): 41104.
[http://dx.doi.org/10.1038/srep41104] [PMID: 28112240]

[65]   Pi J, Shen L, Shen H, *et al.* Mannosylated graphene oxide as macrophage-targeted delivery system for enhanced intracellular M.tuberculosis killing efficiency. Mater Sci Eng C 2019; 103: 109777.
[http://dx.doi.org/10.1016/j.msec.2019.109777] [PMID: 31349400]

[66]   Subramaniam S, Thomas N, Gustafsson H, Jambhrunkar M, Kidd SP, Prestidge CA. Rifampicin-Loaded Mesoporous Silica Nanoparticles for the Treatment of Intracellular Infections. Antibiotics (Basel) 2019; 8(2): 39.
[http://dx.doi.org/10.3390/antibiotics8020039] [PMID: 30979069]

[67]   Hussain S, Joo J, Kang J, *et al.* Antibiotic-loaded nanoparticles targeted to the site of infection enhance antibacterial efficacy. Nat Biomed Eng 2018; 2(2): 95-103.
[http://dx.doi.org/10.1038/s41551-017-0187-5] [PMID: 29955439]

[68]   Mittal KR, Pharasi N, Sarna B, *et al.* Nanotechnology-based drug delivery for the treatment of CNS disorders. Transl Neurosci 2022; 13(1): 527-46.
[http://dx.doi.org/10.1515/tnsci-2022-0258] [PMID: 36741545]

[69]   Ebrahimi Z, Talaei S, Aghamiri S, Goradel NH, Jafarpour A, Negahdari B. Overcoming the blood–brain barrier in neurodegenerative disorders and brain tumours. IET Nanobiotechnol 2020; 14(6): 441-8.
[http://dx.doi.org/10.1049/iet-nbt.2019.0351] [PMID: 32755952]

[70]   Naqvi S, Panghal A, Flora SJS. Nanotechnology: A Promising Approach for Delivery of Neuroprotective Drugs. Front Neurosci 2020; 14: 494.
[http://dx.doi.org/10.3389/fnins.2020.00494] [PMID: 32581676]

[71]   Ullah S.N.M.N, Afzal O, Altamimi A.S.A, Ather H, Sultana S, Almalki W.H, Bharti P, Sahoo A, *et al.* Nanomedicine in the Management of Alzheimer's Disease: State-of-the-Art. Biomedicines 2023; 11(6): 752.
[http://dx.doi.org/10.3390/biomedicines11060752]

[72]   Naser SS, Singh D, Preetam S, *et al.* Posterity of nanoscience as lipid nanosystems for Alzheimer's disease regression. Mater Today Bio 2023; 21: 100701.
[http://dx.doi.org/10.1016/j.mtbio.2023.100701] [PMID: 37415846]

[73]   Sánchez-López E, Ettcheto M, Egea MA, *et al.* Memantine loaded PLGA PEGylated nanoparticles for Alzheimer's disease: *in vitro* and *in vivo* characterization. J Nanobiotechnology 2018; 16(1): 32.
[http://dx.doi.org/10.1186/s12951-018-0356-z] [PMID: 29587747]

[74]   Vilella A, Belletti D, Sauer AK, *et al.* Reduced plaque size and inflammation in the APP23 mouse model for Alzheimer's disease after chronic application of polymeric nanoparticles for CNS targeted zinc delivery. J Trace Elem Med Biol 2018; 49: 210-21.
[http://dx.doi.org/10.1016/j.jtemb.2017.12.006] [PMID: 29325805]

[75]   Duan L, Li X, Ji R, *et al.* Nanoparticle-Based Drug Delivery Systems: An Inspiring Therapeutic Strategy for Neurodegenerative Diseases. Polymers (Basel) 2023; 15(9): 2196.

[http://dx.doi.org/10.3390/polym15092196] [PMID: 37177342]

[76]   Vliet.E.F.V, Knol.M.J, Schiffelers.R.M, Caiazzo.M, Fens.M.H.A.M, Levodopa-loaded nanoparticles for the treatment of Parkinson's disease. J Control Release 2023; 360: 212-24.
[http://dx.doi.org/10.1016/j.jconrel.2023.03.030]

[77]   Barcia E, Boeva L, García-García L, *et al.* Nanotechnology-based drug delivery of ropinirole for Parkinson's disease. Drug Deliv 2017; 24(1): 1112-23.
[http://dx.doi.org/10.1080/10717544.2017.1359862] [PMID: 28782388]

[78]   Silva S, Almeida A, Vale N. Importance of Nanoparticles for the Delivery of Antiparkinsonian Drugs. Pharmaceutics 2021; 13(4): 508.
[http://dx.doi.org/10.3390/pharmaceutics13040508] [PMID: 33917696]

CHAPTER 8

Recent Advancement in the Epigenetic Mediated Nanotechnology Approach: Potential Implication in Cancer Therapy

Jangmang Chongloi[1], Manish Jaiswal[2], Ankit Srivastava[3], Rangnath Ravi[4], Ranjeet Kumar Nirala[5], Biswajita Pradhan[6] and Kali Prasad Pattanaik[7,*]

[1] *Department of Emergency Medicine, JPNATC AIIMS, New Delhi, Delhi 110029, India*

[2] *Department of Biochemistry, AIIMS, New Delhi, Delhi 110029, India*

[3] *Department of Biotechnology, Motilal Nehru National Institute of Technology, Allahabad, India*

[4] *Shivaji College, University of Delhi, New Delhi, Delhi 110021, India*

[5] *Saheed Raj Guru College of Applied Sciences for Women, University of Delhi, New Delhi, Delhi 110096, India*

[6] *School of Biological Sciences, AIPH University, Bhubaneswar, India*

[7] *National Rice Research Institute, Cuttack, Odisha, India*

Abstract: Cancer is a major cause of morbidity and mortality and exhibits a potential health burden worldwide. Despite significant advancements in chemotherapy, radiotherapy, surgery, and immunotherapy, early-stage diagnosis, lack of a specific biomarker, and recurrence of the disease urgently necessitate an alternative diagnostic approach. Multiple lines of evidence show that epigenetic factors, such as DNA methylation and histone modifications, play a crucial role in modulating the onset and progression of cancer phenotypes. Furthermore, epigenetic drugs in clinical settings are highly efficient and ideal with regard to patient response and reproducibility. However, owing to their toxicity, low solubility, poor bioavailability, and low stability, epigenetic drugs can lead to off-target effects and might fail to induce a long-term response. Recently, it has been observed that the integration of epigenetic drugs with nanoscale delivery systems offers a novel and promising approach not only for the targeted delivery of epigenetic drugs to the tumor site but also enhances the stability and permeability with improved efficacy and reduced systemic toxicity. Leveraging this concept, the current chapter aims to elucidate the recent advancements and futuristic approach of nano-epigenetic drug conjugates and their potential implications in cancer therapy.

Keywords: Cancer, DNA methylation, Histone modification, Nano-epigenetic drug, Nanotechnology.

---
* **Corresponding author Kali Prasad Pattanaik:** National Rice Research Institute, Cuttack, Odisha, India;
E-mail: kaliprasad25@gmail.com

Manoranjan Arakha & Arun Kumar Pradhan (Eds.)

INTRODUCTION

Cancer has become the second largest cause of death, with cardiovascular disease being the leading cause of death globally [1]. The most common cancers are breast, lung, colon, rectum, and prostate cancers. Widespread metastasis is the primary cause of cancer-related death. Metastasis occurs when cancer cells grow rapidly, invade adjoining parts of the body, and spread to other organs. Cancer mortality can be reduced through early detection and treatment. Commonly, treatment of cancers is limited to traditional chemotherapy, radiation therapy, and surgery [2]. The etiology of cancer can be quite complicated as it involves both environmental and hereditary influences.

The application of various nanotechnologies in cancer studies has become a rising field and approach aimed at providing early diagnosis, prediction, different therapeutic treatments, and drug design and delivery. Despite the significant advancements in chemotherapy, radiotherapy, surgery, immunotherapy, and early-stage diagnosis, the lack of a specific biomarker and recurrence of the disease urgently necessitate an alternative diagnostic approach. One of the main applications of nanotechnology in cancer research is the development of diagnostic tools. In this context, alternative diagnostic approaches are urgently needed [3, 4].

One of the main applications of nanotechnology in cancer research is the development of nanomaterials for drug delivery. Nanoparticles can be designed to deliver drugs directly to the cancer cells, thereby reducing the side effects of chemotherapy. Nanoparticles can be used to deliver these drugs directly to cancer cells, increasing their efficacy and reducing their toxicity. Nanoparticles can also be used to study epigenetic modifications in cancer cells. One of the challenges in studying epigenetics is that it is difficult to isolate and analyze the epigenetic marks on DNA. Nanoparticles can be designed to bind specifically to certain epigenetic marks, allowing researchers to isolate and study them. For example, nanoparticles have been designed to bind to histone modification, allowing researchers to study the role of histone modification in cancer development.

Besides all of these applications, different cancer drugs and traditional cancer therapies in the clinical setting are only highly efficient and ideal as far as the patient response and reproducibility are concerned. However, owing to the toxicity, low solubility, poor bioavailability, and low stability associated with certain cancer drugs, it can lead to off-target effects and failure to induce long-term responses. Recently, it has been observed that the integration of epigenetic drugs with a nanoscale delivery system offers a novel and promising approach not only in the delivery of the drugs to the tumor site but also for enhancing stability and permeability with minimum side effects [5]. To address these limitations, the integration of epigenetic drugs with nanoscale delivery systems offers a novel and promising approach to cancer therapy. Leveraging this concept, the current chapter aims to provide an overview of recent advances and future perspectives on nano-epigenetic drug conjugates and their potential impact on cancer therapy.

EPIGENETICS PLAYS A MAJOR ROLE IN CANCER

Conrad Hall Waddington first coined the term epigenetics in the early 1940s and referred to it as a molecular mechanism that influences changes in gene expression but does not involve changes in its genetic code. Epigenetics is also understood as the study of heritable changes in gene expression that arise without any changes in DNA sequence [6]. Epigenetic modifications can alter the manner in which genes are expressed, potentially leading to the development of cancer. Multiple lines of evidence show that epigenetic factors, such as DNA methylation and histone modifications, play a crucial role in modulating the onset and progression of cancer phenotype Fig. (**1**) [7]. Non-coding RNAs, such as small RNAs, also play an important role in the regulation of gene expression by controlling mRNA translation. Regions that target miRNAs are often associated with carcinogenesis. Post-translation modifications of histone, such as methylation, are crucial regulators of chromatin structure and gene expression and, therefore have been a major target of epigenetic therapies [8].

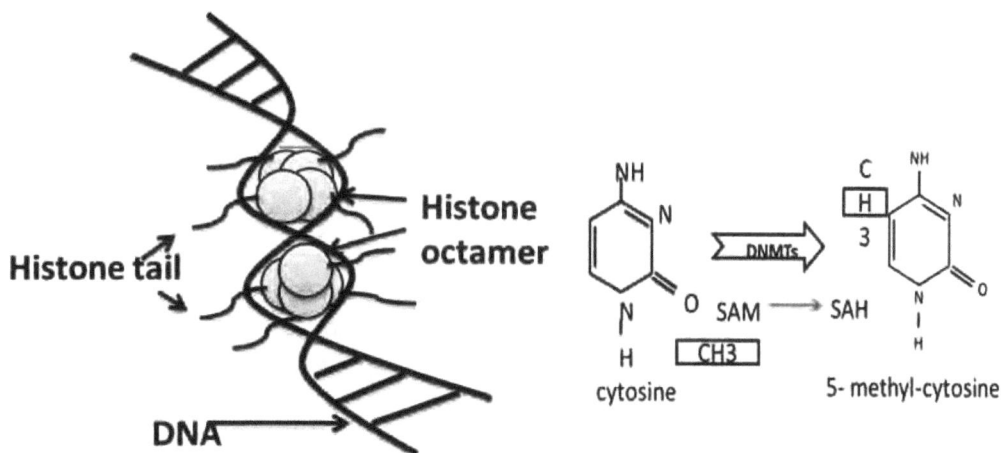

Fig. (1). Histone modification.

Epigenetic drugs are a promising class of new cancer therapeutics that target epigenetic modifications in cancer cells. These drugs work by modifying epigenetic marks on DNA, altering the expression of genes, and potentially leading to cancer cell death. Thus, epigenetic modifications can be used as biomarkers for cancer diagnosis and prognosis [9, 10].

Epigenetic drugs represent a new class of cancer therapeutics that targets specific molecular pathways involved in the onset and progression of cancer. Non-coding small RNA can be divided into tumor-promoting and tumor-suppressing miRNAs. Non-coding RNAs can either be directly involved in tumorigenesis or indirectly affect tumor development by participating in other epigenetic events [11]. Changes in epigenetic modifications of cancer can lead to the regulation of certain cellular responses, such as cell proliferation, apoptosis, invasion, and senescence. Some of these alterations determine cell function and are involved in oncogenic transformations. Reversing these alterations by drugs can revert the cancer phenotype to normal. Thus, a theory was proposed that epigenetic changes are responsible for tumorigenesis, which is the consequence of the combined action of multiple genetic events occurring in the progression of cancer [12]. Unlike traditional chemotherapy drugs, which target rapidly dividing cells, epigenetic drugs target specific modifications that are associated with cancer development and progression. By targeting these modifications, epigenetic drugs can induce long-term changes in gene expression and reverse cancer phenotypes. Epigenetics play an important role in tumorigenesis through DNA methylation, histone modification, chromatin remodeling, and non-coding RNA regulation. This epigenetic change has provided insight into the mechanisms underlying cancer phenotypes and therefore enables us to investigate and provide potential therapeutic [13, 14].

DNA methylation, along with other epigenetic modifications, plays a critical role in the regulation of gene expression and is centrally implicated in the dysregulation of most cancers. The fundamental basis of iDNMTs is that they target the specific epigenetic changes that revert the state of DNA methylation to activate silenced tumor-suppressor genes and, therefore, prevent cancer growth. Initially, the FDA-approved iDNMT is 5-azacitidine, which finds utility against the disease myelodysplastic syndrome (MDS) and acute myeloid leukaemia (AML). The other iDNMTs, which include Decitabine, DNMT inhibitors, DNMT1, and DNMT3A, have shown targeting different other cancers, and thus the great potential summed up in this approach. There are some more important roles shown by the non-nucleoside analogs of the compound, like hydralazine and procaine, in focusing on targeting DNA methylation in most cancers [15, 16]. The following drugs target DNA methylation (Table **1**):

Table 1. Drugs targeting DNA methylation.

iDNMT Drugs	Cancer Type	iDNMT Class	References
Azacitidine	Pancreatic adenocarcinoma	Nucleoside analogs	[17]
Decitabine	Chronic myelomonocytic leukemia, myelodysplastic syndromes	Nucleoside analogs	[18]
DNMT1	Breast, colorectal, gastric, non-small cell lung, pancreatic	Nucleoside analogs	[19]
DNMT3A	MDS, acute myeloid leukemia	-	[20]
DNMT3B	IFC syndrome, SNPs in breast and lung adenoma	-	[21, 22]
4-deoxyuridine	Zebularine	Nucleoside analogs	[23]
Hydralazine	Lung, colon, and rectum cancer	Nonnucleoside analogs	[24]
Procaine	Colon cancer, leukemia	Nonnucleoside analogs	[25 - 27]
Phthaloyl tryptophan RG108	Prostate cancer	Nonnucleoside analogs	[28]
Prednimustine	Hodgkin lymphoma, breast cancer	Miscellaneous	[29 - 31]

Besides DNA methylation, histone modifications are central to gene regulation and cancer development. To target those modifications, a large number of histone deacetylase inhibitors (iHDACs) have been developed, which ultimately target the structure of the chromatin and gene expression. iHDACs have been and are, to this day, very useful chemotherapeutic agents in the management of different cancers through the activation of apoptosis, cell cycle arrest, and differentiation of cancer cells. For instance, romidepsin and panobinostat have shown great anticancer efficacy in the treatment of lymphomas and leukemias. Now, nanoformulations of this type of drug further enhance its therapeutic potential by making it much more effective and less toxic. The emulsion approach decreases the toxicity significantly, and the rate of biodegradation is very high [16, 32]. Below is a list of the drugs that target histone modifications (Table **2**):

Table 2. Drugs targeting Histone modification.

iHDAC Drugs	Cancer Type	iHDAC Class	References
Apicidin	Cutaneous T-cell lymphoma	Cyclic tetrapeptide	[33]
α-ketoamides	Hematologic cancers	Electrophilic ketones	[34]
Phenylbutyrate	Head and neck cancer	Short-chain fatty acid	[35]
FK228, (Depsipeptides)	Anti-cancer agents	Peripheral T-cell lymphoma	[36]
Depudecin	Anti-angiogenesis	Miscellaneous	[37]

(Table 2) cont.....

iHDAC Drugs	Cancer Type	iHDAC Class	References
Romidepsin FK228	Cutaneous T-cell lymphoma	Cyclic tetrapeptide	[38]
Givinostat	Hodgkin lymphoma	Hydroxamates	[39]
LBH589 Panobinostat	Leukaemia, Hodgkin lymphoma	Hydroxamates	[40]
Mocetinostat MGCD0103	Lymphoid leukemia	Benzamide	[41]
CI-994	Non-small cell lung cancer	Benzamide	[42]
PCI-24781	Sarcoma cell line	Hydroxamates	[43]
PXD101 Belinostat	Mesothelioma	Hydroxamates	[44]
SK-7041	Pancreatic cancer	Hydroxamic	[45]
(SAHA), Vorinostat	Cutaneous T-cell lymphoma	Hydroxamates	[46]
Tasidotin	Solid tumours	Cyclic tetrapeptide	[47]
Trapoxin A	Breast cancer	Cyclic tetrapeptide	[48]
Tubacin	Urothelial cancer cell	Miscellaneous	[49]
Trichostatin A (TSA)	Breast, cervical	Hydroxamates	[50]

NANO-EPIGENETIC APPROACH IN CANCER

Nano-epigenetic drug conjugates integrate epigenetic drugs with nanoscale delivery systems to enhance cancer therapeutic outcomes by improving drug efficacy and specificity. Nano-epigenetics employs nanotechnology to study epigenetic modifications in cancers. Conjugating polymers with weight drug molecules, particularly for cancer treatment, increases the molecule-nan--epigenetic drug and improves its pharmacokinetic disposition in cells. These polymer-drug conjugates, which offer high solubility and stability, promote an enhanced permeability and retention (EPR) effect in cancer cells [51, 52]. Epigenetic drugs show high potential for modulating the sensitivity of tumors to other anticancer regimens and overcoming therapy resistance. However, it is a major challenge in the clinical setting to effectively and rationally combine epigenetic drugs with other therapeutic modalities. The concept of nano-epigenetic drug conjugates involves the design and development of nanocarriers that are capable of delivering epigenetic drugs directly to tumor cells, thereby minimizing off-target effects and maximizing the therapeutic effect [53].

This cutting-edge approach to drug delivery combines epigenetic therapies with targeted nanoparticle delivery. These conjugates selectively target cancer and diseased cells, delivering epigenetic drugs directly to affected cells while minimizing damage to healthy cells [54].

Nano-epigenetic drug conjugates, which use nanoparticles to deliver epigenetic drugs, provide a more effective and less toxic alternative to traditional chemotherapy. Polymeric drug conjugates, such as paclitaxel and doxorubicin combinations, increase drug bioavailability. This technology has the potential to revolutionize cancer treatment and improve patient outcomes. Current clinical trial strategies involve combining epigenetic agents with chemotherapies or targeted therapies [55 - 57].

Cancer is an extremely complex disease characterized by extensive inter- and intra-tumor heterogeneity. In tumor tissues, different tumor cells exhibit different patterns of histone modification in individual genes, indicating that epigenetic heterogeneity occurs at the cellular level.

Nanomedicine represents a promising and advanced strategy in frontier cancer treatment. A major advantage of Nano-epigenetic drug conjugates is their ability to selectively target tumor sites. These nanoscale delivery systems are designed to bind to the surface of tumor cells and deliver epigenetic drugs directly to the specific site of action, reducing toxicity and off-target effects. Nanomedicine therapeutics have proven effective in cancer treatment both *in vitro* and *in vivo*. Advancements in cancer treatments are emerging not only from new therapies but also from new site-specific delivery systems that deliver high drug doses to target sites. Nanotechnology-based agents are convenient for site-specific drug delivery and play a significant role in improving early detection, discovering new biomarkers, and advancing cancer immunotherapy [58, 59]. Nanoparticles facilitate drug delivery through active and passive mechanisms. In the active targeting mechanism, a monoclonal antibody or ligand is attached to the nanoparticle, enabling drug delivery to pathological sites or across biological barriers *via* molecular recognition. In the passive targeting mechanism, nanoparticles are directed to the target site through physicochemical or pharmacological factors, leveraging their inherent size and the unique properties of tumor vasculature, such as the enhanced permeability and retention (EPR) effect and the tumor microenvironment [60, 61]. Nanoparticles smaller than 100 nm can easily move through the capillaries of the reticuloendothelial system and reach the hepatic and spleen macrophages. The drugs enter and accumulate in the macrophages to present their effectiveness. Then, these macrophages work as a defense system in treating hepatic and spleen disease. As in the case of defective lymphatic and vascular systems in carcinoid tumors, the drugs that have entered into the tumor-infected areas after exiting the blood circulation system accumulate for a longer period of time to induce their therapeutic effects.

One of the major challenges in cancer therapy is achieving the desired concentration of therapeutic agents at the tumor site while minimizing damage to normal cells. Therefore, it is crucial to develop agents with significant potential for cancer detection, prevention, and treatment. Integrating epigenetic drugs with nanoscale delivery systems enhances drug stability and permeability. Nanomedicine holds incredible promise for revolutionizing cancer therapeutics and diagnostics by developing innovative biocompatible nanocomposites for drug delivery, representing the most pertinent application of nanoparticles. In recent years, nanocarriers have seen unprecedented application as an emerging class for therapeutic cancer treatment. Notably, two therapeutic nanocarriers, liposomes and albumin nanoparticles, have been approved by the US FDA for clinical use. Liposomal doxorubicin and albumin-bound paclitaxel have demonstrated enhanced permeability and retention-based nanovector applications in breast cancer chemotherapy [62, 63]. Nanocarriers protect drugs from degradation and enhance their solubility and bioavailability, leading to improved drug efficacy and reduced required doses, thereby minimizing potential side effects. By employing both passive and active targeting strategies, nanocarriers can achieve increased intracellular drug concentrations in cancer cells while minimizing toxicity in normal cells. This dual targeting approach simultaneously enhances anticancer effects and reduces systemic toxicity.

MECHANISMS OF ACTION OF NANO-EPIGENETIC DRUG CONJUGATES

Global alteration of epigenetic modification is considered one of the hallmarks of cancer [63] and has always been an ideal potential druggable target for anticancer approaches, mainly because of their plasticity and reversible nature [64]. Targeting epigenetic aberrations in cancer relies mainly on DNMT and HDAC inhibitors [65]. DNA methylation is performed by DNA methyltransferases (DNMTs) using S-adenosyl-L-methionine as the methyl donor group [66]. The proposed hypothesis regarding alterations in DNA methylation first came into existence with the discovery that the cytosine base in DNA can be methylated to become 5-methylcytosine [67]. Cytosine analogs of iDNMT drugs, such as azacytidine, are integrated into DNA and inhibit DNMT in actively replicating cells, which can lead to the loss of methylation marks during DNA replication. The loss of these methylation markers can result in the reactivation of aberrantly silenced tumor suppressor genes, and eventually, their expression and functional activity can be restored [67]. The inhibition of DNMT affects key regulatory pathways, such as apoptosis, cell cycle regulation, and immune modulation. Hypomethylation can lead to re-expression of silenced genes and genomic instability, resulting in long interspersed transposable elements and short interspersed transposable elements [68].

Histone deacetylase inhibitors, such as histone acetylases, histone deacetylases, and histone methyltransferases, are some of the enzymes involved in chromatin remodeling [68]. In addition, various iHDACs have emerged as promising anticancer drugs. HDACs remove acetylation marks from histone tails to create a repressive chromatin environment. This plays an important epigenetic role in inducing and inhibiting cell mobility, cell cycle arrest, apoptosis, and eventually cell death [12]. The effect of epigenetic drug conjugates of iHDACs is mediated mainly when abnormally silenced tumor suppressor genes are re-activated. Currently, four iHDACs have been approved by the USFDA, namely, vorinostat, belinostat, romidepsin, and panobinostat. These drugs have been found to act on class I, II, and IV HDACs and lead to cell cycle arrest, which eventually induces apoptosis [69]. The mechanisms of drug release (Fig. **2**) and interactions with tumor cells are crucial for the effectiveness of nano-epigenetic drug conjugates. The nanocarriers can be designed to release the drugs in response to specific environmental stimuli, such as changes in pH or temperature, or through the action of enzymes present in the tumor microenvironment. This ensures that the drugs are delivered to the tumor cells at the right time and in the right amount, maximizing their therapeutic effects. The interaction between nanocarriers and tumor cells can also be modulated to enhance the uptake and efficacy of drugs, leading to improved treatment outcomes.

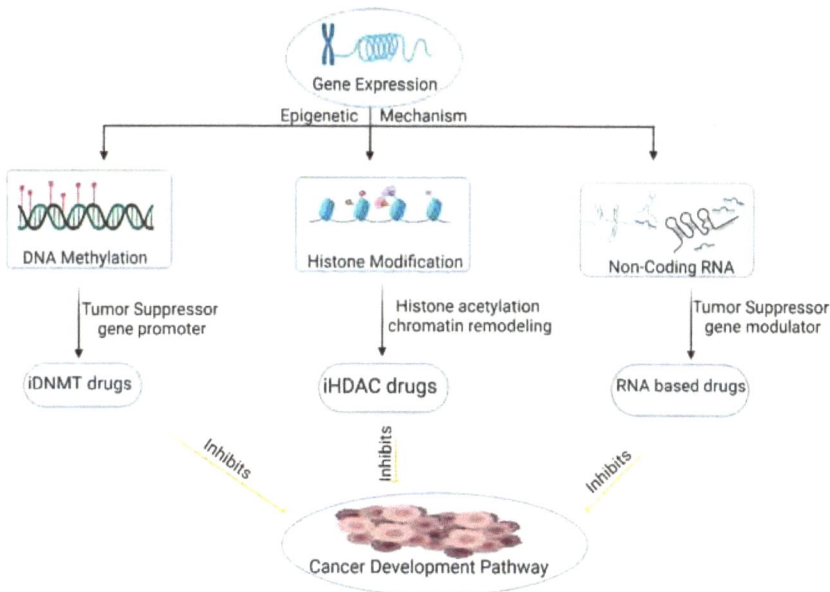

Fig. (2). Mechanism of epigenetic drugs in inhibition of cancer development.

Non-coding RNA are small clusters of RNA that are not naturally translated into a functional protein but have the ability to regulate gene expression at the transcriptional and post-transcriptional levels *via* selective base pairing with their specific target in gene silencing. They play an important epigenetic modulatory role that affects the organization and modification of chromatin [64, 70]. Non-coding RNAs are categorized into two main classes based on their size: short-chain non-coding RNA (< 30 nucleotide chains) and long non-coding RNA (> 200 nucleotides in size). Examples of short-chain ncRNAs include small interfering RNA (siRNA), microRNA (miRNA), and PIWI-interacting RNA (piRNA). Enhancer RNAs and small nucleolar RNA belong to the long non-coding RNAs (lncRNAs) [71]. It has been reported that miRNAs behave as oncogenes by altering tumor-suppressing proteins by modulating the levels of proteins that exhibit the potential of oncogenes. miRNAs can also regulate the expression of more than one RNA. piRNAs are cleaved into individual units, bind to PIWI proteins to induce epigenetic regulation, and act as sequence-specific guides that direct the *de novo* DNA methylation machinery to transposable elements [71]. Long non-coding RNAs can regulate transcription by interfering with the transcriptional machinery, as well as post-transcription, by regulating splicing, mRNA stability, protein translation, and protein stability. An RNA-based drug, MRX34, targets the miR-34 mimic gene and acts as a tumor suppressor gene modulator. Therefore, RNA-based therapeutics conjugated with different conventional chemotherapeutic agents may offer a new approach to treat malignancies [72]. The mechanisms of action of nano-epigenetic drug conjugates highlight the potential of this approach for improving the efficacy and reducing the toxicity of epigenetic drugs in cancer therapy. Further research is needed to fully understand the underlying mechanisms and optimize the design of nanocarriers for the effective and safe delivery of epigenetic drugs in cancer therapy.

CHALLENGES AND LIMITATIONS OF NANO-EPIGENETIC DRUG CONJUGATES

The targeted delivery of therapeutic drugs is a major issue in the clinical setting [73]. The application of conventional therapeutic agents is limited owing to their non-selectivity, side effects, low efficiency, and poor biodistribution. The association of therapeutic agents with nanoparticles is a unique physiochemical and biological property, and designing their pathways for suitable targeting is a promising approach to delivering a wide range of molecules to specific site locations in the body. The use of nanocarriers to deliver epigenetic drugs results in a reduction in side effects. The design of nanocarriers to control drug release significantly reduces the exposure of normal cells to drugs and minimizes their side effects. Nanocarriers can evade the immune system and avoid rapid clearance, leading to improved drug exposure and efficacy. This targeted strategy results in the enhancement of the concentration of the therapeutic agent in cells or tissues, thereby lowering the dose. Therapeutic agents can also be combined with nanoparticles, which can protect them from physiological barriers and improve their bioavailability [64, 74].

One of the major challenges associated with nano-epigenetic drug conjugates is their toxicity. Epigenetic drugs can have a broad range of off-target effects and cause toxicity, leading to undesirable side effects. The integration of these drugs with nanoscale delivery systems can help reduce toxicity and improve therapeutic outcomes; however, further optimization is required to fully overcome this challenge. Another challenge with nano-epigenetic drug conjugates is the solubility and bioavailability of the drugs. Some epigenetic drugs are poorly soluble, which leads to low bioavailability and reduced therapeutic efficacy. Nanoscale delivery systems can help to improve the solubility and bioavailability of these drugs; however, more research is needed to fully address these challenges. The lack of specific biomarkers for their target populations is also a major challenge for the use of nano-epigenetic drug conjugates. The absence of these biomarkers makes it difficult to accurately identify patients who would benefit from these drugs, limiting their clinical applicability [75, 76].

Finally, the challenge associated with nano-epigenetic drug conjugates is the clinical translation and regulatory approval processes. The regulatory approval process can be lengthy and complex, and translation of these drugs from preclinical studies to clinical trials can be challenging. Despite these challenges, there is growing interest in the development and use of nano-epigenetic drug conjugates, and the future outlook for these drugs is promising.

FUTURE PERSPECTIVE OF NANO-EPIGENETIC DRUG CONJUGATES IN CANCER THERAPY

Despite the challenges associated with nano-epigenetic drug conjugates, there is a growing body of evidence demonstrating their potential as a new and innovative approach to cancer therapy. The integration of these drugs with nanoscale delivery systems offers several benefits, including improved stability, permeability, and efficacy, as well as reduced toxicity and off-target effects. In the future, it is likely that nano-epigenetic drug conjugates will play a significant role in cancer treatment, and further research is needed to fully realize their potential [60, 64, 76]. The future of nano-epigenetic drug conjugates in cancer therapy is highly promising, with several emerging trends and research directions that hold great potential for further advancing the field. One of the key areas of focus in the future of nano-epigenetic drug conjugates is the integration of these drugs with other cancer therapies such as chemotherapy, radiotherapy, and immunotherapy. This integration could lead to more effective and synergistic cancer treatments, as different therapies work together to target different aspects of the disease.

Another important area of research is the continued development of advanced and effective methods for synthesizing and formulating nano-epigenetic drug conjugates. This can include the development of new and innovative nanoscale delivery systems, as well as the optimization of existing systems, to improve the stability, bioavailability, and target specificity of drugs. Personalized medicine, or the tailoring of treatments to the unique needs and characteristics of each patient, is an increasingly important trend in the field of cancer therapy. The integration of nano-epigenetic drug conjugates with personalized medicine can lead to the development of highly targeted and effective treatments tailored to the specific needs and characteristics of each patient.

Finally, a number of emerging trends and research directions in the field of nano-epigenetic drug conjugates hold great promise for the future. These can include the development of new epigenetic drugs that target specific cancer-associated pathways, exploration of new nanoscale delivery systems, and investigation of the potential for nanoscale epigenetic drugs to be used in combination with other cancer therapies. Overall, the future of nano-epigenetic drug conjugates in cancer therapy is bright, with many exciting new developments and research directions that hold great potential for further advancement in the field and improving outcomes in cancer patients.

CONCLUSION

In conclusion, nano-epigenetic drug conjugates have emerged as a promising approach for cancer therapy, combining the targeted benefits of epigenetic drugs with the advantages of nanoscale delivery systems. The current literature review highlights recent advancements in nano-epigenetic drug conjugates, including their definition and concept, benefits, mechanisms of action, and challenges and limitations. It also highlights the future perspective of nano-epigenetic drug conjugates in cancer therapy, including their potential for integration with other cancer therapies, advances in synthesis and formulation, and the potential for personalized medicine. Nano-epigenetic drug conjugates have the potential to overcome some limitations of traditional epigenetic drugs, such as toxicity, low solubility, poor bioavailability, and low stability, while enhancing their targeting, stability, and permeability. This approach can lead to a more effective and efficient cancer therapy with fewer side effects and a better response rate. The use of nano-epigenetic drug conjugates in cancer therapy can potentially improve the outcomes of patients with cancer, including longer remission times and reduced toxicity. Additionally, this approach can lead to a more efficient use of healthcare resources and reduce the overall burden on the healthcare system.

Despite promising results in preclinical and clinical studies, there is still much work to be done to fully realize the potential of nano-epigenetic drug conjugates in cancer therapy. Further research is needed to optimize their synthesis, formulation, and delivery and to explore their potential for use in combination with other cancer therapies. Additionally, there is a need for further clinical trials to validate the safety and efficacy of nano-epigenetic drug conjugates and to determine their optimal dosing and administration.

In conclusion, nano-epigenetic drug conjugates represent a promising approach to cancer therapy, offering the potential for targeted and effective treatment with fewer side effects. Further research and development in this field can help to realize their full potential and improve outcomes in cancer patients.

REFERENCES

[1]   Mukherjee A, Madamsetty VS, Paul MK, Mukherjee S. Recent advancements of nanomedicine towards antiangiogenic therapy in cancer. Int J Mol Sci 2020; 21(2): 455.
[http://dx.doi.org/10.3390/ijms21020455]

[2]   Singhal S, Nie S, Wang MD. Nanotechnology Applications in Surgical Oncology. Annu Rev Med. 2010; 61: 359. Available from: /pmc/articles/PMC2913871/
[http://dx.doi.org/10.1146/annurev.med.60.052907.094936]

[3]   Sengupta S, Eavarone D, Capila I, *et al.* Temporal targeting of tumour cells and neovasculature with a nanoscale delivery system. Nature 2005; 436(7050): 568-72.
[http://dx.doi.org/10.1038/nature03794] [PMID: 16049491]

[4]   Prabhu RH, Patravale VB, Joshi MD. Polymeric nanoparticles for targeted treatment in oncology: current insights. Int J Nanomedicine. 2015; 10: 1001. Available from: /pmc/articles/PMC4324541/

[5]   Crucho CIC, Barros MT. Polymeric nanoparticles: A study on the preparation variables and characterization methods. Mater Sci Eng C 2017; 80: 771-84.
[http://dx.doi.org/10.1016/j.msec.2017.06.004] [PMID: 28866227]

[6]   Montalvo-Casimiro M. Available from: www.frontiersin.org

[7]   Dupont C, Armant D, Brenner C. Epigenetics: definition, mechanisms and clinical perspective. Semin Reprod Med 2009; 27(5): 351-7. Available from: https://pubmed.ncbi.nlm.nih.gov/19711245/
[http://dx.doi.org/10.1055/s-0029-1237423] [PMID: 19711245]

[8]   Allis CD, Jenuwein T. The molecular hallmarks of epigenetic control 2016. Available from: https://www.nature.com/articles/nrg.2016.59
[http://dx.doi.org/10.1038/nrg.2016.59]

[9]   Hao X, Luo H, Krawczyk M, Wei W, Wang W, Wang J, *et al.* DNA methylation markers for diagnosis and prognosis of common cancers. Biological Sci 2017; 114(28): 7414-9.
[http://dx.doi.org/10.1073/pnas.1703577114]

[10]   Heyn H, Esteller M. DNA methylation profiling in the clinic: applications and challenges. Nat Rev Genet 2012; 13(10): 679-92. Available from: https://pubmed.ncbi.nlm.nih.gov/22945394/
[http://dx.doi.org/10.1038/nrg3270] [PMID: 22945394]

[11]   Garzon R, Marcucci G, Croce CM. Targeting microRNAs in cancer: rationale, strategies and

challenges. Nat Rev Drug Discov 2010; 9(10): 775-89. Available from: https://pubmed.ncbi.nlm.nih. gov/20885409/
[http://dx.doi.org/10.1038/nrd3179] [PMID: 20885409]

[12]   Baylin SB, Jones PA. Epigenetic Determinants of Cancer. Cold Spring Harb Perspect Biol 2016; 8(9): a019505. Available from: http://cshperspectives.cshlp.org/
[http://dx.doi.org/10.1101/cshperspect.a019505]

[13]   Garofalo M, Croce CM. microRNAs: Master regulators as potential therapeutics in cancer. Annu Rev Pharmacol Toxicol 2011; 51(1): 25-43. Available from: https://pubmed.ncbi.nlm.nih.gov/20809797/
[http://dx.doi.org/10.1146/annurev-pharmtox-010510-100517] [PMID: 20809797]

[14]   Lee YS, Dutta A. MicroRNAs in Cancer. Annu Rev Pathol 2009; 4(1): 199-227. Available from: https://pubmed.ncbi.nlm.nih.gov/18817506/
[http://dx.doi.org/10.1146/annurev.pathol.4.110807.092222] [PMID: 18817506]

[15]   Rodríguez-Paredes M, Esteller M. Cancer epigenetics reaches mainstream oncology. Nat Med 2011; 17: 330–339. Available from: https://www.nature.com/articles/nm.2305
[http://dx.doi.org/10.1038/nm.2305]

[16]   Kaminskas E, Farrell AT, Wang YC, Sridhara R, Pazdur R. FDA drug approval summary: azacitidine (5-azacytidine, Vidaza™) for injectable suspension. The oncologist 2005; 1;10(3): 176-82.

[17]   Gailhouste L, Liew LC, Hatada I, Nakagama H, Ochiya T. Epigenetic reprogramming using 5-azacytidine promotes an anti-cancer response in pancreatic adenocarcinoma cells. Cell Death Dis 2018; 9: 468. Available from: https://www.nature.com/articles/s41419-018-0487-z
[http://dx.doi.org/10.1038/s41419-018-0487-z]

[18]   Saba HI, Wijermans PW. Decitabine in myelodysplastic syndromes. Semin Hematol 2005; 42(3) (Suppl. 2): S23-31.
[http://dx.doi.org/10.1053/j.seminhematol.2005.05.009] [PMID: 16015501]

[19]   Kanai Y, Ushijima S, Nakanishi Y, Sakamoto M, Hirohashi S. Mutation of the DNA methyltransferase (DNMT) 1 gene in human colorectal cancers. Cancer Lett 2003; 192(1): 75-82.
[http://dx.doi.org/10.1016/S0304-3835(02)00689-4] [PMID: 12637155]

[20]   Ley TJ, Ding L, Walter MJ, McLellan MD, Lamprecht T, Larson DE, *et al.* DNMT3A Mutations in Acute Myeloid Leukemia. N Engl J Med. 2010; 363(25): 2424.
[http://dx.doi.org/10.1056/NEJMoa1005143]

[21]   Jiang YL, Rigolet M, Bourc'his D, *et al.* DNMT3B mutations and DNA methylation defect define two types of ICF syndrome. Hum Mutat 2005; 25(1): 56-63.
[http://dx.doi.org/10.1002/humu.20113] [PMID: 15580563]

[22]   Shen XM, Ohno K, Fukudome T, *et al.* Congenital myasthenic syndrome caused by low-expressor fast-channel AChR δ subunit mutation. Neurology 2002; 59(12): 1881-8. Available from: https://pubmed.ncbi.nlm.nih.gov/12499478/
[http://dx.doi.org/10.1212/01.WNL.0000042422.87384.2F] [PMID: 12499478]

[23]   Daifuku R. Pharmacoepigenetics of Novel Nucleoside DNA Methyltransferase Inhibitors. Pharmacoepigenetics. 2019; 425–35.
[http://dx.doi.org/10.1016/B978-0-12-813939-4.00007-3]

[24]   Kaufman DW, Kelly JP, Rosenberg L, Stolley PD, Warshauer ME, Shapiro S. Hydralazine use in relation to cancers of the lung, colon, and rectum. Eur J Clin Pharmacol 1989; 36(3): 259-64. Available from: https://link.springer.com/article/10.1007/BF00558157
[http://dx.doi.org/10.1007/BF00558157] [PMID: 2744066]

[25]   Tada M, Imazeki F, Fukai K, *et al.* Procaine inhibits the proliferation and DNA methylation in human hepatoma cells. Hepatol Int 2007; 1(3): 355-64. Available from: https://pubmed.ncbi.nlm.nih. gov/19669330/
[http://dx.doi.org/10.1007/s12072-007-9014-5] [PMID: 19669330]

[26]   Ali MS, Farah MA, Al-Lohedan HA, Al-Anazi M. Comprehensive exploration of the anticancer activities of procaine and its binding with calf thymus DNA: a multi spectroscopic and molecular modelling study. RSC Adv 2018; 8: 9083-93.
[http://dx.doi.org/10.1039/C7RA13647A]

[27]   Li C, Gao S, Li X, Ma L. Procaine inhibits the proliferation and migration of colon cancer cells through inactivation of the ERK/MAPK/FAK pathways by regulation of RhoA. Oncology research 2018; 26(2): 209.

[28]   Medina-Franco JL, Yoo J, Dueñas-González A. DNA Methyltransferase Inhibitors for Cancer Therapy. Epigenetic Technological Applications. 2015; 265–90.
[http://dx.doi.org/10.1016/B978-0-12-801080-8.00013-2]

[29]   1990. Available from: https://www.ncbi.nlm.nih.gov/books/NBK526160/

[30]   Ideström K, Kimby E, Björkholm M, *et al.* Treatment of chronic lymphocytic leukaemia and well-differentiated lymphocytic lymphoma with continuous low- or intermittent high-dose prednimustine *versus* chlorambucil/prednisolone. Eur J Cancer Clin Oncol 1982; 18(11): 1117-23.
[http://dx.doi.org/10.1016/0277-5379(82)90092-X] [PMID: 6897633]

[31]   Szántó I, Fleischmann T, Eckhardt S. Prednimustine treatment in malignant lymphomas. Oncology 1989; 46(4): 205-7. Available from: https://karger.com/ocl/article/46/4/205/235870/Prednimustine-Treatment-in-Malignant-Lymphomas
[http://dx.doi.org/10.1159/000226716] [PMID: 2662086]

[32]   Kim HJ, Bae SC. Histone deacetylase inhibitors: molecular mechanisms of action and clinical trials as anti-cancer drugs. Am J Transl Res. 2011; 3(2): 166.

[33]   Lee JH, Choy ML, Marks PA. Mechanisms of resistance to histone deacetylase inhibitors. Adv Cancer Res 2012; 116: 39-86.
[http://dx.doi.org/10.1016/B978-0-12-394387-3.00002-1] [PMID: 23088868]

[34]   Wang J, Liang B, Chen Y, Fuk-Woo Chan J, Yuan S, Ye H, *et al.* Wang J, Liang B, Chen Y, Fuk-Woo Chan J, Yuan S, Ye H, *et al.* A new class of α-ketoamide derivatives with potent anticancer and anti-SARS-CoV-2 activities. Eur J Med Chem. 2021; 215: 113267.
[http://dx.doi.org/10.1016/j.ejmech.2021.113267]

[35]   Burkitt K, Ljungman M. Phenylbutyrate interferes with the Fanconi anemia and BRCA pathway and sensitizes head and neck cancer cells to cisplatin. Mol Cancer 2008; 7(1): 24. Available from: https://molecular-cancer.biomedcentral.com/articles/10.1186/1476-4598-7-24
[http://dx.doi.org/10.1186/1476-4598-7-24] [PMID: 18325101]

[36]   Konstantinopoulos PA, Vandoros GP, Papavassiliou AG. FK228 (depsipeptide): a HDAC inhibitor with pleiotropic antitumor activities. Cancer Chemother Pharmacol 2006; 58(5): 711-5. Available from: https://link.springer.com/article/10.1007/s00280-005-0182-5
[http://dx.doi.org/10.1007/s00280-005-0182-5] [PMID: 16435156]

[37]   Kwon HJ, Owa T, Hassig CA, Shimada J, Schreiber SL. Depudecin induces morphological reversion of transformed fibroblasts *via* the inhibition of histone deacetylase. Proc Natl Acad Sci USA. 1998; 95(7): 3356.
[http://dx.doi.org/10.1073/pnas.95.7.3356]

[38]   Saijo K, Katoh T, Shimodaira H, Oda A, Takahashi O, Ishioka C. Romidepsin (FK228) and its analogs directly inhibit phosphatidylinositol 3-kinase activity and potently induce apoptosis as histone deacetylase/phosphatidylinositol 3-kinase dual inhibitors. Cancer Sci. 2012; 103(11): 1994.
[http://dx.doi.org/10.1111/j.1349-7006.2012.02363.x]

[39]   Choulis NH. Dermatological Drugs, Topical Agents, and Cosmetics. Side Effects of Drugs Annual 2014; 36: 203-31.
[http://dx.doi.org/10.1016/B978-0-444-63407-8.00014-9]

[40]   Thomas S, Miller A, Thurn KT, Munster P. Clinical Applications of Histone Deacetylase Inhibitors.

Handbook of Epigenetics: The New Molecular and Medical Genetics. 2011; 597–615.
[http://dx.doi.org/10.1016/B978-0-12-375709-8.00037-X]

[41]   Boumber Y, Younes A, Garcia-Manero G. Mocetinostat (MGCD0103): a review of an isotype-specific histone deacetylase inhibitor. Expert Opin Investig Drugs 2011; 20(6): 823-9.
[http://dx.doi.org/10.1517/13543784.2011.577737] [PMID: 21554162]

[42]   Loprevite M, Tiseo M, Grossi F, Scolaro T, Semino C, Pandolfi A, Favoni R, Ardizzoni A. In vitro study of CI-994, a histone deacetylase inhibitor, in non-small cell lung cancer cell lines. Oncology research 2005; 1;15(1): 39-48.

[43]   Yang C, Choy E, Hornicek FJ, Wood KB, Schwab JH, Liu X, *et al.* Histone deacetylase inhibitor (HDACI) PCI-24781 enhances chemotherapy induced apoptosis in multidrug resistant sarcoma cell lines. Cancer Research 2009; 69(15): 5826-34.
[http://dx.doi.org/10.1158/0008-5472.CAN-08-4703]

[44]   Ramalingam SS, Belani CP, Ruel C, *et al.* Phase II study of belinostat (PXD101), a histone deacetylase inhibitor, for second line therapy of advanced malignant pleural mesothelioma. J Thorac Oncol 2009; 4(1): 97-101.
[http://dx.doi.org/10.1097/JTO.0b013e318191520c] [PMID: 19096314]

[45]   Ryu JK, Lee WJ, Lee KH, *et al.* SK-7041, a new histone deacetylase inhibitor, induces G2-M cell cycle arrest and apoptosis in pancreatic cancer cell lines. Cancer Lett 2006; 237(1): 143-54.
[http://dx.doi.org/10.1016/j.canlet.2005.05.040] [PMID: 16009488]

[46]   Butowski NA, Chang SM. General and neurological complications of targeted therapy. Handb Clin Neurol 2012; 105: 937-45.
[http://dx.doi.org/10.1016/B978-0-444-53502-3.00033-1] [PMID: 22230543]

[47]   Mita AC, Hammond LA, Bonate PL, *et al.* Phase I and pharmacokinetic study of tasidotin hydrochloride (ILX651), a third-generation dolastatin-15 analogue, administered weekly for 3 weeks every 28 days in patients with advanced solid tumors. Clin Cancer Res 2006; 12(17): 5207-15. Available from: https://pubmed.ncbi.nlm.nih.gov/16951240/
[http://dx.doi.org/10.1158/1078-0432.CCR-06-0179] [PMID: 16951240]

[48]   Akbari E, Mousazadeh H, Sabet Z, *et al.* Dual drug delivery of trapoxin A and methotrexate from biocompatible PLGA-PEG polymeric nanoparticles enhanced antitumor activity in breast cancer cell line. J Drug Deliv Sci Technol 2021; 61: 102294.
[http://dx.doi.org/10.1016/j.jddst.2020.102294]

[49]   Rosik L, Niegisch G, Fischer U, Jung M, Schulz WA, Hoffmann MJ. Limited efficacy of specific HDAC6 inhibition in urothelial cancer cells. Cancer Biol Ther. 2014; 15(6): 742.
[http://dx.doi.org/10.4161/cbt.28469]

[50]   Lee JW, Yang DH, Park S, Han HK, Park JW, Kim BY, *et al.* Trichostatin A resistance is facilitated by HIF-1α acetylation in HeLa human cervical cancer cells under normoxic conditions. Oncotarget. 2018; 9(2): 2035.

[51]   Maeda H, Bharate GY, Daruwalla J. Polymeric drugs for efficient tumor-targeted drug delivery based on EPR-effect. Eur J Pharm Biopharm 2009; 71(3): 409-19. Available from: https://pubmed.ncbi.nlm.nih.gov/19070661/
[http://dx.doi.org/10.1016/j.ejpb.2008.11.010] [PMID: 19070661]

[52]   Deb PK, Kokaz SF, Abed SN, Paradkar A, Tekade RK. Pharmaceutical and Biomedical Applications of Polymers. Basic Fundamentals of Drug Delivery. 2019; 203–67.
[http://dx.doi.org/10.1016/B978-0-12-817909-3.00006-6]

[53]   Chen F, Shi Y, Zhang J, Liu Q. Nanoparticle-based Drug Delivery Systems for Targeted Epigenetics Cancer Therapy. Curr Drug Targets 2020; 21(11): 1084-98.
[http://dx.doi.org/10.2174/1389450121666200514222900] [PMID: 32410563]

[54]   Hoppenz P, Els-Heindl S, Beck-Sickinger AG, Beck-Sickinger AG, Hoppenz P, Els-Heindl S.

Peptide-Drug Conjugates and Their Targets in Advanced Cancer Therapies. Front Chem 2020; 8: 571. Available from: www.frontiersin.org [Internet].
[http://dx.doi.org/10.3389/fchem.2020.00571]

[55]   Hasan M, Leak RK, Stratford RE, Zlotos DP, Witt-Enderby PA. Drug conjugates—an emerging approach to treat breast cancer. Pharmacol Res Perspect 2018; 6(4): e00417.
[http://dx.doi.org/10.1002/prp2.417] [PMID: 29983986]

[56]   Mahapatro A, Singh DK. Biodegradable nanoparticles are excellent vehicle for site directed in-vivo delivery of drugs and vaccines 2011.
[http://dx.doi.org/10.1186/1477-3155-9-55]

[57]   Panyam J, Labhasetwar V. Biodegradable nanoparticles for drug and gene delivery to cells and tissue. Adv Drug Deliv Rev 2003; 55(3): 329-47.
[http://dx.doi.org/10.1016/S0169-409X(02)00228-4] [PMID: 12628320]

[58]   Mitchell MJ, Billingsley MM, Haley RM, Wechsler ME, Peppas NA, Langer R. Engineering precision nanoparticles for drug delivery. Nat Rev Drug Discov 2021; 20: 101–124. Available from: https://www.nature.com/articles/s41573-020-0090-8
[http://dx.doi.org/10.1038/s41573-020-0090-8]

[59]   Rodríguez F, Caruana P, De la Fuente N, Español P, Gámez M, Balart J, *et al.* Nano-Based Approved Pharmaceuticals for Cancer Treatment: Present and Future Challenges. Biomolecules 2022; 12(6): 784.
[http://dx.doi.org/10.3390/biom12060784]

[60]   Aghebati-Maleki A, Dolati S, Ahmadi M, Baghbanzadeh A, Asadi M, Fotouhi A, *et al.* Nanoparticles and cancer therapy: Perspectives for application of nanoparticles in the treatment of cancers. J Cell Physio. Wiley-Liss Inc.; 2020. p. 1962–72.

[61]   Ma P, Ma M. Paclitaxel Nano-Delivery Systems: A Comprehensive Review. J Nanomed Nanotechnol 2013; 4(2): 164.
[http://dx.doi.org/10.4172/2157-7439.1000164]

[62]   Misra R, Acharya S, Sahoo SK. Cancer nanotechnology: application of nanotechnology in cancer therapy. Drug Discov Today 2010; 15(19-20): 842-50.
[http://dx.doi.org/10.1016/j.drudis.2010.08.006] [PMID: 20727417]

[63]   Yoo CB, Jones PA. Epigenetic therapy of cancer: past, present and future. Nat Rev Drug Discov 2006; 5(1): 37-50. Available from: https://pubmed.ncbi.nlm.nih.gov/16485345/
[http://dx.doi.org/10.1038/nrd1930] [PMID: 16485345]

[64]   Roberti A, Valdes AF, Torrecillas R, Fraga MF, Fernandez AF. Epigenetics in cancer therapy and nanomedicine. Clin Epigenet 2019; 11: 81.
[http://dx.doi.org/10.1186/s13148-019-0675-4]

[65]   Zhu B, Reinberg D. Epigenetic inheritance: Uncontested? Cell Res 2011; 21(3): 435-41.
[http://dx.doi.org/10.1038/cr.2011.26] [PMID: 21321606]

[66]   Hoffmann I, Roatsch M, Schmitt ML, Carlino L, Pippel M, Sippl W, *et al.* The role of histone demethylases in cancer therapy. Mol Oncol 2012; 6(6): 683-703.
[http://dx.doi.org/10.1016/j.molonc.2012.07.004]

[67]   Herman JG, Baylin SB. Gene silencing in cancer in association with promoter hypermethylation. N Engl J Med 2003; 349(21): 2042-54. Available from: https://pubmed.ncbi.nlm.nih.gov/14627790/
[http://dx.doi.org/10.1056/NEJMra023075] [PMID: 14627790]

[68]   Cang S, Ma Y, Liu D. New clinical developments in histone deacetylase inhibitors for epigenetic therapy of cancer. J Hematol Oncol 2009; 2 (22). Available from: http://www.jhoonline.org/content/2/1/22
[http://dx.doi.org/10.1186/1756-8722-2-22]

[69]   Xue K, Gu JJ, Zhang Q, *et al.* Vorinostat, a histone deacetylase (HDAC) inhibitor, promotes cell cycle

arrest and re-sensitizes rituximab- and chemo-resistant lymphoma cells to chemotherapy agents. J Cancer Res Clin Oncol 2016; 142(2): 379-87. Available from: https://link.springer.com/article/10.1007/s00432-015-2026-y
[http://dx.doi.org/10.1007/s00432-015-2026-y] [PMID: 26314218]

[70]   Kanwal R, Gupta K, Gupta S. Cancer epigenetics: an introduction. Methods Mol Biol 2015; 1238: 3-25.
[http://dx.doi.org/10.1007/978-1-4939-1804-1_1] [PMID: 25421652]

[71]   Ferreira HJ, Esteller M. Non-coding RNAs, epigenetics, and cancer: tying it all together. Cancer Metastasis Rev 2018; 37: 55-73.
[http://dx.doi.org/10.1007/s10555-017-9715-8]

[72]   Chakraborty C, Sharma AR, Sharma G, Doss CGP, Lee SS. Therapeutic miRNA and siRNA: Moving from Bench to Clinic as Next Generation Medicine. Mol Ther Nucleic Acids 2017; 8: 132-43.
[http://dx.doi.org/10.1016/j.omtn.2017.06.005] [PMID: 28918016]

[73]   Attia MF, Anton N, Wallyn J, Omran Z, Vandamme TF, Correspondence MF, *et al.* An overview of active and passive targeting strategies to improve the nanocarriers efficiency to tumour sites 2019. Available from: https://academic.oup.com/jpp/article/71/8/1185/6122081
[http://dx.doi.org/10.1111/jphp.13098]

[74]   Feinberg AP, Koldobskiy MA, Göndör A. Epigenetic modulators, modifiers and mediators in cancer aetiology and progression. Nat Rev Genet 2016; 17: 284-99.
[http://dx.doi.org/10.1038/nrg.2016.13]

[75]   Cramer SA, Adjei IM, Labhasetwar V. Advancements in the delivery of epigenetic drugs. Expert Opin Drug Deliv 2015; 12(9): 1501-12.
[http://dx.doi.org/10.1517/17425247.2015.1021678] [PMID: 25739728]

[76]   Lee JJ, Yazan LS, Abdullah CAC. A review on current nanomaterials and their drug conjugate for targeted breast cancer treatment. Int J Nanomed 2017; 12.
[http://dx.doi.org/10.2147/IJN.S127329]

CHAPTER 9

Artificial Intelligence and Nanotechnology-Integrated Recent Applications in Early Lung Cancer Detection and Therapy

Ananya Bhattacharjee[1], Nayan Ranjan Ghosh Biswas[2], Abhishek Bhattacharjee[3,*], R. Murugan[1], Ranjit Prasad Swain[4] and Ram Kumar Sahu[5]

[1] *Bio-Medical Imaging Laboratory (BIOMIL), National Institute of Technology, Silchar-788010, Assam, India*

[2] *Department of Pharmaceutical Sciences, Dibrugarh University, Dibrugarh-786004, Assam, India*

[3] *Department of Pharmaceutical Sciences, Assam University (A Central University), Silchar-788011, Assam, India*

[4] *Department of Pharmaceutics, GITAM School of Pharmacy, GITAM (Deemed to be University), Visakhapatnam-530045, Andhra Pradesh, India*

[5] *Department of Pharmaceutical Sciences, Hemvati Nandan Bahuguna University (A Central University), Garhwal-249161, Uttarakhand, India*

Abstract: Globally, lung cancer is a leading cause of mortality resulting from mutating genes and altered pathologies. Early-stage detection of such lethal disease is very critical, and subsequent nanomaterial-based personalized drug delivery systems can help reduce death rates when metastasized tumors have not spread to other regions of the body. Artificial intelligence and nanotechnology, collectively known as nanoinformatics, have advanced challenging scopes in nanotherapeutics. Artificial intelligence-based computer-aided diagnosis systems have huge potential in performing early theranostics by integrating machine learning and deep learning techniques in lung cancer detection. These models have a high degree of accuracy, specificity, and sensitivity due to artificial intelligence-based image processing and segmentation techniques. The emergence of widespread applications of nanotechnology in the design of site-specific novel drug delivery systems is crucial in lung cancer therapy as they can target cancerous cells. Optimizing nanomaterials' properties have proven to be beneficial in therapy in accordance with the interaction of drugs with biological fluid and the immune system. Embedded nanocarriers in carbon nanotubes, polymeric micelles, and liposomes possess great pharmacokinetic and pharmacodynamic potential in the management of the healthcare of lung cancer patients.

* **Corresponding author Abhishek Bhattacharjee:** Department of Pharmaceutical Sciences, Assam University (A Central University), Silchar-788011, Assam, India; E-mail: abhishek.bhattacharjee@aus.ac.in

Keywords: Artificial   intelligence,   Lung   cancer,   Nanotechnology, Nanoinformatics, Nanotherapeutics.

INTRODUCTION

Cancer is a set of diseases that are distinguished from one another by the abnormal and uncontrolled proliferation of cells originating from a variety of sources, and it can affect a number of organs throughout the body. Cancer is the primary reason for mortality in many parts of the world [1]. Malignancy, the name of cancer, is the unchecked or abnormal proliferation of cells that weakens our immune system. Human cells divide by a process known as mitosis *via* cell growth and multiplication, followed by a programmed death, which replaces the old ones with the newer ones. When this orderly apoptotic process fails, there is unceasing growth and multiplication of cells, which ultimately results in the development of a lump or tumor, which can either develop into cancer or sometimes remains as a noncancerous mass (benign tumor). Cancer is basically a disorder that develops due to the genetic mutation in the cells upon their exposure to various noxious stimuli, viz. chemicals, fumes and smoke, radiations, viruses, *etc.* It is observed that about 200 different types of cancer are present. It can be classified according to the organ in the body in which it first appears (for example, cancer of the brain, cancer of the lungs, cancer of the cervix, *etc.*). Another way to categorize cancer is by the type of cell in which it first manifests. There are only a few main categories of cancer, and each is characterized by its cellular origin. They are as follows [2]:

Carcinoma

A form of cancer known as epithelial carcinoma that starts in the cells that line or cover internal organs, such as the skin or tissue of the esophagus, kidney, pancreas, prostate gland, head, neck, *etc.* Some of the subtypes of cancer comprise adenocarcinoma, basal cell carcinoma, transitional cell carcinoma, and squamous cell carcinoma.

Sarcoma

It starts in mesenchymal connective tissues such as bones, cartilage, blood vessels, muscle, adipose tissues, *etc.*

Leukaemia

This particular form of cancer attacks the white blood cells in the body. It develops in the tissues responsible for producing blood cells, such as bone marrow.

Lymphoma and Myeloma

Myeloma, on the other hand, is a type of cancer that affects the plasma cells, whereas lymphoma affects the lymphatic system.

Lung cancer is the only form of the many types of prevalent malignancies estimated to be responsible for nearly 1.80 million deaths worldwide in 2020-21. Lung cancer is also the second most common form of malignancy, accounting for 20% of all cancer-related deaths.

Lung Carcinoma

Among the several kinds of lung cancer, the most frequent are non-small cell lung cancer (NSCLC) and small cell lung cancer (SCLC). Of the total number of lung cancer cases, NSCLC accounts for 85 percent, while SCLC makes up the remaining 15 percent. NSCLC can be further divided into three different subtypes: Adenocarcinoma, squamous cell carcinoma, and giant cell carcinoma. The most common subtype is adenocarcinoma, which typically develops in the lungs' outer layers [3]. The middle of the lungs is the most common location for the development of squamous cell carcinoma, which is commonly related to smoking. The less prevalent form of lung cancer, known as large cell carcinoma, can arise in any area of the lungs. On the other hand, the SCLC can be further subdivided into three subtypes, each of which has a distinctive therapeutic profile and prognosis. These subtypes are the small cell carcinoma, the mixed small cell and large cell carcinoma, and the squamous cell carcinoma. Regardless of the subtype, lung cancer typically starts as abnormal cells that grow out of control, eventually forming a tumor [4]. Early lung cancer diagnosis enables the development of effective treatment plans. However, most cases of lung cancer are detected at a more advanced level. Detecting lung cancer nodules manually from the image modalities with the naked eye is a laborious task [5]. The requirement for a second opinion as a result of a scarcity of healthcare experts has a substantial influence on the procedure [6]. Thus, it appears that early identification of lung disorders is crucial for avoiding lung cancer. Artificial intelligence, which includes deep learning (DL) and machine learning (ML) techniques, which are incorporated into computer-aided diagnosis systems, plays a crucial role in the early diagnosis of lung cancer [7]. Conventional healthcare management has limitations, but predictive approaches such as deep learning (DL) algorithms can assist in circumventing them. The adoption of DL-based detection might minimize invasive procedures, hence improving the efficacy and sustainability of present healthcare practices [8]. In addition, radiomics is the process of extracting data that can be obtained from medical imaging. This data is then used in oncology to enhance diagnostic capabilities and clinical decision-making to enable precision medicine [9]. Both lung nodule segmentation and classification are part of an automated lung cancer diagnosis system. The deep learning-based automated systems provide an edge over the performance of conventional detection techniques.

Pathology of Lung Cancer

There are many different subtypes of lung cancer, and the pathological aspects of the disease can differ significantly depending on the stage of the disease and other circumstances. The process through which lung cancer develops is a complicated one that is influenced by a wide variety of hereditary and environmental factors. The inhalation of tobacco smoke is one of the most prevalent etiological factors of lung cancer. The smoke from cigarettes comprises various cancer-causing agents that have the ability to cause DNA damage in lung cells and ultimately result in mutations that can encourage the occurrence of cancer. Additional risk factors for lung cancer include having a history of the disease in one's family, having certain genetic abnormalities, being exposed to radon gas and asbestos, and breathing polluted air. Once the genetic mutations occur, they can activate oncogenes or inactivate tumor suppressor genes, leading to the uncontrolled growth and division of cells that characterize cancer. This may lead to the growth of tumors in the tissue of the lungs, which can then metastasize or spread to other regions of the body by invading neighboring tissues, organs, and blood vessels [10]. Because of a meager diagnosis at the very onset, most lung cancer patients become aware at the latter stage when it is not preferred to go with the surgical treatment. Hence, in order to outspread the survival expectancy and improve life quality, systemic chemotherapy and various other approaches such as hormonal therapy, radiation therapy, and a combination of nanotechnology with artificial intelligence for theranostics purposes are under continuous use and investigation and trial [11].

Artificial Intelligence in Lung Cancer Detection

The ability to establish successful treatment strategies for lung cancer is made possible by early detection of the disease. However, the majority of cases of lung cancer are detected at a more advanced level. Detecting lung cancer nodules manually from the image modalities with the naked eye is a laborious task [5]. The requirement for a second opinion as a result of a scarcity of healthcare experts has a substantial influence on the procedure [6]. Thus, it appears that early identification of lung disorders is crucial for avoiding lung cancer. Artificial intelligence, including machine learning (ML) and deep learning (DL) techniques integrated with computer-aided diagnosis systems, plays an important part in the early detection of lung cancer [7]. Conventional healthcare management has limitations, but predictive approaches such as deep learning algorithms can assist in circumventing them. The adoption of DL-based detection might minimize invasive procedures, hence improving the efficacy and sustainability of present healthcare practices [8]. Additionally, radiomics is the process of extracting data that may be extracted from medical imaging. This data is then used in oncology to enhance diagnostic capabilities and clinical decision-making in order to provide precision medicine [9]. Both lung nodule segmentation and classification are part of an automated lung cancer diagnosis system. The deep learning-based automated systems provide an edge over the performance of conventional detection techniques.

Nanotechnology

The topic of drug delivery is one where nanotechnology has shown immense potential. Nanoparticles generally range between 1-100 nanometres in size, with certain exceptions in some cases. These can be designed to have particular qualities, including dimensions or size, form, shape, surface charge, composition, and other specialized properties, to correspond with the treatment approach and strategy [12]. These characteristics can be tailored to target particular body tissues or cell types while also increasing the solubility, stability, and bioavailability of medications. Nanoparticles have novel possibilities for medication targeting and delivery due to their small size, enabling them to connect with biological processes at the cellular and molecular levels. Patients with compromised respiratory systems, such as those with COPD or interstitial lung disease, are especially vulnerable to the adverse effects and challenges of conventional chemotherapy and surgery. In such cases, alternative treatment options may be considered in the form of nanotechnology. In general, the use of nanotechnology in drug delivery has the potential to enhance the efficiency and safety of existing medications, in addition to making it possible to provide medications that were previously impossible to administer [13]. Several nanotechnologies are being developed for lung cancer detection and therapy. Liposomes, dendrimers, polymeric nanoparticles, gold nanoparticles, quantum dots, carbon nanotubes, nanoparticle-based immunotherapy, and nano-biosensors are some of the novel approaches that are being used for this purpose [14].

Nano-informatics

Nano-informatics is the application of information science and nanotechnology. In the context of lung cancer, nano-informatics can play a significant role in developing and optimizing nanotechnology-based diagnostic and therapeutic approaches [15]. By using advanced computational techniques, the design of nanoparticles can be done that can specifically target cancer cells, deliver drugs directly to them, and monitor their effects in real-time, which can lead to better efficacy of treatment with fewer side effects. Another application of nanoinformatics in lung cancer is in the analysis of large-scale data sets generated by high-throughput techniques such as genomics, proteomics, and metabolomics [16]. By using machine learning algorithms and other data analysis tools, researchers can identify patterns and biomarkers that can help diagnose lung cancer at an earlier stage and predict its prognosis. This can lead to more personalized treatment approaches tailored to the individual patient.

AI, ML, and DL-based nanotechnology have a lot of potential for lung cancer treatment and early diagnosis. The development of nanosensors is one of the most promising uses of AI in lung cancer, which can detect cancer biomarkers in the breath or blood of patients [17]. These nanosensors can detect changes in the concentration of certain molecules that are indicative of lung cancer, such as volatile organic compounds (VOCs). The data collected by the nanosensors can then be analyzed using artificial intelligence algorithms to identify patterns and make a diagnosis. In addition to early detection, these techniques can also be used to develop more targeted and effective treatments for lung cancer [18].

Hyphenation of AI with Nanotechnology

In order to improve medication discovery and delivery, numerous issues in nanotechnology are constantly being addressed. Designing nano-systems, nano-computing, and AI techniques in scheming principles of nanoscale simulation is the significant refrain, with stress on curtailing computational time and effectually estimating, predicting, and simulating system parameters. Drug delivery necessitates the utilization of a range of approaches, such as formulation, manufacturing procedures, storage of bioactive chemicals, and transportation to the locations that need to be treated in order to achieve the desired level of therapeutic impact. However, the majority of AI approaches are utilized for the evaluation and interpretation of biological and genetic data. Recently, the design, characterization, and production of drug delivery nanosystems have been optimized through the application of AI methodology. With this in mind, suitable techniques for AI-based optimization for combination drug delivery using different NP categories are examined to increase drug localization at the cancer site. This is done in an effort to reduce the risk of adverse drug reactions. In the treatment of cancer, oncologists often need to deal with the unique genetic characteristics of their patients, which might involve a wide variety of molecular irregularities. Various factors, including pharmacological characteristics and biological parameters, are taken into account for drug administration when treating a cancer patient. When combined with patient data such as genetic information, metabolic information, proteomic information, histological information, and therapy history, the numerous molecular descriptors unique to drugs represent huge databases that make it difficult to optimize drug mixtures for specific patients [19]. Fig. (**1**) depicts a pictorial representation of neural network (AI)-enabled detection of cancer biomarkers by PEGylated quantum dots [19]. Table **1** explains the influence of AI on drug development and the depiction of its effects [19].

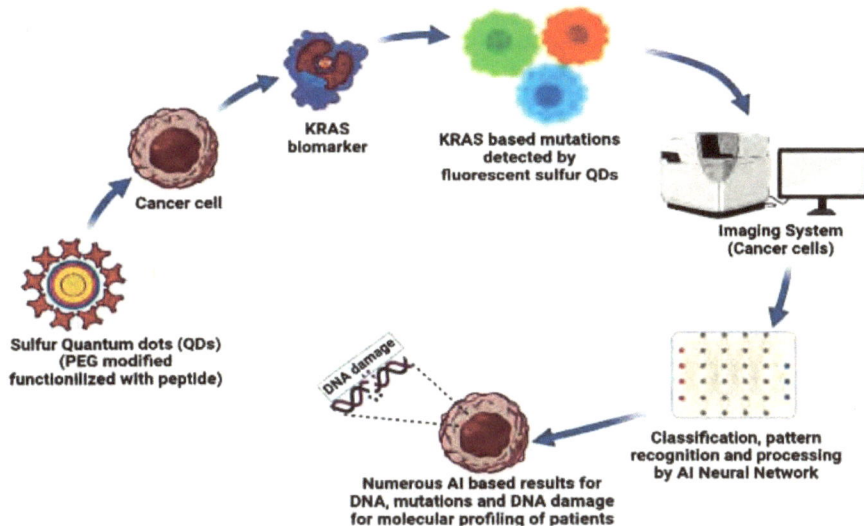

Fig. (1). A visual representation of a potential method for sensing biomarkers that is supported by AI. [Adopted from open access article [19].

Nanotherapeutics

Nanotherapeutics refers to the application of novel drug delivery systems in the production of pharmacologic and therapeutic interventions for various xenobiotics that are supposed to be transported across the biological membrane, thereby bringing a change in their disposition across the biological compartments of our body [20]. In nanotherapy, nanoparticles or materials on the nanoscale are utilized to improve the delivery, efficacy, and stability of medications, as well as to enable targeted drug delivery to particular tissues or cells. The primary objective of nanotherapeutics is to develop next-generation treatments with improved safety, biocompatibility, bioavailability, therapeutic activity, and bioadherence by reducing the dose [21].

Several nanocarriers are being developed for lung cancer therapy and detection, which include liposomes, quantum dots, dendrimers, nano-biosensors, carbon nanotubes, gold nanoparticles, NP-based immunotherapy, polymeric nanoparticles, *etc.* (Fig. **2**) depicts some of the nanocarriers used for theranostics purposes in lung carcinoma [22].

Table 1. The impact of AI on developing new drugs and depiction of existing ones (Adopted from open access article [19]).

Current AI Approaches	Area under Focus	Advantages	Challenges
AI-based API system. Passive AI System. Artificial Neural Network (ANN).	Absorption, distribution, metabolism, and excretion (ADME). Molecular entity features of the drugs. Predicting the synergy of anticancer drugs.	Predicting extravasation in tumor tissue. AI tools can integrate information from multiple sources and simplify the process of experimentation and *in silico* system simulations, resulting in enhanced drug delivery systems. These AI models may assist in adding novel molecular component traits, as well as features of molecules that are already well known, in order to anticipate effective therapy based on a molecule's bioavailability or local concentration for better therapeutic outcomes. In order to predict and quantify the synergism of anticancer medications, an ANN with a back-propagation technique was built. The results helped determine the drug concentrations and cytotoxicity.	Significant challenges in the evaluation of molecular, pharmacokinetics, and patient information. The availability of diverse datasets provides information on various molecules, which are considered to be in the range of 10^{60} molecules. If such datasets are not available, AI models will provide biased information. Although the model performed well in determining optimal compositions and presented the maximum synergistic effect, the model's capability with more complex information could be evaluated.

Liposomes

In the past few decades, there has been a growth in the importance of liposomes as an efficient drug delivery mechanism for anti-cancer medications. The extraordinary biocompatibility of liposomes has led to an upsurge in research on them over the course of the past decade, which has resulted in the development of a great deal of innovative formulations, including archaeosomes, virosomes, temperature-responsive liposomes, cationic liposomes, ethosomes, and niosomes. Liposomes are single- or multiple-bilayer nanocarriers that can be created from lipids that are natural or manufactured. Phospholipid-based liposomal vesicles were first created by Banham and colleagues in 1965, and they were quickly recognized as having potential as drug delivery vehicles for chemotherapeutic drugs. These are classed as large (500-1000 Å) and tiny (300-500 Å) unilamellar vesicles, as well as multilamellar vesicles, which are formed of various concentric phospholipid bilayers (1.0-5.0 m). Liposomal versions of some medications used as approved lung cancer treatments are constantly going through extensive research. Several medications that have been permitted for use as lung cancer treatments are constantly being modified into liposomal formulations. The following drugs have been developed into liposomal formulations for targeting various carcinomas: docetaxel (DTX), doxorubicin (DOX), cisplatin (CPPD), etoposide (ETP), irinotecan (IRI) paclitaxel (PTX), vinorelbine (VNB), and erlotinib. In addition, researchers are also looking into the effectiveness of tretinoin and DOX-conjugated liposomes as chemotherapeutics for malignancies that have metastasized to the lungs.

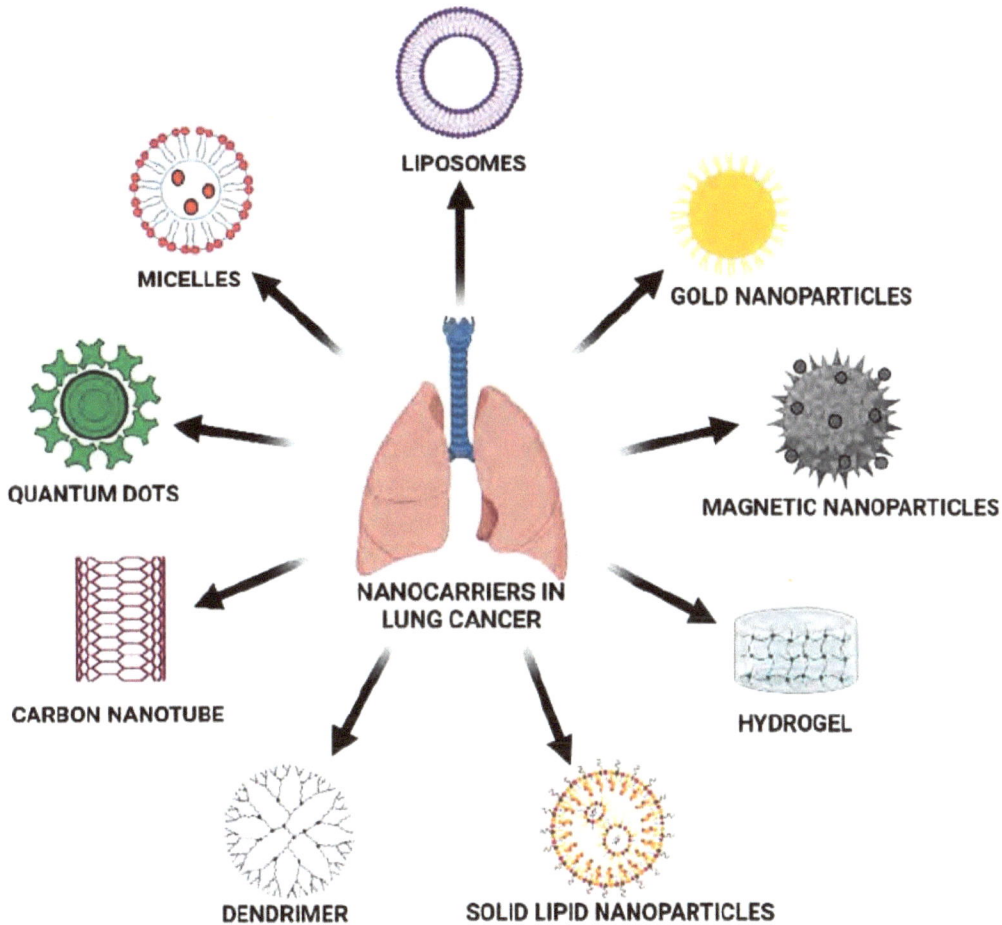

LIPOSOMES

MICELLES

GOLD NANOPARTICLES

QUANTUM DOTS

MAGNETIC NANOPARTICLES

NANOCARRIERS IN LUNG CANCER

CARBON NANOTUBE

HYDROGEL

DENDRIMER   **SOLID LIPID NANOPARTICLES**

Fig. (2). Several different drug delivery techniques that utilize nanocarriers. [Figure adopted from open access article [22]].

Polymeric Nanoparticles

Polymeric nanoparticles can be synthesized from biocompatible materials and can be used to deliver drugs to specific tissues or cells in which the drugs are either matrixed or encapsulated in various polymers. In order to increase the drug's affinity and stability, these polymeric NPs are used to bind to the ligand and target particular tumor cells or tissues. They can also be employed for the nonspecific passive targeting of cancer cells. There is a high demand for polymeric nanoparticles for both active and passive approaches to target cancer cells. In order to synthesize polymeric nanoparticles, a variety of polymers can be used. Polymeric nanoparticles often make use of polymers such as poly(lactide-c--glycolide) (PLGA), poly (lactic acid) (PLA), and a few additional polymers such as poly(glycolic acid) (PGA), poly(-caprolactone) (PCL), and poly(acrylic acid), amongst others. When used as a treatment for lung cancer, polymeric nanoparticles have proven to be effective thus far. The use of taxanes-loaded PEG-PLA (polyethylene-glycol-modified polylactic acid) nanoparticles in the A549 lung tumor xenograft model greatly increased the efficacy of chemotherapy as well as radiation therapy in both *in vitro* and *in vivo* conditions.

Dendrimers

Dendrimers are branched, synthetic polymers that can be used as drug delivery vehicles. They are robust, clearly defined monodispersed nanostructures with the potential for high functionality, having a diameter range of 2–10 nm. They have a three-dimensional, highly branching architecture. The several functional groups that can be joined to the globular or semi-globular arrangement of dendrimer branches include silicon groups (silicon-based dendrimers), peptides, carbohydrates (glycodendrimers), *etc.* Dendrimers' three-dimensional structure can be fragmented down into three parts: the core, or point where dendrons attach; the inner shells, or dendrons that surround the core; and the outer surface, which contains polyvalent attachment sites for potentially reactive groups. Strategies such as acylation or PEGylation of the active terminal surface and creating dendrimers with protein receptors could be developed to improve bioavailability and target-specific distribution, respectively. Due to their prolonged circulatory system distribution time, decreased accumulation in numerous bodily organs, and low toxicity, PEGylated Dendrimers have caught the attention of formulation experts. PEG-dendrimers can be made by combining PEG and poly(ethylene oxide) (PEO). When examined in a separate experiment, Dendrimer tailored with PEG showed encouraging potential as a drug delivery technique for aerosols. In addition, PEGylated doxorubicin dendrimers created an extended time of drug residence in cancer tissues in rats when they were administered through inhalation. This resulted in an effective reduction in the amount of carcinoma load.

Carbon Nanotubes

Carbon nanotubes or CNTs are hollow, lipophilic tubes made of carbon atoms that can range in diameter from 4 nm to 100 mm and are dependent on how the graphene molecules are organized. CNTs, which belong to the fullerene family, resemble a single sheet of graphene that has been folded up. There are two distinct varieties of nanotubes, which are referred to as single-walled and double-walled, respectively. The first method involves rolling a single sheet of graphene into a cylinder-like shape, whilst the second method involves rolling numerous concentric layers of graphene into a cylindrical shape. Both methods provide the same end result. To specifically target epidermal growth factor receptors (EGFR), single-walled carbon nanotubes (SWCNTs) were the ones to be integrated with the anticancer medication cisplatin. The results showed that the medication worked better against squamous carcinoma cells that expressed a lot of EGFR. In addition, due to its nanoscale size, it demonstrated greater effectiveness in stopping cancer as opposed to merely addressing the issue of cancer. Docetaxel-loaded CNTs have been shown to be more successful than free docetaxel in the treatment of PC3 and S180 (murine sarcoma) cell lines, according to research conducted *in vitro* and in preclinical settings. In yet another study, the dynamics of PEGylated nanotubes in combination with a chemotherapeutic drug were investigated. This combination enhanced both the cellular absorption and the cytotoxic profile in non-small cell lung cancer. According to the results of the cytotoxicity investigation, PEGylated conjugates killed mitochondria more frequently than the unconjugated moiety.

Gold Nanoparticles

Gold nanoparticles (AuNPs), which were just recently recognized as a potential delivery method for delivering and releasing medications into a wide variety of cell types, have been proven to be capable of administering not only pharmaceuticals but also genetic materials, proteins, and even tiny molecules. Due to the fact that they possess a variety of appealing characteristics, they are an excellent instrument for the delivery of chemotherapeutic and diagnostic drugs. The fact that the inside section of these particles is risk-free, biocompatible, and inert is the primary benefit of utilizing it as the foundation for carriers. The diameter of an AuNP can range anywhere from 5 nm to 200 nm, and their production can be facilitated by controlled dispersion. A range of medicinal compounds or biomacromolecules can be conjugated (covalently or non-covalently) with AuNPs' highly changeable and multivalent surface configurations. The noble metal gold has also been extensively studied for its medicinal uses, which include its use in the sensitive identification, detection, and classification of several lung tumor types. To specifically target overexpressed

EGFR tumor cells, fluorescently tagged antibodies have been added to the surfaces of the gold nanocluster. This arrangement makes it feasible to use nanoparticles based on gold in magnetic resonance imaging (MRI) as well as in the treatment of cancer.

Quantum Dots for Imaging

Quantum dots (QDs) are semiconductor nano-crystals that emit fluorescent light when excited by light or electricity. Recently, nanofabrication was used to create QDs, which are atom-like nanosized colloidal particles. These QDs are regarded as promising lung cancer treatments. QDs can be improved to become more soluble and biocompatible, which transforms them into an improved fluorescence probe. The core of the substance is made up of a crystalline metalloid that is shielded by a cap or shell, which increases the bioavailability of the material as a whole. Some of the distinguishing features of quantum dots (QDs) include their extensive absorption spectra, severe photobleaching, and photostable states. Due to the narrow wavelength range of the emission spectrum, this nanoparticle system displays repeating fluorescence and excitation cycles. Because of these properties, QDs have been a popular topic of discussion, particularly for clinical imaging of tumor tissue specimens taken from cancer patients. Such imaging is used to identify the kind and stage of the disease, as well as to devise a treatment strategy and forecast the patient's prognosis.

Nano-biosensors for Tumor Detection

Nano-biosensors are a type of biosensor that utilizes nanotechnology to detect and analyze biological molecules and processes at the nanoscale. Biosensors are devices that use biological molecules or biological processes to detect and measure various substances in a sample. Nanotechnology allows for the miniaturization of biosensors, enabling highly sensitive and specific detection of biological analytes. Nano-biosensors typically consist of a transducer, which converts a biological signal into an electrical or optical signal, and a recognition element, which binds to the target molecule or biological process. The recognition element can be a protein, DNA, RNA, or other biomolecules and is often immobilized on the surface of the transducer using nanotechnology.

The capability of nano-biosensors to detect lung cancer biomarkers in exhaled breath or blood samples with high sensitivity and specificity positions them as a potentially useful tool for the early diagnosis and ongoing surveillance of lung cancer. From a study conducted, it was evident that a lung cancer biomarker that could be detected using a nano-biosensor is the protein carcinoembryonic antigen (CEA), which is often elevated in lung cancer patients. Researchers have developed nano-biosensors based on carbon nanotubes, gold nanoparticles, and other nanomaterials to detect CEA with high sensitivity and selectivity. In addition to biomarker detection, nano-biosensors can also be used for imaging lung cancer cells and monitoring the response of lung cancer cells to treatment. For example, researchers have developed fluorescent nano-biosensors that can detect changes in pH or oxygen levels in lung cancer cells, which can be used to monitor the effectiveness of chemotherapy or radiation therapy.

In general, nano-biosensors have a great deal of promise for enhancing both the diagnosis and treatment of lung cancer, and the research that is now being conducted in this area is likely to result in the development of novel and cutting-edge strategies for the early identification of cancer and the personalization of treatment.

ARTIFICIAL INTELLIGENCE IN LUNG CANCER DETECTION

This section discusses the various ML and DL segmentation and classification approaches used for automatic lung cancer diagnosis. Oncology detection is now frequently carried out using a variety of imaging modalities, including X-rays, CT scans, MRIs, PET/CT scans, and digital images from histopathology. In this section, the related work is broadly divided into two categories: binary and multi-class lung cancer classification. Also, the various image modalities used by different researchers in the past are mentioned separately. Moreover, the database containing each imaging modality is also mentioned on which deep learning approaches are applied.

Artificial Intelligence-based Lung Cancer Detection for Different Imaging Modalities

CT Images

This subsection discusses the binary and multi-class classification of lung cancer through CT images.

In lung CT binary classification, a method for detecting lung nodules was proposed in which adversarial attack samples and synthetic nodules are involved to increase the model's generalizability and robustness [23]. For the binary categorization of lung CT scans, a cross-residual convolutional neural network (CNN) was applied, and it attained an accuracy of roughly 92% [24]. A CNN model based on maximum intensity projection for an automatically detecting lung cancer system was proposed. As input images, a number of slabs with thicknesses ranging from 5 mm to 15 mm were sent into the system. The model was shown to have a sensitivity of 93% [25]. For the purpose of determining the severity classification of lung cancer based on CT images, an optimized deep belief network was presented. The accuracy of the model was 96% [26]. In order to perform binary classification of the benign and malignant nodules that were present in the CT scans, a gradient boosting machine was utilized, which obtained an AUC of 96.87% and an accuracy of 93.78% [27].

In multi-class classification, a deep joint model-based segmentation algorithm was proposed, followed by a grid scheme for identifying the nodules. Features such as texton features, perimeter, area, short axis, and long axis were extracted, which were later fed to a Shuffled Shepard Optimized Sine-Cosine multi-class classifier. This classifier classified solid nodules, part solid, and ground-glass opacity (GGO) nodules effectively with 83.7% specificity, 92% accuracy, 81.5% F1-score, and 93% sensitivity [28]. An advanced CNN model was employed for the multi-class categorization of normal, malignant, and benign (NMB) classes of lung CT images, which obtained 97% accuracy [29]. An improved efficient net model was introduced to differentiate NMB CT image classes. The model earned perfect scores in terms of accuracy, precision, specificity, recall, and F1 score [30]. The multi-class classification of four non-small cell lung cancer subtypes was carried out by incorporating a combined approach of majority voting followed by the Bayesian prior method and achieved 81.2% accuracy [31]. Fig. (3) depicts three sample images of benign, malignant, and normal CT image datasets [32].

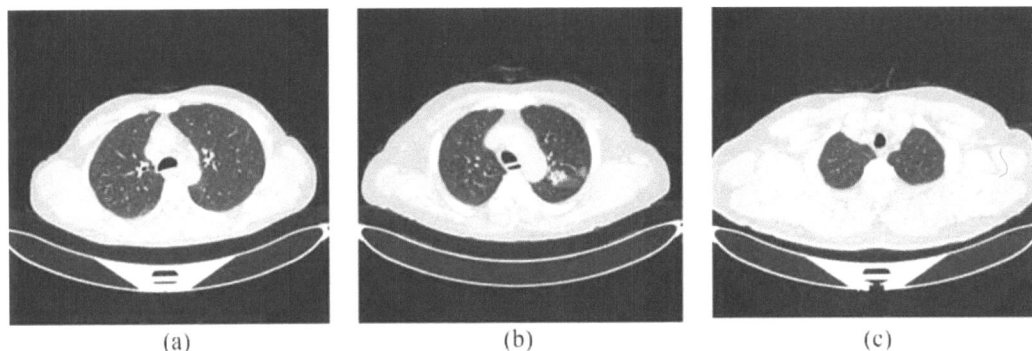

(a)   (b)   (c)

Fig. (3). Lung cancer CT dataset [32]. (a) Benign (b) Malignant (c) Normal.

X-ray Images

The healthy and cancerous lungs using X-ray images were distinguished using several image processing techniques such as cropping, scaling, and median filtering followed by Haar wavelet transform-based feature extraction. Later, these features were fed to a probabilistic neural network, which obtained an accuracy of 80% [33]. Multiple lung diseases, including lung cancer, were detected using a deep CNN network, which achieved a 96.76% AUC value [34]. A Dense Net 121 model was proposed for three classes of lung X-ray images: malignant, normal, and non-nodule. The model obtained 74.43% mean accuracy, 75% mean specificity, and 75% recall [35]. The lung nodules were identified using a grayscale histogram and local binary pattern features followed by an optimized neuro-fuzzy classifier. The model obtained 97% accuracy, 98% AUC, 99% specificity, and 85% recall [36]. A deep learning approach for multi-class classification of various lung diseases such as pneumonia, tuberculosis, lung cancer, and COVID-19 was proposed, which achieved accuracy, F1-score, precision, recall, and AUC of 94%, 95%, 97%, 94%, and 99.82%, respectively [37]. Four deep architectures were proposed for multi-class classification of lung diseases such as lung cancer, COVID-19, and pneumonia. The highest accuracy achieved was 98.05% accuracy and recall [38]. An artificial neural network was used for lung tumor boundary detection. The input images were first made noise-free, and then image enhancement was performed, followed by lung segmentation. The model achieved 100% accuracy [39]. Fig. (**4**) shows an X-ray dataset of Cancerous and healthy subjects [40].

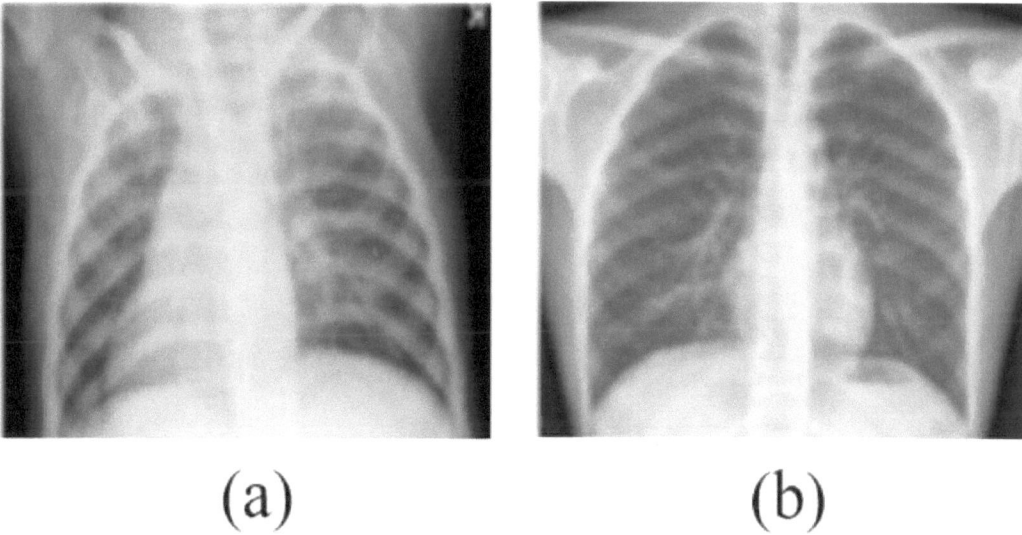

(a)   (b)

Fig. (4). X-ray dataset [40]. (a) Cancerous Lungs (b) Healthy Lungs.

MRI

Using MRI, CT, and ultrasound data, a PSO and genetic optimization-based SVM classifier was presented for the purpose of detecting lung cancer. According to a study [41], the accuracy of the model was 89.5%. A method for detecting lung nodules, which is based on deep learning and makes use of a quicker R-convolution neural network (CNN) constructed with transfer learning, has been developed. After that, a technique for reducing false positives (FPs) based on anatomical characteristics is developed in order to minimize the number of FPs while retaining the number of real nodules [42]. In order to rebuild MRI images for the purpose of detecting lung cancer, a CNN-based technique was utilized, and the results showed an accuracy of 89.2% [43]. A salient detection network, also known as an SDN, is established in order to facilitate thoracic MRI nodule identification. According to another study [44], an encoding-decoding block is developed, which is responsible for the production of the salient detection map. This is done in order to effectively combine multi-level feature maps and boost the rate of small nodule recognition [44]. Fig. (5) displays the MRI dataset for lung cancer [42].

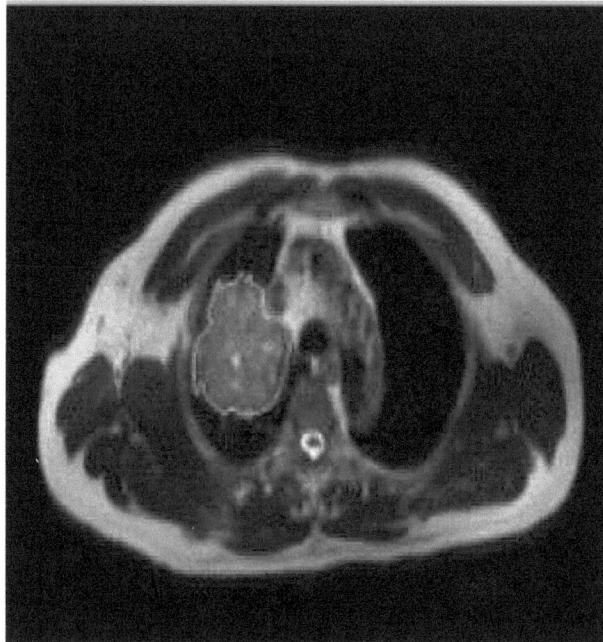

Fig. (5). Lung cancer MRI dataset [42].

PET/CT

An ANN model was employed for automated lung cancer identification using ultralow dose PET/CT images and achieved 98.9% AUC [45]. CLAHE and Wiener filtering were used to reduce the artifacts caused by fluctuations in contrast and noise. Using morphological operators, the pulmonary Region of Interest (ROI) was derived from PET/CT images. The computed fuzzy C means clustering approach was then used to classify the retrieved Haralick characteristics into normal and pathological PET/CT image categories [46]. A CNN was utilized to locate and segment individual regions of the lungs, mediastinum, and malignancies using a variety of different fusion techniques. The detection accuracy and dice score obtained were 99% and 63.85%, respectively [47]. An

SVM-based lung cancer identification approach was proposed using multiple features from multi-modal PET/CT images [48]. It was recommended to use a VGG-SegNet algorithm for nodule delineation. This model's results were then put into an SVM-RBF decoder for lung nodule classification in PET/CT images, which resulted in a precision of 97.83% [49]. It was proposed to use an integrated regional network that is skilled in extracting pulmonary nodule features from the PET/CT images, and this system earned a dice score of 84% [50]. In order to segment lung lesions in PET/CT images using the U-Net design, some shot learning was incorporated as a learning strategy. The IOU score obtained was 44.48% after 5-fold cross-validation [51]. A VGG model integrated with SVM and Random Forest was carried out to classify multi-classes. The highest accuracy obtained was 98.70% [52]. A unique and effective multi-class classification system for lung cancer stages based on tumor, node, and metastases (TNM) criteria was proposed by incorporating a 3D deep CNN and lung cancer stage classification network. The accuracy obtained for the T, N, and M stages were 96.23%, 97.63%, and 96.92%, respectively [53]. Fig. (**6**) depicts an annotated PET/CT dataset [53].

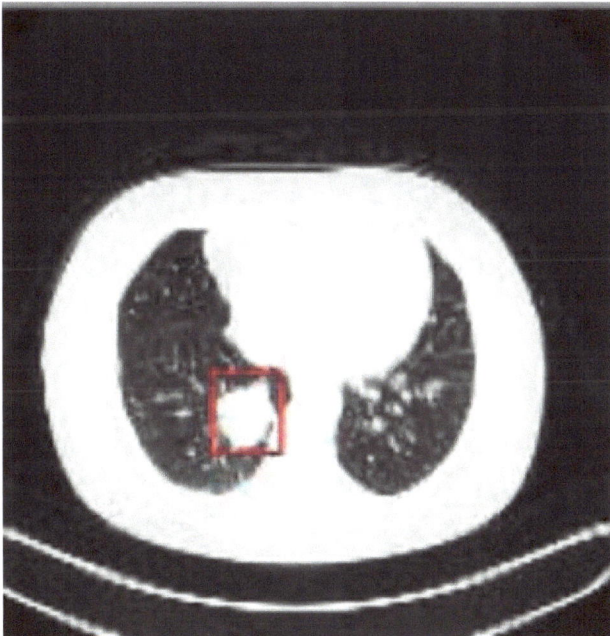

Fig. (6). Lung cancer PET/CT dataset [53].

Histopathology Modality

The histopathological images were run through a CNN model in order to perform multi-class classification of benign, adenocarcinoma, and squamous cell carcinoma. It had an accuracy of 96.11% throughout training and 97.2% during validation [54]. An Inception v3 was used for the binary classification of adenocarcinoma and SCC and achieved an AUC of 97% [55]. An atrous fusing module and CNN were integrated by combining Inception V2 and ResNet and replacing 8 ResNet 18 blocks with the inception module. The histopathological images were divided into cancerous and normal patches [56]. A comparison of ResNet and VGG 16 was performed on input image patches. VGG 16 models performed better than the ResNet model and obtained 83.3% and 75.41% AUC and accuracy, respectively [57]. A CNN-based SCC and adenocarcinoma classification model was proposed, and 97% detection accuracy was achieved [58]. A deep ResNet framework was used for the classification of lung cancer histopathological images and achieved an AUC of 79% [59]. Fig. (7) depicts a sample image of a lung cancer histopathological dataset containing adenocarcinoma, squamous cell carcinoma, and benign tissue classes [60].

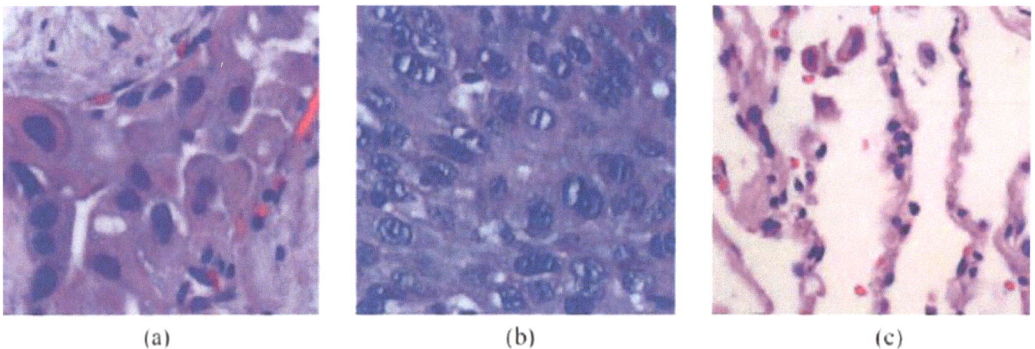

(a)   (b)   (c)

Fig. (7). Lung cancer histopathological dataset [60] (**a**) Adenocarcinoma (**b**) Squamous Cell Carcinoma (**c**) Benign.

In summary, Fig. (8) shows the general schematic representation of lung cancer classification through ML and DL approaches on different image modalities such as CT, X-ray, MRI, PET/CT, and histopathological images. Depending on the image modality, the input image is fed to the segmentation model, followed by feature extraction and ML classifier. The segmentation model isolates the nodule from other parts of the lung input images. Several features, such as area, circumference, perimeter, circularity, and so on, may be extracted from the detected nodule. ML classifiers such as SVM, Random Forest, Decision Tree, and so on may be used for classification purposes. The classification can be binary or multi-class. Binary classification is the classification of any two classes, such as benign or malignant, adenocarcinoma or squamous cell carcinoma, and so on. The multi-class classification can be based on the classification of more than two classes, such as benign, malignant or normal, adenocarcinoma, squamous cell carcinoma, or benign tissue. In this way, a computerized automated diagnosis system can help radiologists detect lung cancer in its early stages.

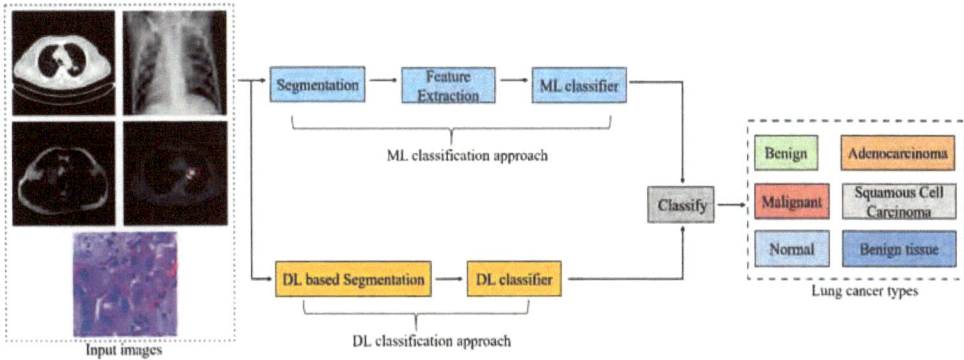

Fig. (8). Schematic representation of lung cancer classification using ML and DL approaches on different types of input images.

NANOMATERIAL-BASED PERSONALIZED DRUG DELIVERY SYSTEM

Even though a large number of synthetic and naturally occurring components were found, identified, and screened on lung cancer cell lines, additional study is still required in order to have a comprehensive understanding of how these substances function. According to a study [61], it is absolutely necessary to develop unique drug delivery strategies that may either passively or actively target cancer cells in order to overcome the problems that are connected with conventional treatment. Drug administration to the lungs has presented a lot of intriguing opportunities for boosting therapeutic efficacy due to the increased surface area of the alveolar region. The dearth of mechanistic knowledge of the cellular processes that underlie tumor heterogeneity has sparked a great deal of interest in developing the most effective therapeutic approaches and treatment regimens. By showcasing desirable qualities like the prolonged retention of therapeutic agents in lungs, resulting in the reduction of dose and an increase in patient compliance, the development of nano-based delivery systems over the past two decades has managed to open up a variety of possibilities for clinical therapeutics and increased bioavailability [62]. The conception and use of personalized new drug delivery systems that make use of nanomaterials for the treatment and detection of cancer have been covered in this section of the chapter [63].

Targeting Strategies of Nanocarriers

The strategies for targeting nanomaterials to deliver chemotherapeutic agents to the cancerous tissue are commonly categorized into three sets:

Passive Targeting

When compared to the typical healthy condition, passive targeting strategies increase the rate of infiltration in the vasculature in the cancer microenvironment *via* the EPR phenomenon. Targeting ability often rises 20% to 30% compared to healthy cells when the permeability and retention are improved. The majority of the time, tumors develop in unfavorable environments like hypoxia or inflammation, and quickly expanding malignant tumors obstruct the nearby environment's blood or new vessel growth. The tumor stroma may benefit from this selectively improved nanosystems' permeation and such improvement in the permeation of newly formed capillaries may also increase retention time. Nevertheless, lymphatic drainage may eliminate such tiny molecules and shorten the duration that the medicine is retained in the body [64]. As a result, lymphatic drainage needs to be regulated. Physical and chemical characteristics such as particle size, charge, and superficial chemistry are known to headway passive targeting to the cancer based on the EPR effect. It can be hard for a pharmacological molecule with a certain level of therapeutic activity to bind to all of its target cells. By creating more distinct, targeted nanocarriers that successfully attach to tumor cells following the irruption process, the fundamental issue of unequal targeting of tumor cells may be resolved, thereby extending the retention durations and improving the pharmacokinetic profile [65].

Receptor-Based Active Targeting

By modifying the nanocarrier's surface, the drug that is specifically targeted to the tumor cell is delivered by receptor-mediated endocytosis. By altering the surface of nanocarriers, substances like albumin, folic acid, hyaluronic acid, herceptin, transferrin, and RGD (arginine-glycine-aspartic acid) act as ligands to the specific receptors overexpressed in the tumors and have increased the effectiveness of tumor targeting. These are a few of the receptor-based targeting strategies discussed in this segment that have been developed by numerous researchers to deliver novel nanocarriers to specific sites while maximizing tumor internalization, cell uptake, and penetration [66]. Fig. (**9**) demonstrates a pictorial representation of various receptors overexpressed in lung carcinoma and their binding with specific ligands [66].

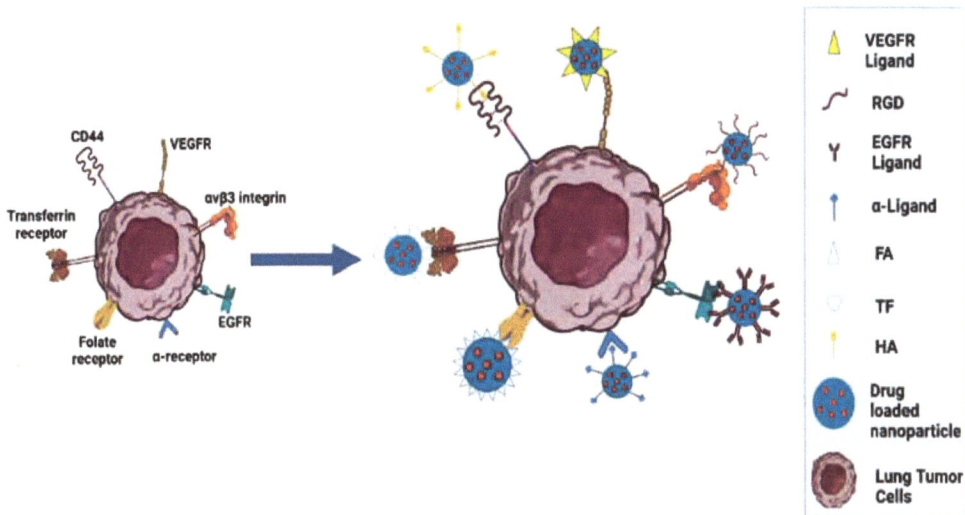

Fig. (9). Various ligand-receptor targeting strategies for delivering anti-cancer nanotherapeutics to some of the receptors commonly overexpressed in lung carcinoma. [Figure adopted from open access article [66]].

EGFR Based Targeting

Herceptin may be a choice for EGFR targeting. Herceptin can quickly identify lung adenocarcinomas with overexpressed HER2, a member of the EGFR family of receptors. Paclitaxel (PTX) and docetaxel (DTX) were administered through herceptin-modified nanocarriers in order to prevent drug loss before they approached the cells and organs that were supposed to receive them. These modifications have boosted cellular absorption, enhanced drug release, and demonstrated cytotoxic behavior [67].

Transferrin Receptor-Based Targeting

According to research studies, apo-transferrin and paclitaxel nanocarrier were coupled to deliver the medication to a particular target site. Another example of this kind of active targeting is the modification of the surface of a (PTX) paclitaxel nanocarrier that is linked with transferrin surface protein. This modification in the past has provided better tumor suppression [68]. The selective tumor-targeting peptide XQ1 (HAIYPRHGGGF) was found to fix to transferrin receptors in HeLa and A549 lung cancer cell lines, which enhanced the antineoplastic activity of camptothecin nanocarriers [69].

Folate Receptor Based Targeting

The folate receptor (FR) is a promising target for targeted therapy due to its selective expression on malignant tissues. Currently, folate conjugates and monoclonal antibodies (mAbs) are the two main FR-targeted therapy strategies that are in the clinical development stage. Folic acid is the preferred ligand for FR targeting [70]. Folic acid-tailored temozolomide (TMZ)-loaded chitosan nanoparticles are an example of a novel use of active targeting by FR. These nanoparticles are designed to specifically target cancer of the lungs. In comparison to unbound TMZ and CS-TMZ-NP (nanoparticles devoid of folic acid), CS-TMZ-FLA-NP had the highest antiproliferative effect against cancer linked with lungs while causing the least amount of damage to healthy tissue [71].

Targeting on the Basis of the Vascular Endothelial Growth Factor (VEGF) Receptor

According to immunohistochemical analysis, oral administration of bovine lactoferrin (BLF) has been shown to reduce the overexpression of VEGF in lung carcinoma, which is associated with angiogenesis and vascularization. The progression, penetration, and metastasis of the tumor are all significantly influenced by these pathways. BLF is important for effectively treating the carcinoma since it has the potential to prevent angiogenesis [72].

Targeting on the Basis of the Luteinizing Hormone-Releasing (LHRN) Receptor

Inhalation of LHRH receptor-fabricated nanostructured lipid carriers (LHRH-NLC), which were used in the study, resulted in the effective delivery of paclitaxel to the lungs. The findings demonstrated that nanostructured lipid paclitaxel conjugated to luteinizing hormone had 16 folds better apoptotic features than the usual medication and was specifically localized in tumor locales as paralleled to non-targeted NLC nanoparticles [73].

CD44 Receptor-Based Targeting

Lapatinib nanocarriers decorated with hyaluronic acid (HA) are utilized to actively target the overexpressed CD44 receptor on cancer cells. HA-modified paclitaxel-loaded nanocarrier has been coupled to bind to CD44, a cell surface receptor that is upregulated in a few particular kinds of lung cancer and is found to be a very effective targeting strategy. For tumor-specific targeting, this intercellular adhesion molecule interacts with a receptor for HA, thereby bringing tumor cell death [74].

Integrin Adhesion Molecule αvβ3-Based Targeting

The integrin attachment protein αvβ3, which is overexpressed on lung tumor cells, can bind with RGD (Arginyl glycylaspartic acid), a ligand unique to integrin receptor, and can be decorated on the surface of nanocrystals, which can subsequently be internalized by receptor-mediated endocytosis. RGD-modified nanocapsules with paclitaxel drug (RGD-PTX NC) can be made using the Schiff base reaction to achieve $\alpha v \beta 3$-based targeting [75].

Targeting that is Sensitive to Stimuli Based on a System

A stimuli-responsive system-based approach to cancer treatment may be efficient and targeted. There are significant differences in intracellular cytosolic contents and extracellular microenvironmental elements in tumor cells, including hypoxia in the tissue, pH, temperature, and glutathione concentrations. It may be possible to create these nanocarriers with chemotherapeutic medications inside of them using a technology that triggers the release of the drugs from the nanocarriers in response to a stimulus specific to the microenvironment of cancer cells [76]. Some of the prime stimuli-responsive systems for delivering chemotherapeutic agents have been discussed in this portion:

pH-Responsive Systems

Researchers have been looking into the possibility of using the low pH of the tumor microenvironment (4.5-6.5 pH) and the pH of the lysosomal environment (5.5-6.8 pH) to their advantage in order to directly carry medication into the mitochondria, nucleus, and cytoplasm of the intracellular compartment or to the extracellular cancer surroundings. A spray-dried powder inhalation system was developed based on a pH-sensitive peptide and mannitol in order to increase drug payloads to low-pH cancer sites to maximize nucleic acid transfection and maintain stability in broncho-alveolar lavage fluid. This method was intended to increase drug payloads to low-pH cancer sites in order to ensure maximum nucleic acid transfection [77].

Temperature-Responsive Systems

It has been established that modifying the medicine rate of release in a thermally dependent manner at a particular region can be highly advantageous in the therapy of cancer. This is due to the fact that hyperthermia happens to be one of the hallmarks of tumorous tissues. It is also possible to modify the surface of nanocrystals by employing thermos-responsive polymers. Researchers developed thermo-responsive cellulose nanocrystal grafted PNIPAM [poly(N-isopropyl acrylamide)] brushes in order to encapsulate 5-fluorouracil. When the temperature was increased from 25 °C to 37-40 °C, PNIPAM generated a significant increase in the amount of medication that was released from the system [78].

Enzyme-Responsive Systems (ERS)

Due to the fact that certain enzymes continue to be plentiful close to the cancer site, ERS can be exploited to facilitate the controlled delivery of medications [79]. Formulation containing doxorubicin-loaded hybrid liposomal system incorporating phospholipids and poloxamer 188 was discovered by Tagami *et al.* They found that this system coupled to the PLA2 (phospholipase A2) enzyme that is activated in cancer cells. The drug release was eight times greater when the PLA2 enzyme was present. ERS can successfully transport pharmaceutical nanocrystals *via* the pulmonary route for targeted chemotherapy [80].

Redox-Responsive Systems

Redox-sensitive nanocarrier systems hold plenty of promise and can be beneficial for treating cancer. Anti-cancer medications can be delivered to cancer cells with precision using these stimuli-responsive techniques. It is possible to transport nanoparticle systems to pulmonary cancer cells by taking advantage of the fact that tumors, or more specifically, cancer cells, have an elevated level of intracellular ROS [81]. In an investigation that was carried out by Chen and his colleagues, a liposome-based nanosystem that contains cholesterol, soya PC (phosphatidylcholine), and a redox-cognizant cationic oligopeptide lipid with proton sponge effect was designed for the simultaneous delivery of PTX and anti-survival siRNA. This system was designed for the simultaneous delivery of PTX and anti-survival siRNA. The medication and siRNA were quickly released from the liposomes into the cytosolic environment through endo-lysosomal leakage, which lowered the expression of the survival gene and increased the induction of apoptosis [82].

Magnetic Field-Responsive Systems

In addition, magnetic fields have been employed as an external trigger to deliver pharmaceutical compounds to the cytosolic or externally regionalized malignant neoplasms in the lungs. Using PEG-polyethylenimine-DOPA (dopamine), Park *et al.* formulated a magnetite nanocrystal system with magnetically driven intracellular transference of loaded siRNA into cancer cells, which has been proven to have a stronger anti-cancer effect than its traditional equivalents [83].

Prevention of Spreading of Metastasized Tumors to other Regions of the Body

The dissemination of cancer from its primary site to different locations on the body is known as metastasis. For the purpose of generating new tumors, cancer cells may splinter off from the parent tumor and move through the bloodstream or lymphatic system to other organs or tissues [84].

Treatment for cancer is notoriously difficult, particularly in cases where the disease has already spread to other areas of the body. The fact that metastatic cancer is responsible for the vast majority of cancer-related fatalities means that it will likely always remain a significant barrier. During the process of metastasis, cancer cells go through a series of processes that are collectively referred to as the "metastatic cascade". These steps present an opportunity for the development of anti-metastatic therapeutic techniques [85]. The current conventional medicine and diagnostic procedures for metastasis have their own set of pitfalls and hurdles. In this input, we discuss in depth the possible developments that nanocarrier-aided approaches can bring to the diagnosis and management of metastatic disease, either on their own or in combination with existing conventional techniques [86].

Numerous studies have demonstrated the effectiveness of targeting the primary cancer site with nanoparticles for a variety of strategies, including inducing apoptosis in tumor cells, targeting cancer stem cells (CSCs), impeding the epithelial to mesenchymal transition (EMT), modifying the tumor microenvironment (TME), or eliciting immune responses [87]. For example, preclinical research on the treatment of solid tumor metastasis with the bioactive sphingolipid ceramide has shown some encouraging results [88]. Both primary lung cancer and lung cancer that spread to other parts of the body can benefit from the anti-cancer properties of a natural drug known as Podophyllotoxin (PPT). Layered double hydroxides (LDH) nanodelivery systems have the potential to be utilized to capture podophyllotoxins, and this technique has increased performance while exhibiting lesser cytotoxicity to noncancerous cells [89]. In a different work, Liu *et al.* created ideal-sized gold nanoclusters protected by cationic bovine serum albumin (CBSA) and developed nanoparticles from hyaluronic acid (HA) for the delivery of drugs to breast cancer. In subcutaneous breast cancer, this formulation demonstrated greater allocation. The team later added nitric oxide (NO) donor to this formulation to increase tumor blood supply, tumor penetration, and ultimately tumor accumulation [90]. In addition, nanoparticles laden with indocyanine green (ICG) and paclitaxel (PTX) have the potential to greatly suppress the growth of primary tumors and stop the spread of cancer to the lungs.

Therefore, techniques utilizing nanotechnology offer an opportunity for the development of new therapies with the ability to halt or prevent later stages in the metastatic cascade by targeting and destroying numerous components in the metastatic cascade [91].

Interaction of Drug with Biological Fluids

Anticancer drugs can interact with biological fluids in several ways, which can affect their distribution, metabolism, and excretion in the body. The interaction between anticancer drugs and biological fluids can influence the efficacy and toxicity of the drugs. One of the key factors that affect the interaction of anticancer drugs with biological fluids is the pH of the fluid. Many anticancer drugs are weak acids or bases, and their solubility and ionization state depend on the pH of the fluid they are in. For example, some drugs, such as methotrexate, are more soluble in acidic fluids, while others, like etoposide, are more soluble in alkaline fluids. The presence of plasma proteins can also affect the interaction of anticancer drugs with biological fluids. Some drugs bind to plasma proteins, which can reduce their concentration in the free form and alter their distribution and elimination [92]. On the other hand, some drugs, such as paclitaxel, can bind to albumin in the plasma, which can enhance their solubility and stability. The rate of drug metabolism and elimination can also be affected by biological fluids. Some anticancer drugs are metabolized by enzymes in the liver or kidneys and then eliminated from the body through urine or feces. The presence of other drugs, food, or endogenous compounds in the biological fluids can affect the activity of these enzymes and alter the rate of drug metabolism and elimination [93]. As nanomaterials' role in physiology and medicine expands, the number of instances in which they will come into contact with biological systems will undoubtedly increase. A couple of aspects of the interactions between nanomaterials, proteins, and biological fluids have been discussed in the following section:

Protein Binding

Blood and all of the proteins that are contained within it are the first tissues with which a nanoparticle can interface after it has been delivered systemically. This is true whether the nanoparticle is being used for treatment, imaging a tumor, or assisting in the creation of a diagnosis. When diagnostic nanomaterials are utilized *in vitro* or *ex vivo* to evaluate samples of biological fluid, they will, in a similar fashion, come into touch with complicated protein combinations. When a surface has dimensions on the nanoscale, protein adsorption on a substrate is a far more complicated process than when the surface has proportions that are bigger. Protein interactions with materials at the nano and macro scales are thus quantitatively and qualitatively distinct. The interaction of nanoparticles with biological fluids causes the nanoparticles to become coated with proteins, which in turn causes a change in the surface charge and characteristics of the nanoparticles. Because of a rise in hydrodynamic size or aggregation, as well as a change in surface charge and zeta potential, this biological coating can become less effective over time. This might happen if certain conditions are met. Albumin, immunoglobulin, fibrinogen, apo-lipoproteins, and complement cascade proteins are the ones that bind the most strongly and competitively to polymeric nanoparticles, liposomes, iron oxide nanoparticles, and carbon nanotubes [94].

Nano Bio-fluid Interaction

The ubiquitous biofluids, which are responsible for mediating the flow of energy and materials and consist of water, dissolved trivial molecules, and ions, are necessary for the processes that take place in living organisms. The interaction between nano bio-fluids is the one that has received the least amount of research compared to other nano-bio interactions. There is a need for a bit more research into the mechanisms by which molecules of water and oxygen influence the functionality of nanomedicine in biofluids. At the interface between nanoscale biofluids and the outside world, the fundamental driving force is bonding with hydrogen [95].

It has been discovered that sodium gadolinium fluoride (NaGdF4) NPs with PAA caps exhibit better relaxivity for enhanced biological imaging than those with PEI and PEG caps. This was discovered by comparing the three types of caps. This is because PAA and water molecules form strong hydrogen bonds, which improve the relaxivity of the NaGdF4 NPs for better bioimaging [96].

In a nutshell, the collaboration of anticancer drugs with biological fluids is a complex and dynamic process that can affect the pharmacokinetics and pharmacodynamics of these drugs. Understanding these interactions is essential for optimizing the efficacy and safety of anticancer therapy.

PHARMACOKINETICS AND PHARMACODYNAMICS POTENTIAL NANOCARRIERS

Researchers are able to increase the rate of dissolution in a formulation by nanosizing it, which in turn boosts the amount of medicine that is absorbed and increases its bioavailability. When applied to certain tissues, nanoparticles are able to be taken up selectively, which allows for increased tissue selectivity to be accomplished through their application. Nanoparticles can be utilized to offer better protection or limit renal clearance in order to prolong the pharmacological effect of drugs that have a short half-life or that are easily degraded, such as small peptides and nucleic acids. This can be accomplished in order to prolong the pharmacological effect. Nanoparticles, however, may also enhance the release of the drug to specific tissues, leading to additional adverse effects. There are frequent differences between the pharmacokinetic and pharmacodynamic characteristics of the parent medicine and the drug encapsulated in the nanoparticles. In order to comprehend and anticipate the effectiveness and adverse effects of nanoparticles, it is crucial to monitor their pharmacokinetics (PK) and bio-distribution (BD). Their PK profiles are principally influenced by the chemical and physical features of the nanoparticles, such as their size, charge, and surface chemistry [97].

In this chapter, since liposomes, polymeric micelles, and carbon nanotubes are the nanoparticles for drug administration that have been examined and characterized the most in relevance to lung carcinoma, they make up the majority of the instances in this segment. However, similar principles may be applied to different kinds of nanoparticles, such as hybrid and silica-based ones. The PK study entails monitoring the concentration of the medication in all major tissues after its delivery and continuing to do so over the course of time up until the elimination phase. It is absolutely necessary to monitor the concentration of the drug over a lengthy period of time (often three half-lives) in order to provide a comprehensive description of the behavior of the medication or nanoparticles *in vivo*. Several different programs can be utilized in order to fit the PK profile in the blood, which allows for the acquisition of crucial PK parameters that quantitatively explain how the body controls the medication or nanoparticles. During a PK study, some of the most important metrics to analyze are the maximum concentration (C_{max}), the half-life ($t_{1/2}$), the clearance (Cl), the area under the curve (AUC), and the mean resident time (MRT).

When a pharmacological formulation remains in circulation for a longer period of time, it typically possesses the following characteristics: an increased $t_{1/2}$, a lower Cl, a greater AUC, and a higher MRT. It is common for a formulation to exhibit the following characteristics if it is rapidly removed from the body: a short $t_{1/2}$, a high Cl, a short AUC, and a short MRT. The behavior or profile of a drug or nanoparticle can not only be described using PK data but it can also be predicted using this data. PK data are utilized extensively to define the dose and dosing regimen, as well as the toxicity profile, in order to maintain a desirable blood concentration for optimized therapeutic agents with a low risk of adverse events. This is done in order to maintain a desirable blood concentration for optimally effective therapeutic agents [98].

CONCLUSION

The field of nanomedicine is advancing at a rapid rate, and a wide variety of drug carriers are currently being evaluated to determine how effective they are at delivering pharmaceuticals locally and to specific tumor locations of interest. In addition, merging hybrid techniques with nanotherapies is always being tested to improve the effectiveness of cancer patients' treatment regimes. However, the typical medication development and delivery processes provide a number of difficulties for these approaches. Understanding pharmacological synergies is important, but establishing specific patient profiles based on their individual molecular signatures is now crucial to the efficacy of targeted drug delivery. More clinical indicators are also needed to reduce treatment failures and the likelihood of cancer recurrence in order to improve cancer treatments. In this situation, smart computational models are capable of properly processing complex data and delivering precise outcomes. AI is vital in building a road map to evaluate real-time monitoring of drug delivery protocols, categorizing patients based on their genomic signatures, and offering precise information on treatment response due to its established capabilities with clinical imaging. This includes but is not limited to, helping to advance image-guided drug delivery. In spite of the fact that numerous approaches have been investigated in the pharmaceutical and nanotechnology sectors over the course of the years, there has not been a great deal of study that has specifically focused on the use of AI for targeted medication delivery. In this chapter, the purpose of the study was to demonstrate how the application of AI may assist in conquering some of the drawbacks of manufacturing strategies and its likely effect on patient profiling, the detection of cancer biomarkers, and the imaging of nanoparticles in order to improve the outcomes of nanotechnology-based therapeutics. By using significant findings from the study, further research could result in the creation of intelligent systems for biomarker recognition and nanoparticle monitoring and analysis.

REFERENCES

[1]   Schabath MB, Cote ML. Cancer progress and priorities: lung cancer. Cancer Epidemiol Biomarkers Prev 2019; 28(10): 1563-79.
[http://dx.doi.org/10.1158/1055-9965.EPI-19-0221] [PMID: 31575553]

[2]   Zhang PW, Chen L, Huang T, Zhang N, Kong XY, Cai YD. Classifying ten types of major cancers based on reverse phase protein array profiles. PLoS One 2015; 10(3): e0123147.
[http://dx.doi.org/10.1371/journal.pone.0123147] [PMID: 25822500]

[3]   Thai AA, Solomon BJ, Sequist LV, Gainor JF, Heist RS. Lung cancer. Lancet 2021; 398(10299): 535-54.
[http://dx.doi.org/10.1016/S0140-6736(21)00312-3] [PMID: 34273294]

[4]   Wang S, Dong L, Wang X, Wang X. Classification of pathological types of lung cancer from CT images by deep residual neural networks with transfer learning strategy. Open Med (Wars) 2020; 15(1): 190-7.
[http://dx.doi.org/10.1515/med-2020-0028] [PMID: 32190744]

[5]   Liu S, Wang Y, Yang X, *et al.* Deep learning in medical ultrasound analysis: a review. Engineering (Beijing) 2019; 5(2): 261-75.
[http://dx.doi.org/10.1016/j.eng.2018.11.020]

[6]   Mollura DJ, Culp MP, Pollack E, *et al.* Artificial intelligence in low-and middle-income countries: innovating global health radiology. Radiology 2020; 297(3): 513-20.
[http://dx.doi.org/10.1148/radiol.2020201434] [PMID: 33021895]

[7]   Painuli D, Bhardwaj S, köse U. Recent advancement in cancer diagnosis using machine learning and

deep learning techniques: A comprehensive review. Comput Biol Med 2022; 146: 105580.
[http://dx.doi.org/10.1016/j.compbiomed.2022.105580] [PMID: 35551012]

[8] Bohr A, Memarzadeh K. The rise of artificial intelligence in healthcare applications. Artificial Intelligence in healthcare. Elsevier 2020; pp. 25-60.
[http://dx.doi.org/10.1016/B978-0-12-818438-7.00002-2]

[9] Shur JD, Doran SJ, Kumar S, *et al.* Radiomics in oncology: a practical guide. Radiographics 2021; 41(6): 1717-32.
[http://dx.doi.org/10.1148/rg.2021210037] [PMID: 34597235]

[10] Inamura K. Lung cancer: understanding its molecular pathology and the 2015 WHO classification. Front Oncol 2017; 7: 193.
[http://dx.doi.org/10.3389/fonc.2017.00193] [PMID: 28894699]

[11] Zhang H, Meng D, Cai S, *et al.* The application of artificial intelligence in lung cancer: a narrative review. Transl Cancer Res 2021; 10(5): 2478-87.
[http://dx.doi.org/10.21037/tcr-20-3398] [PMID: 35116562]

[12] Sanvicens N, Marco MP. Multifunctional nanoparticles – properties and prospects for their use in human medicine. Trends Biotechnol 2008; 26(8): 425-33.
[http://dx.doi.org/10.1016/j.tibtech.2008.04.005] [PMID: 18514941]

[13] Dhamija E, Meena P, Ramalingam V, Sahoo R, Rastogi S, Thulkar S. Chemotherapy-induced pulmonary complications in cancer: Significance of clinicoradiological correlation. Indian J Radiol Imaging 2020; 30(1): 20-6.
[http://dx.doi.org/10.4103/ijri.IJRI_178_19] [PMID: 32476746]

[14] Carrasco-Esteban E, Domínguez-Rullán JA, Barrionuevo-Castillo P, *et al.* Current role of nanoparticles in the treatment of lung cancer. J Clin Transl Res 2021; 7(2): 140-55.
[PMID: 34104817]

[15] Sharma N, Sharma M, Sajid Jamal QM, Kamal MA, Akhtar S. Nanoinformatics and biomolecular nanomodeling: a novel move en route for effective cancer treatment. Environ Sci Pollut Res Int 2020; 27(16): 19127-41.
[http://dx.doi.org/10.1007/s11356-019-05152-8] [PMID: 31025282]

[16] Lal CV, Bhandari V, Ambalavanan N. Genomics, microbiomics, proteomics, and metabolomics in bronchopulmonary dysplasia. Seminars in perinatology. Elsevier 2018; Vol. 42: pp. 425-31.
[http://dx.doi.org/10.1053/j.semperi.2018.09.004]

[17] Chiu HY, Chao HS, Chen YM. Application of artificial intelligence in lung cancer. Cancers (Basel) 2022; 14(6): 1370.
[http://dx.doi.org/10.3390/cancers14061370] [PMID: 35326521]

[18] Gharra A, Broza YY, Yu G, *et al.* Exhaled breath diagnostics of lung and gastric cancers in China using nanosensors. Cancer Commun (Lond) 2020; 40(6): 273-8.
[http://dx.doi.org/10.1002/cac2.12030] [PMID: 32459390]

[19] Das KP, J C. Nanoparticles and convergence of artificial intelligence for targeted drug delivery for cancer therapy: Current progress and challenges. Frontiers in Medical Technology 2023; 4: 1067144.
[http://dx.doi.org/10.3389/fmedt.2022.1067144] [PMID: 36688144]

[20] Prasad M, Lambe UP, Brar B, *et al.* Nanotherapeutics: An insight into healthcare and multi-dimensional applications in medical sector of the modern world. Biomed Pharmacother 2018; 97: 1521-37.
[http://dx.doi.org/10.1016/j.biopha.2017.11.026] [PMID: 29793315]

[21] Wei X, Song M, Jiang G, *et al.* Progress in advanced nanotherapeutics for enhanced photodynamic immunotherapy of tumor. Theranostics 2022; 12(12): 5272-98.
[http://dx.doi.org/10.7150/thno.73566] [PMID: 35910806]

[22] Sharma A, Shambhwani D, Pandey S, *et al.* Advances in lung cancer treatment using nanomedicines.

ACS Omega 2023; 8(1): 10-41.
[http://dx.doi.org/10.1021/acsomega.2c04078] [PMID: 36643475]

[23]   Liu S, Setio AAA, Ghesu FC, *et al.* No surprises: Training robust lung nodule detection for low-dose CT scans by augmenting with adversarial attacks. IEEE Trans Med Imaging 2021; 40(1): 335-45.
[http://dx.doi.org/10.1109/TMI.2020.3026261] [PMID: 32966215]

[24]   Lyu J, Bi X, Ling SH. Multi-level cross residual network for lung nodule classification. Sensors (Basel) 2020; 20(10): 2837.
[http://dx.doi.org/10.3390/s20102837] [PMID: 32429401]

[25]   Zheng S, Guo J, Cui X, Veldhuis RNJ, Oudkerk M, van Ooijen PMA. Automatic pulmonary nodule detection in CT scans using convolutional neural networks based on maximum intensity projection. IEEE Trans Med Imaging 2020; 39(3): 797-805.
[http://dx.doi.org/10.1109/TMI.2019.2935553] [PMID: 31425026]

[26]   Shanid M, Anitha A. Adaptive optimisation driven deep belief networks for lung cancer detection and severity level classification. Int J Bio-inspired Comput 2021; 18(2): 114-21.
[http://dx.doi.org/10.1504/IJBIC.2021.118101]

[27]   Zhang G, Yang Z, Gong L, Jiang S, Wang L. Classification of benign and malignant lung nodules from CT images based on hybrid features. Phys Med Biol 2019; 64(12): 125011.
[http://dx.doi.org/10.1088/1361-6560/ab2544] [PMID: 31141794]

[28]   Chinniah P, Maram B, Velrajkumar P, Vidyadhari C. DeepJoint Segmentation-based Lung Segmentation and Hybrid Optimization-Enabled Deep Learning for Lung Nodule Classification. Int J Pattern Recognit Artif Intell 2022; 36(13): 2252021.
[http://dx.doi.org/10.1142/S0218001422520218]

[29]   Reddy S, Bhuvaneshwari V, Tikariha AK, Amballa YS, Raj BSS. Automatic pulmonary nodule detection in ct scans using xception, resnet50 and advanced convolutional neural networks models. Int Res J EngTechnol 2022; 9(4): 3226-37.

[30]   Bhattacharjee A, Shankar K, Murugan R, Goel T. A powerful transfer learning technique for multiclass classification of lung cancer ct images. 2022 International Conference on Engineering and Emerging Technologies (ICEET) 2022; 1-6.
[http://dx.doi.org/10.1109/ICEET56468.2022.10007294]

[31]   Li J, Song F, Zhang P, *et al.* A multi-classification model for non-small cell lung cancer subtypes based on independent subtask learning. Med Phys 2022; 49(11): 6960-74.
[http://dx.doi.org/10.1002/mp.15808] [PMID: 35715882]

[32]   Kareem HF. The IQ-OTH/NCCD lung cancer dataset, version 1, retrieved on 14th November, 2021. Available from: https://wwwkagglecom/hamdallak/the-iqothnccd-lung-cancer-dataset/metadata.

[33]   Wulan TD, Kurniastuti I, Nerisafitra P. Lung Cancer Classification in X-Ray Images Using Probabilistic Neural Network. 2021 International Conference on Computer Science, Information Technology, and Electrical Engineering (ICOMITEE) 2021; 35-9.
[http://dx.doi.org/10.1109/ICOMITEE53461.2021.9650087]

[34]   Blais MA, Akhloufi MA. Deep learning and binary relevance classification of multiple diseases using chest X-ray images. In: 2021 43rd Annual International Conference of the IEEE Engineering in Medicine & Biology Society (EMBC). IEEE; 2021. p. 2794-7.
[http://dx.doi.org/10.1109/EMBC46164.2021.9629846]

[35]   Ausawalaithong W, Thirach A, Marukatat S, Wilaiprasitporn T. Automatic lung cancer prediction from chest X-ray images using the deep learning approach. In: 2018 11th Biomedical Engineering International Conference (BMEiCON). IEEE; 2018. p. 1-5.
[http://dx.doi.org/10.1109/BMEiCON.2018.8609997]

[36]   Varela-Santos S, Melin P. A new modular neural network approach with fuzzy response integration for lung disease classification based on multiple objective feature optimization in chest X-ray images.

Expert Syst Appl 2021; 168: 114361.
[http://dx.doi.org/10.1016/j.eswa.2020.114361]

[37]   Alshmrani GMM, Ni Q, Jiang R, Pervaiz H, Elshennawy NM. A deep learning architecture for multi-class lung diseases classification using chest X-ray (CXR) images. Alex Eng J 2023; 64: 923-35.
[http://dx.doi.org/10.1016/j.aej.2022.10.053]

[38]   Ibrahim DM, Elshennawy NM, Sarhan AM. Deep-chest: Multi-classification deep learning model for diagnosing COVID-19, pneumonia, and lung cancer chest diseases. Comput Biol Med 2021; 132: 104348.
[http://dx.doi.org/10.1016/j.compbiomed.2021.104348] [PMID: 33774272]

[39]   Pandiangan T, Bali I, Silalahi ARJ. Early lung cancer detection using artificial neural network. Atom Indonesia 2019; 45(1): 9-15.
[http://dx.doi.org/10.17146/aij.2019.860]

[40]   Malik H, Anees T, Mui-zzud-din . BDCNet: multi-classification convolutional neural network model for classification of COVID-19, pneumonia, and lung cancer from chest radiographs. Multimedia Syst 2022; 28(3): 815-29.
[http://dx.doi.org/10.1007/s00530-021-00878-3] [PMID: 35068705]

[41]   Asuntha A, Singh N, Srinivasan A. PSO, genetic optimization and SVM algorithm used for lung cancer detection. J Chem Pharm Res 2016; 8(6): 351-9.

[42]   Li Y, Zhang L, Chen H, Yang N. Lung nodule detection with deep learning in 3D thoracic MR images. IEEE Access 2019; 7: 37822-32.
[http://dx.doi.org/10.1109/ACCESS.2019.2905574]

[43]   Bai Y, Li D, Duan Q, Chen X. Analysis of high-resolution reconstruction of medical images based on deep convolutional neural networks in lung cancer diagnostics. Comput Methods Programs Biomed 2022; 217: 106592.
[http://dx.doi.org/10.1016/j.cmpb.2021.106592] [PMID: 35172253]

[44]   Zhang L, Li Y, Wu W, Chen H, Peng Y. Salient detection network for lung nodule detection in 3D Thoracic MRI Images. Biomed Signal Process Control 2021; 66: 102404.
[http://dx.doi.org/10.1016/j.bspc.2020.102404]

[45]   Schwyzer M, Ferraro DA, Muehlematter UJ, *et al.* Automated detection of lung cancer at ultralow dose PET/CT by deep neural networks – Initial results. Lung Cancer 2018; 126: 170-3.
[http://dx.doi.org/10.1016/j.lungcan.2018.11.001] [PMID: 30527183]

[46]   Punithavathy K, Ramya M, Poobal S. Analysis of statistical texture features for automatic lung cancer detection in PET/CT images. 2015 International Conference on Robotics, Automation, Control and Embedded Systems (RACE) 2015; 1-5.
[http://dx.doi.org/10.1109/RACE.2015.7097244]

[47]   Kumar A, Fulham M, Feng D, Kim J. Co-learning feature fusion maps from PET-CT images of lung cancer. IEEE Trans Med Imaging 2020; 39(1): 204-17.
[http://dx.doi.org/10.1109/TMI.2019.2923601] [PMID: 31217099]

[48]   Guo N, Yen RF, El Fakhri G, Li Q. SVM based lung cancer diagnosis using multiple image features in PET/CT. 2015 IEEE Nuclear Science Symposium and Medical Imaging Conference (NSS/MIC) 2015; 1-4.
[http://dx.doi.org/10.1109/NSSMIC.2015.7582234]

[49]   Khan MA, Rajinikanth V, Satapathy SC, *et al.* VGG19 network assisted joint segmentation and classification of lung nodules in CT images. Diagnostics (Basel) 2021; 11(12): 2208.
[http://dx.doi.org/10.3390/diagnostics11122208] [PMID: 34943443]

[50]   Lei Y, Wang T, Jeong JJ, *et al.* Automated lung tumor delineation on positron emission tomography/computed tomography *via* a hybrid regional network. Med Phys 2023; 50(1): 274-83.
[http://dx.doi.org/10.1002/mp.16001] [PMID: 36203393]

[51] Protonotarios NE, Katsamenis I, Sykiotis S, *et al.* A few-shot U-Net deep learning model for lung cancer lesion segmentation *via* PET/CT imaging. Biomed Phys Eng Express 2022; 8(2): 025019.
[http://dx.doi.org/10.1088/2057-1976/ac53bd] [PMID: 35144242]

[52] Saikia T, Kumar R, Kumar D, Singh KK. An automatic lung nodule classification system based on hybrid transfer learning approach. SN Computer Science 2022; 3(4): 272.
[http://dx.doi.org/10.1007/s42979-022-01167-0]

[53] Tyagi S, Talbar SN. LCSCNet: A multi-level approach for lung cancer stage classification using 3D dense convolutional neural networks with concurrent squeeze-and-excitation module. Biomed Signal Process Control 2023; 80: 104391.
[http://dx.doi.org/10.1016/j.bspc.2022.104391]

[54] Hatuwal BK, Thapa HC. Lung cancer detection using convolutional neural network on histopathological images. Int J Comput Trends Tech 2020; 68(10): 21-4.
[http://dx.doi.org/10.14445/22312803/IJCTT-V68I10P104]

[55] Coudray N, Ocampo PS, Sakellaropoulos T, *et al.* Classification and mutation prediction from non–small cell lung cancer histopathology images using deep learning. Nat Med 2018; 24(10): 1559-67.
[http://dx.doi.org/10.1038/s41591-018-0177-5] [PMID: 30224757]

[56] Li Z, Zhang J, Tan T, *et al.* Deep learning methods for lung cancer segmentation in whole-slide histopathology images—the acdc@ lunghp challenge 2019. IEEE J Biomed Health Inform 2021; 25(2): 429-40.
[http://dx.doi.org/10.1109/JBHI.2020.3039741] [PMID: 33216724]

[57] ˇSari´c M, Russo M, Stella M, Sikora M. CNN-based method for lung cancer detection in whole slide histopathology images. In: 2019 4th International Conference on Smart and Sustainable Technologies (SpliTech). IEEE; 2019. p. 1-4.

[58] Mangal S, Chaurasia A, Khajanchi A. Convolution neural networks for diagnosing colon and lung cancer histopathological images. arXiv preprint arXiv:200903878. 2020.

[59] Wu Z, Wang L, Li C, *et al.* DeepLRHE: a deep convolutional neural network framework to evaluate the risk of lung cancer recurrence and metastasis from histopathology images. Front Genet 2020; 11: 768.
[http://dx.doi.org/10.3389/fgene.2020.00768] [PMID: 33193560]

[60] LARXEL. Lung and colon cancer histopathological images dataset, version 1, retrieved on 20th march, 2023. Available from: https://wwwkagglecom/datasets/andrewmvd/lung-and-colon-cancer histopathological-images 2020.

[61] Mukherjee A, Paul M, Mukherjee S. Recent progress in the theranostics application of nanomedicine in lung cancer. Cancers (Basel) 2019; 11(5): 597.
[http://dx.doi.org/10.3390/cancers11050597] [PMID: 31035440]

[62] Xue Y, Gao Y, Meng F, Luo L. Recent progress of nanotechnology-based theranostic systems in cancer treatments. Cancer Biol Med 2021; 18(2): 336-51.
[http://dx.doi.org/10.20892/j.issn.2095-3941.2020.0510] [PMID: 33861527]

[63] Doroudian M, Azhdari MH, Goodarzi N, O'Sullivan D, Donnelly SC. Smart nanotherapeutics and lung cancer. Pharmaceutics 2021; 13(11): 1972.
[http://dx.doi.org/10.3390/pharmaceutics13111972] [PMID: 34834387]

[64] Attia MF, Anton N, Wallyn J, Omran Z, Vandamme TF. An overview of active and passive targeting strategies to improve the nanocarriers efficiency to tumour sites. J Pharm Pharmacol 2019; 71(8): 1185-98.
[http://dx.doi.org/10.1111/jphp.13098] [PMID: 31049986]

[65] Subhan MA, Yalamarty SSK, Filipczak N, Parveen F, Torchilin VP. Recent advances in tumor targeting *via* EPR effect for cancer treatment. J Pers Med 2021; 11(6): 571.

[http://dx.doi.org/10.3390/jpm11060571] [PMID: 34207137]

[66]   Wang J, Zhou T, Liu Y, Chen S, Yu Z. Application of nanoparticles in the treatment of lung cancer with emphasis on receptors. Front Pharmacol 2022; 12: 781425.
[http://dx.doi.org/10.3389/fphar.2021.781425] [PMID: 35082668]

[67]   Habban Akhter M, Sateesh Madhav N, Ahmad J. Epidermal growth factor receptor based active targeting: a paradigm shift towards advance tumor therapy. Artificial Cells, Nanomedicine, and Biotechnology. 2018; 46(sup2): 1188-98.
[http://dx.doi.org/10.1080/21691401.2018.1481863]

[68]   Riaz MK, Zhang X, Wong KH, *et al.* Pulmonary delivery of transferrin receptors targeting peptide surface-functionalized liposomes augments the chemotherapeutic effect of quercetin in lung cancer therapy. Int J Nanomedicine 2019; 14: 2879-902.
[http://dx.doi.org/10.2147/IJN.S192219] [PMID: 31118613]

[69]   Lu L, Xu Q, Wang J, Wu S, Luo Z, Lu W. Drug nanocrystals for active tumor-targeted drug delivery. Pharmaceutics 2022; 14(4): 797.
[http://dx.doi.org/10.3390/pharmaceutics14040797] [PMID: 35456631]

[70]   Fernández M, Javaid F, Chudasama V. Advances in targeting the folate receptor in the treatment/imaging of cancers. Chem Sci (Camb) 2018; 9(4): 790-810.
[http://dx.doi.org/10.1039/C7SC04004K]

[71]   Li K, Liang N, Yang H, Liu H, Li S. Temozolomide encapsulated and folic acid decorated chitosan nanoparticles for lung tumor targeting: improving therapeutic efficacy both *in vitro* and *in vivo*. Oncotarget 2017; 8(67): 111318-32.
[http://dx.doi.org/10.18632/oncotarget.22791] [PMID: 29340056]

[72]   Ayuningtyas NF, Chea C, Ando T, *et al.* Bovine Lactoferrin Suppresses Tumor Angiogenesis through NF-κB Pathway Inhibition by Binding to TRAF6. Pharmaceutics 2023; 15(1): 165.
[http://dx.doi.org/10.3390/pharmaceutics15010165] [PMID: 36678795]

[73]   Li X, Taratula O, Taratula O, Schumann C, Minko T. LHRH-targeted drug delivery systems for cancer therapy. Mini Rev Med Chem 2017; 17(3): 258-67.
[http://dx.doi.org/10.2174/1389557516666161013111155] [PMID: 27739358]

[74]   Kesharwani P, Chadar R, Sheikh A, Rizg WY, Safhi AY. CD44-targeted nanocarrier for cancer therapy. Front Pharmacol 2022; 12: 800481.
[http://dx.doi.org/10.3389/fphar.2021.800481] [PMID: 35431911]

[75]   Huang Z, Lv F, Wang J, *et al.* RGD-modified PEGylated paclitaxel nanocrystals with enhanced stability and tumor-targeting capability. Int J Pharm 2019; 556: 217-25.
[http://dx.doi.org/10.1016/j.ijpharm.2018.12.023] [PMID: 30557679]

[76]   Lin X, Wu J, Liu Y, Lin N, Hu J, Zhang B. Stimuli-responsive drug delivery systems for the diagnosis and therapy of lung cancer. Molecules 2022; 27(3): 948.
[http://dx.doi.org/10.3390/molecules27030948] [PMID: 35164213]

[77]   AlSawaftah NM, Awad NS, Pitt WG, Husseini GA. pH-responsive nanocarriers in cancer therapy. Polymers (Basel) 2022; 14(5): 936.
[http://dx.doi.org/10.3390/polym14050936] [PMID: 35267759]

[78]   Zhao X, Bai J, Yang W. Stimuli-responsive nanocarriers for therapeutic applications in cancer. Cancer Biol Med 2021; 18(2): 319-35.
[http://dx.doi.org/10.20892/j.issn.2095-3941.2020.0496] [PMID: 33764711]

[79]   Li M, Zhao G, Su WK, Shuai Q. Enzyme-responsive nanoparticles for anti-tumor drug delivery. Front Chem 2020; 8: 647.
[http://dx.doi.org/10.3389/fchem.2020.00647] [PMID: 32850662]

[80]   Tagami T, Kubota M, Ozeki T. Effective remote loading of doxorubicin into DPPC/poloxamer 188 hybrid liposome to retain thermosensitive property and the assessment of carrier-based acute

cytotoxicity for pulmonary administration. J Pharm Sci 2015; 104(11): 3824-32.
[http://dx.doi.org/10.1002/jps.24593] [PMID: 26228287]

[81]   Tan S, Wang G. Redox-responsive and pH-sensitive nanoparticles enhanced stability and anticancer ability of erlotinib to treat lung cancer *in vivo*. Drug Des Devel Ther 2017; 11: 3519-29.
[http://dx.doi.org/10.2147/DDDT.S151422] [PMID: 29263650]

[82]   Lim S-J, Hong S-S, Choi JY, Kim JO, Lee M-K, Kim SH. Development of paclitaxel-loaded liposomal nanocarrier stabilized by triglyceride incorporation. Int J Nanomedicine 2016; 11: 4465-77.
[http://dx.doi.org/10.2147/IJN.S113723] [PMID: 27660440]

[83]   Wong J, Prout J, Seifalian A. Magnetic nanoparticles: new perspectives in drug delivery. Curr Pharm Des 2017; 23(20): 2908-17.
[PMID: 28215155]

[84]   Seyfried TN, Huysentruyt LC. On the origin of cancer metastasis. Critical Reviews™ in Oncogenesis. 2013; 18, 1-2.
[http://dx.doi.org/10.1615/CritRevOncog.v18.i1-2.40]

[85]   Dujon AM, Capp JP, Brown JS, *et al.* Is there one key step in the metastatic cascade? Cancers (Basel) 2021; 13(15): 3693.
[http://dx.doi.org/10.3390/cancers13153693] [PMID: 34359593]

[86]   Mu Q, Wang H, Zhang M. Nanoparticles for imaging and treatment of metastatic breast cancer. Expert Opin Drug Deliv 2017; 14(1): 123-36.
[http://dx.doi.org/10.1080/17425247.2016.1208650] [PMID: 27401941]

[87]   Gupta P, Saraff M, Gahtori R, *et al.* Phytomedicines targeting cancer stem cells: Therapeutic opportunities and prospects for pharmaceutical development. Pharmaceuticals (Basel) 2021; 14(7): 676.
[http://dx.doi.org/10.3390/ph14070676] [PMID: 34358102]

[88]   Ogretmen B. Sphingolipid metabolism in cancer signalling and therapy. Nat Rev Cancer 2018; 18(1): 33-50.
[http://dx.doi.org/10.1038/nrc.2017.96] [PMID: 29147025]

[89]   Choi SJ, Choy JH. Layered double hydroxide nanoparticles as target-specific delivery carriers: uptake mechanism and toxicity. Nanomedicine (Lond) 2011; 6(5): 803-14.
[http://dx.doi.org/10.2217/nnm.11.86] [PMID: 21793673]

[90]   Liu R, Hu C, Yang Y, Zhang J, Gao H. Theranostic nanoparticles with tumor-specific enzyme-triggered size reduction and drug release to perform photothermal therapy for breast cancer treatment. Acta Pharm Sin B 2019; 9(2): 410-20.
[http://dx.doi.org/10.1016/j.apsb.2018.09.001] [PMID: 30976492]

[91]   Dianat-Moghadam H, Azizi M, Eslami-S Z, *et al.* The role of circulating tumor cells in the metastatic cascade: biology, technical challenges, and clinical relevance. Cancers (Basel) 2020; 12(4): 867.
[http://dx.doi.org/10.3390/cancers12040867] [PMID: 32260071]

[92]   Aur´ıa-Soro C, Nesma T, Juanes-Velasco P, Landeira-Vi˜nuela A, Fidalgo Gomez H, Acebes-Fernandez V, *et al.* Interactions of nanoparticles and biosystems: microenvironment of nanoparticles and biomolecules in nanomedicine. Nanomaterials. 2019; 9(10): 1365.

[93]   Zhang XQ, Xu X, Bertrand N, Pridgen E, Swami A, Farokhzad OC. Interactions of nanomaterials and biological systems: Implications to personalized nanomedicine. Adv Drug Deliv Rev 2012; 64(13): 1363-84.
[http://dx.doi.org/10.1016/j.addr.2012.08.005] [PMID: 22917779]

[94]   Wang X, Li S, Wang S, Zheng S, Chen Z, Song H. Protein binding nanoparticles as an integrated platform for cancer diagnosis and treatment. Adv Sci (Weinh) 2022; 9(29): 2202453.
[http://dx.doi.org/10.1002/advs.202202453] [PMID: 35981878]

[95]   Wang Y, Cai R, Chen C. The nano–bio interactions of nanomedicines: Understanding the biochemical

driving forces and redox reactions. Acc Chem Res 2019; 52(6): 1507-18.
[http://dx.doi.org/10.1021/acs.accounts.9b00126] [PMID: 31149804]

[96]   Zheng XY, Zhao K, Tang J, *et al.* Gd-dots with strong ligand–water interaction for ultrasensitive magnetic resonance renography. ACS Nano 2017; 11(4): 3642-50.
[http://dx.doi.org/10.1021/acsnano.6b07959] [PMID: 28350963]

[97]   Abdifetah O, Na-Bangchang K. Pharmacokinetic studies of nanoparticles as a delivery system for conventional drugs and herb-derived compounds for cancer therapy: a systematic review. Int J Nanomedicine 2019; 14: 5659-77.
[http://dx.doi.org/10.2147/IJN.S213229] [PMID: 31632004]

[98]   Li SD, Huang L. Pharmacokinetics and biodistribution of nanoparticles. Mol Pharm 2008; 5(4): 496-504.
[http://dx.doi.org/10.1021/mp800049w] [PMID: 18611037]

CHAPTER 10

Heavy Metal Adsorption by Plant Phenolic Polymer: Lignin Nanoparticles

Prolita Pattanayak[1], Bikash Chandra Behera[2], Snehalata Pradhan[3] and **Arun Kumar Pradhan[1,*]**

[1] *Centre for Biotechnology, Siksha 'O' Anusandhan (Deemed to be University), Bhubaneswar, Odisha 751030, India*

[2] *School of Biological Sciences, National Institute of Science Education and Research, Bhubaneswar, Odisha 752050, India*

[3] *Samanta Chandra Sekhar Autonomous College, Puri, Odisha 752001, India*

Abstract: Due to the extreme toxicity of heavy metal ions, water pollution caused by them is currently a major concern. The search for novel, economically viable adsorbents derived from biomass has been intensified recently. Heavy metals are causing a growing number of pollution incidents, leading to significant harm to the health of human beings and also affecting the aquatic environment. Therefore, there is a need for effective and efficient methods for removing high atomic weight metal ions. Lignin is a complex biopolymer that has tremendous scope for usage as a material for the development and production of biodegradable products. Herein, novel hybrid nanoparticles can be prepared. Owing to its distinct physicochemical properties, crosslinked phenol structure, and availability, lignin can be used to possibly generate broad ranges of sorbents, especially those that adsorb heavy metal ions. Thus, the objective of this chapter is to explore the different categories of modifications in the process of lignin's transformation into sophisticated adsorbents for heavy metal ions. Additionally, lignin-derived adsorbents are able to offer significant ecological advantages because of their biocompatibility, stability, and abundance in the plant world. The development of novel lignin-based adsorbents with improved heavy metal ion adsorbent performance may be facilitated by the information in this chapter.

Keywords: Adsorption, Biomass, Composite, Heavy metal, Lignin nanoparticle.

INTRODUCTION

A class of metals identified as heavy metals and metalloids includes those with atomic numbers greater than 20 and densities greater than 5000 kg/m^3. Even in trace amounts, they are harmful to humans despite the fact that some of the heavy

---

* **Corresponding author Arun Kumar Pradhan:** Centre for Biotechnology, Siksha 'O' Anusandhan (Deemed to be University), Bhubaneswar, Odisha 751003, India; E-mail: arunpradhan@soa.ac.in

Manoranjan Arakha & Arun Kumar Pradhan (Eds.)

metals and metalloids are naturally occurring elements with biological functions [1]. Some examples are cadmium, iron, arsenic, lead, nickel, mercury, chromium, zinc, silver, *etc.* In an environmental context, they can be defined as metal (and metalloid) species that are typically bio-accumulative, cannot easily be decomposed by natural factors, and have the ability to cause cancer [2].

Concern about heavy metals' potential for harming humans and other life forms began with the hazardous activity of these substances on organisms that had been observed as early as the late 19th century [2]. Following that, techniques for accurately determining the accumulation of heavy metals and tracking their distribution, occurrence, and future have been developed. Water pollution and environmental degradation pose the biggest threats to the biodiversity of sources of drinking, agricultural, and industrial water worldwide. One significant class of harmful substances found in water sources and often produced by human industrial activity on earth are heavy metals. Elements like As, Cd, Pb, and Hg are unnecessary for plants and animals, and at any amount of exposure, they are hazardous to all biological processes. Because of this consequence, even in extremely low quantities, they may be harmful to species. Heavy metal toxicity is correlated with their oxidation states. For instance, Cr(III) appears to be less harmful than Cr(VI), which is extremely toxic [3].

The bioavailability of heavy metals is a crucial factor in toxicity assessments. Higher bioavailable metal fractions provide more danger to animals and plants. The total concentration of a particular metal that enters cells, rather than its total amount in the environment, determines how harmful it is. Various heavy metal cleaning methods, including physical, chemical, electrical, and biological methods, have been proposed [4]. It is frequently difficult to decide on the best decontamination technique because of this. Sometimes more than one strategy was required for effective results. Chemical precipitation, ion exchange, coagulation, electrochemical methods, and adsorption are among the most applied technologies [5].

Bioaccumulation is a significant feature of their pollution since harmful heavy metals build up in human tissues through the food chain. According to the duration and intensity of exposure, heavy metal exposure in humans can result in a variety of disorders [6, 7]. Short-term exposure to heavy metals can harm the kidney, liver, and lungs, among other organs, as well as the gastrointestinal, cardiovascular, and central neurological systems. On the other hand, exposure for a long time has been linked to a number of degenerative conditions and may raise the chance of developing certain cancers. Hexavalent chromium, for example, can cause apoptosis of human lung fibroblasts, making them hazardous. Several heavy metals, including Pd, As, Cd, and Hg, can also cause cancer [8 - 10]. The majority of industrial waste include numerous high-risk metals, and the present equipment and technology for treating pollutants are partially efficient [11]. It is necessary to build enhanced systems of purification to treat heavy metal ions in industrial regions.

Furthermore, a number of techniques have been accessed to highlight the shortcomings and potential to eliminate heavy metals, and their combination has been identified, which provides a potential way for developing effective hybrid technologies. Since it is efficient, inexpensive, and friendly to the environment, adsorption is frequently used in the treatment of wastewater (Table **1**). It is important to find the optimum adsorbent, one that has strong adsorptive performance and minimal environmental effect and that can be simply separated from wastewater.

Table 1. The benefits and drawbacks of the adsorption method.

Benefits	Drawbacks
Flexibility, ease of use, cheap cost, and excellent effectiveness against heavy metal contaminants	It removes heavy metals from water through hydroxide precipitation by forming metal hydroxides.
It produces excellent wastewater and effluent treatment.	The discharge of precipitated hydroxide into the environment is a major drawback.
Adsorbents can be regenerated and utilized again using appropriate desorption techniques.	During adsorption, it has a bit of a reversible property.
Low concentrations of high atomic mass metals are removed from effluent using adsorption.	Adsorption might not remove heavy metals efficiently from the inorganic wastes hence precipitation using chemicals is used alternatively.

LIGNIN POTENTIAL AS HEAVY METAL ABSORBER

Carbonaceous materials have extensive utility as adsorbents for the water treatment industry because of their excellent chemical stability and high specific surface area [12]. Activated carbons, for instance, are made using fossil fuels like coal and oil, as well as nanoscale carbons like carbon nanotubes and graphene. However, producing and renewing activated carbon is an expensive and energy-intensive process [13]. To replace the expensive activated carbons, a renewable organic biomass material that is inexpensive, ecologically friendly, and contains a significant proportion of carbon could potentially be used [14]. The majority of cellulose, including hemicellulose and lignin, consisting of carbonyl, hydroxyl, and carboxyl groups on their surfaces, are found in biosorbents obtained from plant residues. Surface functional groups have a major role in the potential of biosorbents made from plant wastes [15]. These include a broad variety of adsorbates, such as inorganic ions of heavy metal. A naturally occurring biopolymer of carbohydrates is Lignin. It is a three-dimensional networked structure consisting of lignocellulose biomass. It contains chemical groups such as methoxy, carboxy, hydroxy, phenolic, and aliphatic groups. It is made in biorefinery industries on a massive scale and finds applications in adhesives, adsorbents, dispersants, and plasticizers. Due to lignin's lower affinity to heavy metal ions, it typically has a poor capacity for adsorption [16]. Lignin undergoes functionalization with sulfonic and amine groups to improve the extraction of heavy metals from water [17, 18]. To efficiently remove high atomic weight metallic ions from an aqueous solution, it was altered using various materials, including lignin-grafted nanocomposites [19] or lignin-based nanotraps [18], chitosan-lignin composites, lignin-based hydrogels, and chitin-lignin hybrids [20]. Additionally, it co-polymerized as a result of better adsorption activity of the changed functional groups with increased surface area, diffusion, and dispersion [17].

Lignin has attracted significant interest for adsorbing various heavy metals in recent years. A variety of properties, including abundance, affordability, and biodegradability, as well as the availability of numerous -OH groups within its macromolecules, contribute to its effectiveness as an adsorbent [21]. However, lignin's heterogeneous chemical structure restricts its industrial applicability since it has a relatively poor capacity for heavy metal adsorption [22]. Hence, it needs to be modified before utilization, including by chemical grafting processes and modification with other compounds [23]. Furthermore, nanoparticles synthesized from lignin function as absorbents owing to their huge specific surface area, porous structure, and environmentally friendly nature. In order to create new lignin-based biosorbents that boost lignin's adsorption loading for ions of heavy metals, understanding the determinants of lignin's adsorption capability is essential [24].

A variety of lignin surface modifications with novel functionalizations, such as amines, carboxyl, nitrates, and sulphonic groups, have been described as high-rate metal absorbers. Copper, cadmium, lead, zinc, nickel, and mercury are the primary metals that have been recorded. Different heavy metals have varying affinities for lignin, depending on the sources of lignin, molecular weight, functional groups, and extraction techniques [25]. Typically, three types of lignin to be utilized as heavy metal absorbers have been reported, i.e., functionalized polymeric, the hydrogel form, and lignin in nanoparticle/composite [21].

This study discusses methods for eliminating heavy metals using various lignin-based biopolymers in aqueous solutions. The regeneration of different adsorbents using desorption reagents post-desorption-adsorption cycles is also covered. This also involves reusing the used adsorbents in order to reduce toxicity and secondary pollution.

PRODUCTION OF LIGNIN-BASED MATERIALS AND FUNCTIONAL LIGNIN

Significant progress has been made in various areas pertaining to the concept of bio-refinery and the biotransformation of natural polysaccharides derived from biorenewable sources to create a range of industry-relevant biomaterials and other value-added products [26]. Chemical alterations are necessary, nevertheless, in order to use biomaterials in complex functional products [27]. However, certain products made of lignin are used to eliminate hazardous ions of metals from water. To investigate the metal adsorption processes for possible commercialization, new lignin-based materials must be developed. To further adapt lignin-derived materials for heavy metals and metalloid removal, grafting, cross-linking, hybridization, and copolymerization should be used as lignin modification techniques. The ideal pH range must also be expanded, and the selectivity, adsorption capability, and recyclable nature of these materials have to be improved (Table **2**).

Table 2. Percentage composition of monomeric units sourced from various pulp sources [28, 29].

Lignin Source	Monomeric Unit (%)		
	Coniferyl Alcohol (G)	P-coumaryl Alcohol (H)	Sinapyl Alcohol (S)
Softwood	90-95	0.5-3.5	0-1
Hardwood	25-50	-	50-75
Pulp	25-50	10-25	25-50

Modified Lignin – Chemical and Physical Modification

Lignin may be modified to create a polymer that can adsorb metals through grafting nitrogen-, oxygen - and sulfur-functional groups. NMR spectra may be used to determine the chemical structure, particularly the active sites, in order to optimize the circumstances for the ensuing change [24, 30].

Carboxylic and hydroxyl groups in lignin are oxygenated functional groups that occur naturally [24]. Ethers and hydroxyl groups, which contain oxygen, may work as a reducing agent for recovering metals or provide a free electron pair to chelate metal ions [31]. The dispersion and adsorption capabilities of lignin can be enhanced by an increase in the number of oxygenated groups. Li and colleagues used sulfuric acid at various concentrations to pulp lignin and introduced a very acidic carboxyl group in a portion of the aromatic ring's side chain [32]. The method produced a novel substance known as lignin-biochar (LS-BC3), where the adsorption loading of lead (II) ions is 679 mg/g, demonstrating that the procedure sticks to the model of pseudo-second-order kinetics. Moreover, the research also shows a relationship between maximum adsorption capacity and carboxylic group availability. Similar to this, two-stage synthetic methods were used to produce lignin adsorbents from biomass [33]. Initially, two fractions were obtained by an alkali-oxygen pretreatment: soluble and solid. Precipitation with nitric acid of concentration 0.25 mol/L separates the soluble fraction from the water-soluble lignin. On the other hand, enzymatic hydrolysis was used to extract lignin from the solid residues. These two biosorbents have lead (II) sorption capacities of 263 and 91 mg g^{-1}, which is considered remarkable [33].

Lignin Nanoparticles

Lignin-derived nanoparticles can further boost adsorption capacity because of greater specific surface area and porosity. However, it is still difficult to produce high-yield nanoparticles using an environmentally friendly process. This challenge can be addressed by combining ultrasonic treatment with improved solvent shifting to produce lignin nanoparticles. The dispersibility, form regularity, and yield of lignin nanoparticles may all be improved by ultrasonic treatment, which is crucial for further adsorbing metallic ions of high atomic mass [34, 35]. These modified lignin derivatives have an expected adsorption capacity of 500 mg/g, which is 5 times higher than grafted lignin, and by using physical and chemical adsorption techniques, heavy metals can be removed efficiently [36]. Moreover, the adsorption period of these absorbent materials can be significantly reduced, and their reusability can be increased.

Rice straw lignin was utilized by Shweta and Jha to produce biodegradable nanocomposites of carboxymethylated tetraethoxysilane (CML-T) [37]. The surface was then changed using carboxymethylation, yielding particles with an average diameter of between 160 and 560 nm. In CML-T nanocomposites, methyl and hydroxyl groups provide binding sites for the removal of metal ions. This substance has an adsorption capacity of 71% for nickel ions and 82% for cadmium ions. Furthermore, Geng and colleagues produced magnetic lignosulfonate (MLS) material by ultrasound-assisted synthesis, which involved only one step [38]. The material demonstrated a capacity to absorb various pollutants like chromium ions and Rodamian B. Novel hybrid nanoparticles were produced by Zhang and colleagues who employed epichlorohydrin [39]. It acted as a crosslinker between numerous amino-functionalized magnetic nanoparticles and carboxymethylated lignin. Additionally, Wang and colleagues produced a novel substance known as bentonite-supported organosolv lignin-stabilized zero-valent iron nanoparticles [40]. It exhibited a phenomenal elimination capability of

100% for Cr(VI) for 50 mg L^{-1} at pH 3.0, with a contact period of 30 min and a concentration of 1 g L^{-1} of adsorbent. The authors did not provide the maximum adsorption capacity.

Lignin-based Composites

These are another type of absorbent material that could potentially be used to remove heavy metal ions [24]. Compared to pure materials, hybrid materials with lignin and organic or inorganic precursors have better physicochemical qualities. It is now more appropriate to use lignin-derived materials for the cleanup of metals from water-based solutions [41] because of these increased characteristics. By dispersing various amounts of anionic polyelectrolyte polymer lignosulfonate to a solution of oxidized graphene (GO), one form of hybrid composites may be produced hydrothermally. These materials have a variety of porous structures, a large specific surface area, and plenty of sites for adsorption [42]. After being fabricated in a column-packed apparatus for ongoing treatment of effluent, the lignosulfonate-modified graphene hydrogel (LS-GH) composite demonstrated a good adsorption capacity of 1308 mg g^{-1} for Pb(II) within the initial 40 minutes. The equilibrium studies are fitted to the Langmuir mechanism. In addition, LS-GH is regenerable and reusable for up to 10 adsorption-desorption cycles.

Additionally, lignin-based hydrogels can transport inorganic nanoparticles in a manner that is ecologically secure. A ferrous sulfide (FeS) nanoparticle@lignin hydrogel composite was created by Liu and colleagues through a 12-hour contact between ferrous sulfate and lignin hydrogel [43]. It demonstrated enhanced environmental and mechanical durability and successfully removed cadmium with an increased adsorption ability of 833 g/kg in about 30 minutes. The mechanisms that largely account for enhanced adsorption capacity are hydrogel swelling (0.6%), sorption of nanoparticles (2%), lignin complexation (13%), and chemical precipitation of cadmium sulfide (CdS) (84%). The Langmuir adsorption model best captured the kinetic energy adsorption when it was changed to pseudo-second order kinetics. Additionally, in the process of titanate/TiO2 nanomaterials synthesis, lignin was used as an additive by Fu *et al*. The product made from lignin extracted from wood chips and titanate nanotubes demonstrated high removal capability on Cd^{2+}, Pb^{2+}, and Cu^{2+}. The highest adsorption capacities shown are 678 mg g^{-1}, 309 mg g^{-1}, and 258 mg g^{-1}, respectively, over a broad range of pH (2–7). The literature reported metal adsorption capability of a few lignin-based composite materials.

Lignin-based Hydrogels

Biopolymers prepared from hydrogels are regarded as a better ecologically responsible and long-lasting substitute for the purpose of metal remediation in aqueous solutions [44]. Lignin is a good option to be included in basic hydrogel formulations because of its amazing qualities of biocompatibility, safety, biodegradability, and promotion of sustainability [24]. To generate novel functional adsorbents, lignin-based hydrogels can be combined with natural polymers, synthetic polymers, and/or nanomaterial polymers [45 - 47]. They have shown enhanced morphological, physicochemical, reusability, and water-loving characteristics. These adsorbents are either natural or synthetic polymeric materials [48]. Crosslinking, either physically or chemically, leads to the generation of their three-dimensional structure.

Hydrogels are usually distinguished on the basis of their flexibility, resistance to compression, elasticity, and, most significantly, by the fact that they are extremely hydrophilic materials that absorb high amounts of water without getting rid of their insolubility [49]. The shape and structure of the hydrogel have a major role in the permeation of aqueous and soluble contaminants across the chains of the polymer. The elements that make up the structural framework of polymer contain polar functional groups such as hydroxyl, -COOH, -SO3H, and -CONH, which are susceptible to changes in media based on pH, strength of ions, or temperature [49 - 51]. According to studies, these groups enable ion exchange under either ion exchange or metal coordination under ideal circumstances, which favors the metal adsorption in solution [46, 51].

Removing inorganic pollutants from water-based solutions, which is thought to be the most pertinent possible application for the present study, is not the sole usage of hydrogels. Its applications might be found in a variety of fields, including biomedicine and agriculture [24, 28].

Lignin-based Biochars

Lignin-based biochar is another potential substance for treating heavy metal pollution [52]. In the absence of oxygen or with minimal oxygen present, biomass is thermally pyrolyzed, hydrothermally carbonized, or gasified to generate biochar [52]. The raw material that makes up this substance has some characteristics, such as agricultural leftovers, municipal waste matter, manure, or wood chips, as well as various techniques of decomposition using heat employed to create it, which have a significant impact on its efficacy [32, 53]. Additionally, claims have been made suggesting that adding lignin to the biochar production results in friendly biochar adsorbents with outstanding characteristics for removing metal. This is because lignin is stable, biocompatible, and abundant.

This biochar was developed by pyrolyzing it at various temperatures while in an argon environment [54]. They noticed that biochar produced at elevated temperatures had less aliphatic content and more aromatic structure. The capacity to absorb Pb(II) is significantly impacted by the rise in pyrolysis temperature. The biochar based on lignin demonstrated a noteworthy capacity for adsorbing Pb(II), measuring 1003 mg g^{-1}, in contrast to the initial raw material of 519 mg g^{-1} [54]. The production of more oxygen-containing groups occurs during hydrothermal carbonization at lower temperatures (150–250 °C) [32]. Wang and colleagues used hydrothermal carbonization followed by chemical activation, utilizing ZnCl$_2$ or H$_3$PO$_4$, to produce a biocarbon from residues rich in lignin. For the adsorption of metals, the material offers more readily available active binding sites. The authors discovered that acid groups were present in the activated biocarbon by analyzing the zeta potential. The ability of the substance to eliminate lead, copper, and cadmium ions from extremely concentrated (600 mg L$^{-1)}$ metal solutions was also established [55, 56].

Lignin-based Adsorbents for Removing Individual Heavy Metal Ion

Findings from the current study demonstrated that esterification, amination, grafting copolymerization, and compounding may be used to create lignin-based adsorbents. This bioadsorbent is found to be effective in adsorption with several heavy metals (Fig. **1**). Physical, chemical, and micromorphological characteristics of these adsorbents will vary depending on the different production techniques and reactants, which will eventually impact how well lignin-based adsorbents adsorb. Furthermore, as per the hypothesis of hard acid soft base reactions, soft acids react with soft bases and hard acids with hard bases, whereas the intermediate ones can react with both sides. As a result, the affinity for ions of heavy metal will vary depending on the various active functional groups present in adsorbents derived from lignin.

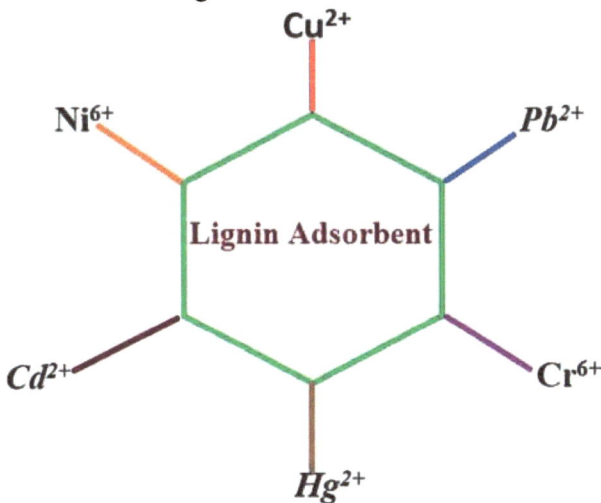

Fig. (1). Lignin bioadsorbent for different heavy metal adsorption.

Adsorption of Cu²⁺

A frequent heavy metal contaminant in industrial effluent is Cu^{2+}. Cu^{2+} consistency in the human body can reach certain levels that result in severe hemolysis and anemia. Because of this, removing Cu^{2+} from wastewater is an essential social issue [57].

One of the key factors affecting the way adsorbents work during adsorption is specific surface area. For example, nanocapsules from lignin were developed that could be adjusted in size utilizing lignosulfonate as the raw material, modified by a process called modification, and using interfacial-initiated microemulsion polymerization [32]. Additionally, altered lignin showed reactivity towards branched polyethylenimine (PEI), having a spherical particle configuration, a specific surface area of 36.445 m²/g, and a 100 nm size. The adsorption loading of lignin-based nanocapsules was a maximum of 56.3 mg/g at 298 K and pH 7, and the adsorption of Cu^{2+} by them in the batch process followed a linear model of pseudo-second-order kinetics and Langmuir linear-model. Using lignin as well as montmorillonite, super-adsorbed hydrogels were created to adsorb Cu^{2+} (Lignin--PAAc/montmorillonite hydrogel) [51]. The crosslinking agent used in this study was N, N′-methylenebis (acrylamide), to which acrylic acid was added to create a polyacrylic acid network. This network structure made it easier to trap Cu^{2+} by exposing the functional groups present in acrylic acid, montmorillonite, and lignin. The adsorption process was studied using batch adsorption experimental studies on Cu^{2+} and followed the Freundlich model. At pH 6, the capacity of Cu^{2+} for adsorption was 74.35 mgg⁻¹. In addition, employing NaCl solution treatment for 5 cycles, the efficiency of Cu-loaded hydrogel to remove Cu^{2+} remained greater than 80%, showing the hydrogel's high degree of cyclability.

Lignin and $MgO\text{-}SiO_2$ inorganic oxide were employed to create a hybrid adsorbent at the micron scale with an uneven surface and 214 m²/g of specific surface area [58]. Additionally, the interior of the adsorbent's pores was suitable for the absorption of Cu^{2+} and exhibited a 2.8 nm pore size and a Cu^{2+} diameter of 0.14 nm. The adsorption response to copper (II) ions was, therefore, quick. The highest adsorption loading achieved was 83.98 mgg⁻¹ under conditions of 5 g/L adsorbent dose, pH 5, for 60 minutes at room temperature. Additionally, the lignin and inorganic oxides contributed to the adsorption process. The computed figures for ΔG and ΔH (ΔG = 38.91 KJ/mol and ΔH = 168.7 KJ/mol, respectively) were determined to be negative. This indicates that the adsorption was characterized as spontaneous and endothermic.

However, lignin has some ability to adsorb Cu^{2+}. In addition to changing organic molecules, the presence of inorganic components also significantly altered the structure and functionality of materials derived from lignin. The convenience for the adsorbing metal ions was made possible by the interwoven network topology, large specific surface area, and various pore structures. It further demonstrates that lignin had potential uses for eliminating Cu^{2+}. These findings also demonstrate that lignin-based adsorption materials have promising adsorption potential for heavy metal ions.

Adsorption of Pb^{2+}

Lead ions are typically found in large amounts in soil and pose a serious threat to industrial effluent [56]. Due to its hazardous nature and difficulty in degrading, even small amounts of lead can be harmful to a human being [59]. Excessive lead consumption can result in malfunction, organ problems, and even cancer [56]. Therefore, it is crucial to create a system capable of efficiently removing Pb^{2+} from effluent.

Ogunsile and Bamgboye used sulfonated resinified soda lignin (SRL) and soda lignin to create a new lignin gel [60]. The study explored the influence of temperature, initial concentration, pH, reaction duration, and dosage of adsorbent on the efficiency of adsorbing Pb2+ using SRL. Results indicated that Pb^{2+} was removed at pH 6 with an efficiency of 96.34%. The SRL's G, H, and S values were 4.6 KJmol^{-1}, 42.39 KJmol^{-1}, and 157.5 Jmol^{-1}, respectively. It implies that adsorption spontaneously took place as an endothermic reaction. A new method for understanding adsorption called the "1-n synergistic adsorption theory" was introduced [61]. The authors employed KL to run experiments on adsorbing lead(II) ions, along with characterizing the interaction between adsorbates. Since a single molecule of adsorbate may interact only with one active site, cooperative adsorption previously concentrated on the 1-1 case. One adsorbent's interactions with several sites were the main emphasis of the hypothesis. Additionally, the KL adsorption was a multivalent binding process, allowing for the practical application of the theory. Under neutral circumstances, KL with carboxyl and phenolic hydroxyl groups had a Pb^{2+} adsorption capability of 49.8 mg/g.

Adsorption of Cr^{6+}

Chromium (Cr) is a type of trace element that our bodies need. Even a small amount can greatly help with healthy growth. It has proven to be beneficial for human development and blood sugar management. The human body will still suffer if it consumes too much chromium [62]. The two most common states of chromium are trivalent and hexavalent. Nearly little harm is caused by Cr^{3+} to the human body. However, Cr^{6+} is highly toxic and readily accumulates in the human body, interfering with the body's normal metabolism [63]. Chemical reactions allow for the transformation of Cr^{6+} and Cr3+. As a result, in addition to adsorption, valence transformation may also be used to remove Cr^{6+} from wastewater.

Adsorption of Hg^{2+}

Mercury (Hg) is a widely recognized harmful heavy metal and is not essential for the human body. When water contains Hg^{2+}, it can enter our bodies through the food we eat, potentially causing damage to our brain, liver, and neurological systems. Mercury-related water contamination is an issue. As a result, the significance of Hg^{2+} removal from water increases.

The total surface area per unit mass and microscopic structure of adsorption materials derived from lignin vary over time, which also affects how well they adsorb substances. Modified lignin-bentonite (AL) and sodium alginate were used for making novel green composite microspheres called SA/ML-BT [64]. These microspheres were effective at adsorbing Hg^{2+}. The SA/ML-BT they made were 2.8mm in diameter. These spheres were porous with many internal folds and a smooth plane; their total surface area per unit mass was 5.98 m^2g^{-1}. Absorption adhered to both the non-linear Freundlich model and the non-linear pseudo-second-order kinetics. Additionally, batch and column adsorption tests were carried out [64]. They discovered that the batch and column adsorption tests at 298 K and pH 3-7 had maximum adsorption loadings of 24.4 and 42.1 mg/g, respectively. It suggested that Hg^{2+} adsorption in that work was suitable for the column experiment. Even after undergoing three cycles of desorption using a 0.1 M HCl desorption agent, the efficiency of removal of mercury (II) ions by these composites stayed consistently higher than 80%. Furthermore, SA/ML-BT demonstrated recyclability.

Likewise, functionalized organic solvent lignin (OLDTC) containing dithiocarbamate groups was prepared by starting with organic solvent lignin and then adding triethylenetetramine and CS_2 [65]. Its specific surface area is 4.5 cm^2g^{-1} and it has an irregular surface. While the S content in OLDTC was from 0.66% to 16.1%, the N content was 12.9%. Because OLDTC has dithiocarbamate groups and is porous, it effectively adsorbs Hg^{2+}. The maximal adsorption loading is 210 mg/g at pH 6 and 298 K.

The incorporation of functional groups during the modification of chemical reaction is primarily responsible for the adsorption of Hg2+ chemically, where functional groups, including sulfur and nitrogen, have a significant effect. To increase the adsorption impact of adsorbents and expose additional functional groups through reactivity, it is important to explore lignin-based adsorbents.

Adsorption of Cd^{2+}

Wastewater from industries contains large amounts of Cd^{2+}. Given its chemical properties, cadmium finds extensive use in the production of electroplating, dyes, alloys and batteries. It is a non-essential component for humans and is extremely hazardous to the human body. A particular quantity of cadmium can harm the liver, kidneys, and gastrointestinal tract, soften bones, and induce liver and kidney disease. Thus, it is vital to remove Cd^{2+} from effluent to successfully mitigate the hazard to human health.

By utilizing lignin as a raw material, 1,2,4-triazole-3-thiol-modified lignin-based adsorbent was created using alkyne hydrothiolation that was triggered by UV light [66]. Under conditions of several ions coexisting, this type of material exhibits the ability to adsorb Cd^{2+}. Functional groups in LBA, including -N = CH- and -NH-, may interact with Cd^{2+}. The greatest adsorption loading of 87.4 mgg^{-1} was at pH 6, temperature of 298 K, and 100 rpm. Upon investigation of LBA's selectivity in mixed solutions, it was observed that the respective values for Pb^{2+}, Cd^{2+}, and Cu^{2+} were 0.38, 2.06, and 0.05. The combined solution demonstrated significant selectivity for Cd^{2+}. LBA showed high repeatability after five rounds by utilizing 0.1 M HCl as a desorption agent, with only a 15.9% reduction in adsorption capacity.

Although the removal of Cd^{2+} is clearly impacted by the lignin-based adsorbent, there is still more to learn about adsorption loading. Regarding the cyclic nature of the adsorbent, the desorption agent with acidic properties proves to be better suited for the Cd2+ adsorption and desorption cycles. Adsorption loading of lignin-derived adsorbent continues to function at a certain level even after multiple cycles. According to the aforementioned studies, lignin-based adsorbents perform less well at adsorbing Cd^{2+} compared to other heavy metal ions like Pb^{2+} and Cu^{2+}. This suggests that the functional groups that are reactive (such as those that contain oxygen, sulfur, and nitrogen) possess relatively weaker affinity for these specific high atomic mass metallic ions. Further research on ways to use adsorbents based on lignin for enhancing the adsorption of Co^{2+} adsorbents is necessary, nevertheless, Cd^{2+} is hazardous and challenging to eliminate from the environment.

CONCLUSION AND VIEWPOINTS

Following cellulose, lignin is one of the most common biopolymers available, and its potential as a substantial raw material has garnered attention in recent years for making eco-friendly and sustainable products. Numerous studies in the literature indicate how lignin may absorb heavy metals from water. Because of the existence of multiple hydroxyl groups in the macromolecular structure, active adsorption sites are thought to be responsible for this feature. However, altering the lignin's functional groups makes metal and metalloid adsorption more favorable. The metal ions removal and wastewater treatment are two areas in which lignin-derived products, such as lignin nanoparticles, modified lignin, and lignin-based hydrogels and composites, have demonstrated promising results. These environmentally friendly and sustainable products offer a useful tool to improve environmental preservation.

At present, lignin is used in the development of adsorbents for a variety of purposes. By modifying, compounding, doping, *etc.*, it can introduce functional groups, including nitrogen, oxygen, or sulfur, increasing the adsorbent's active sites and enhancing its ability to adsorb heavy metal ions. The evident variations between different forms of lignin significantly impact the adsorption loading of the adsorbent when they are prepared. As a result, selecting an appropriate raw material is crucial for creating lignin-derived adsorbent. Lignin, nonetheless, has the potential and challenges in developing adsorbents. Secondary environmental contamination can occur from the preparation procedure, the toxic nature of the substances utilized to alter lignin, and the elimination of metallic ions with high atomic weight. For instance, the usage of altered materials like formaldehyde, glutaraldehyde, and carbon disulfide can affect the environment if they are not treated.

Based on a systematic review of published works, additional research is needed to fully understand the alteration of lignin structure and the application in the creation of novel materials for capturing high atomic mass metals and semimetals. Functionalization techniques ought to receive extra attention when we want to increase the selectivity of novel materials. Only a few research have been done on the removal of anionic species, although several publications have described the exclusion of metal cations from aqueous solutions. It is vital to look at how effectively sorbent materials can remove certain harmful metal and metalloid ions from water since they are present in an aqueous solution as anions. To enhance the novel lignin-based adsorbents' adsorption properties, biodegradability, and recyclability, further research should be done on the employment of lignin for fabricating inorganic/organic hybrid materials. Making high-performance, cost-effective materials with superior adsorption efficiency and selectivity out of this biopolymer is still very challenging.

The investigation of lignin-based adsorption materials has advanced due to ongoing research into these materials. However, it was discovered throughout the study process that the altered materials were not ecologically friendly. Therefore, we should focus on streamlining and environmentally friendly adsorbent preparation in our forthcoming study. Similarly, we could increase the recyclability of adsorbents by utilizing the properties of the chemical reactions between metal ions and adsorbents to produce adsorbents with strong cycling capabilities. But as research progresses, it's also important to pay attention to the varied uses of lignin-derived adsorbents. For instance, the high atomic weight metal ion-loaded adsorbents are recycled. The characteristics of these ions are used to create fluorescence-detecting materials, lignin-based flame-retardants and supercapacitors, and other multipurpose materials. In addition to helping to resolve the problem of wastewater pollution and achieve environmental protection, this type of multifunctional study also makes it easier to use lignin resources in a way that adds value.

REFERENCES

[1]   Sosa-Rodríguez FS, Vazquez-Arenas J, Peña PP, *et al.* Spatial distribution, mobility and potential health risks of arsenic and lead concentrations in semiarid fine top-soils of Durango City, Mexico. Catena 2020; 190: 104540.
[http://dx.doi.org/10.1016/j.catena.2020.104540]

[2]   Fei Y, Hu YH. Recent progress in removal of heavy metals from wastewater: A comprehensive review. Chemosphere 2023; 335: 139077.
[http://dx.doi.org/10.1016/j.chemosphere.2023.139077] [PMID: 37263507]

[3]   Enniya I, Rghioui L, Jourani A. Adsorption of hexavalent chromium in aqueous solution on activated carbon prepared from apple peels. Sustain Chem Pharm 2018; 7: 9-16.
[http://dx.doi.org/10.1016/j.scp.2017.11.003]

[4]   Caliman, FlorentinaAnca, *et al.* Soil and groundwater cleanup: benefits and limits of emerging technologies. Clean Technologies and Environmental Policy 13, 2011; 241-268.

[5]   Sadeghifar H, Ragauskas A. Lignin as a bioactive polymer and heavy metal absorber- an overview. Chemosphere 2022; 309(Pt 1): 136564.
[http://dx.doi.org/10.1016/j.chemosphere.2022.136564] [PMID: 36155017]

[6]   Diarra I, Prasad S. The current state of heavy metal pollution in Pacific Island Countries: a review. Appl Spectrosc Rev 2021; 56(1): 27-51.
[http://dx.doi.org/10.1080/05704928.2020.1719130]

[7]   Volesky B, Holan ZR. Biosorption of heavy metals. Biotechnol Prog 1995; 11(3): 235-50.
[http://dx.doi.org/10.1021/bp00033a001] [PMID: 7619394]

[8]   Jaishankar, Monisha, *et al.* Toxicity, mechanism and health effects of some heavy metals. Interdisciplinary toxicology 7.2, 2014: 60.
[http://dx.doi.org/10.2478/intox-2014-0009]

[9]   Khanam R, Kumar A, Nayak AK, *et al.* Metal(loid)s (As, Hg, Se, Pb and Cd) in paddy soil: Bioavailability and potential risk to human health. Sci Total Environ 2020; 699: 134330.
[http://dx.doi.org/10.1016/j.scitotenv.2019.134330] [PMID: 31522043]

[10]   Yin, Jie, *et al.* Impact of environmental factors on gastric cancer: a review of the scientific evidence, human prevention and adaptation. journal of environmental sciences 89; 2020: 65-79.
[http://dx.doi.org/10.1016/j.jes.2019.09.025]

[11]   Valle H, Sánchez J, Rivas BL. Poly(*N* -vinylpyrrolidone- *co* -2-acrylamido-2-methylpropanesulfonate sodium): Synthesis, characterization, and its potential application for the removal of metal ions from aqueous solution. J Appl Polym Sci 2015; 132(2): app.41272.
[http://dx.doi.org/10.1002/app.41272]

[12] Chen, Suhong, *et al.* Equilibrium and kinetic studies of methyl orange and methyl violet adsorption on activated carbon derived from Phragmitesaustralis. Desalination 2010; 252 (1-3): 149-156.
[http://dx.doi.org/10.1016/j.desal.2009.10.010]

[13] Yahya, MohdAdib, *et al.* A brief review on activated carbon derived from agriculture by-product. AIP Conference Proceedings 2018; 1972 (1).
[http://dx.doi.org/10.1063/1.5041244]

[14] Li F, Wang X, Yuan T, Sun R. A lignosulfonate-modified graphene hydrogel with ultrahigh adsorption capacity for Pb(II) removal. J Mater Chem A Mater Energy Sustain 2016; 4(30): 11888-96.
[http://dx.doi.org/10.1039/C6TA03779H]

[15] Redha A. Ali. Removal of heavy metals from aqueous media by biosorption. *Arab.* J Basic Appl Sci 2020; 27(1): 183-93.
[http://dx.doi.org/10.1080/25765299.2020.1756177]

[16] Lora JH, Glasser WG. Recent industrial applications of lignin: a sustainable alternative to nonrenewable materials. J Polym Environ 2002; 10(1/2): 39-48.
[http://dx.doi.org/10.1023/A:1021070006895]

[17] Ge H, Hua T, Chen X. Selective adsorption of lead on grafted and crosslinked chitosan nanoparticles prepared by using Pb2+ as template. J Hazard Mater 2016; 308: 225-32.
[http://dx.doi.org/10.1016/j.jhazmat.2016.01.042] [PMID: 26844403]

[18] Xiao D, Ding W, Zhang J, Ge Y, Wu Z, Li Z. Fabrication of a versatile lignin-based nano-trap for heavy metal ion capture and bacterial inhibition. Chem Eng J 2019; 358: 310-20.
[http://dx.doi.org/10.1016/j.cej.2018.10.037]

[19] Li Z, Chen J, Ge Y. Removal of lead ion and oil droplet from aqueous solution by lignin-grafted carbon nanotubes. Chem Eng J 2017; 308: 809-17.
[http://dx.doi.org/10.1016/j.cej.2016.09.126]

[20] Nair V, Panigrahy A, Vinu R. Development of novel chitosan-lignin composites for adsorption of dyes and metal ions from wastewater. Chem Eng J 2014; 254: 491-502.
[http://dx.doi.org/10.1016/j.cej.2014.05.045]

[21] Santander P, Butter B, Oyarce E, Yáñez M, Xiao L-P, Sánchez J. Lignin-based adsorbent materials for metal ion removal from wastewater: A review. Ind Crops Prod 2021; 167: 113510.
[http://dx.doi.org/10.1016/j.indcrop.2021.113510]

[22] Abdolali A, Guo WS, Ngo HH, Chen SS, Nguyen NC, Tung KL. Typical lignocellulosic wastes and by-products for biosorption process in water and wastewater treatment: A critical review. Bioresour Technol 2014; 160: 57-66.
[http://dx.doi.org/10.1016/j.biortech.2013.12.037] [PMID: 24405653]

[23] Eraghi Kazzaz A, Hosseinpour Feizi Z, Fatehi P. Grafting strategies for hydroxy groups of lignin for producing materials. Green Chem 2019; 21(21): 5714-52.
[http://dx.doi.org/10.1039/C9GC02598G]

[24] Ge Y, Li Z. Application of lignin and its derivatives in adsorption of heavy metal ions in water: a review. ACS Sustain Chem& Eng 2018; 6(5): 7181-92.
[http://dx.doi.org/10.1021/acssuschemeng.8b01345]

[25] Sadeghifar H, Ragauskas A. Perspective on technical lignin fractionation. ACS Sustain Chem& Eng 2020; 8(22): 8086-101.
[http://dx.doi.org/10.1021/acssuschemeng.0c01348]

[26] Arevalo-Gallegos A, Ahmad Z, Asgher M, Parra-Saldivar R, Iqbal HMN. Lignocellulose: A sustainable material to produce value-added products with a zero waste approach—A review. Int J Biol Macromol 2017; 99: 308-18.
[http://dx.doi.org/10.1016/j.ijbiomac.2017.02.097] [PMID: 28254573]

[27]   Dax D, Xu C, Långvik O, Hemming J, Backman P, Willför S. Synthesis of SET-LRP-induced galactoglucomannan-diblock copolymers. J Polym Sci A Polym Chem 2013; 51(23): 5100-10.
[http://dx.doi.org/10.1002/pola.26942]

[28]   Thakur VK, Thakur MK, Raghavan P, Kessler MR. Progress in green polymer composites from lignin for multifunctional applications: a review. ACS Sustain Chem& Eng 2014; 2(5): 1072-92.
[http://dx.doi.org/10.1021/sc500087z]

[29]   Wang Q, Zheng C, Shen Z, *et al.* Polyethyleneimine and carbon disulfide co-modified alkaline lignin for removal of Pb2 + ions from water. Chem Eng J 2019; 359: 265-74.
[http://dx.doi.org/10.1016/j.cej.2018.11.130]

[30]   Jiang X, Liu J, Du X, Hu Z, Chang H, Jameel H. X., Z. Hu, HM Chang and H. Jameel, Phenolation to improve lignin reactivity toward thermosets application. ACS Sustain Chem& Eng 2018; 6(4): 5504-12.
[http://dx.doi.org/10.1021/acssuschemeng.8b00369]

[31]   Supanchaiyamat N, Jetsrisuparb K, Knijnenburg JTN, Tsang DCW, Hunt AJ. Lignin materials for adsorption: Current trend, perspectives and opportunities. Bioresour Technol 2019; 272: 570-81.
[http://dx.doi.org/10.1016/j.biortech.2018.09.139] [PMID: 30352730]

[32]   Li Y, Wang F, Miao Y, *et al.* A lignin-biochar with high oxygen-containing groups for adsorbing lead ion prepared by simultaneous oxidization and carbonization. Bioresour Technol 2020; 307: 123165.
[http://dx.doi.org/10.1016/j.biortech.2020.123165] [PMID: 32203865]

[33]   Song K, Chu Q, Hu J, *et al.* Two-stage alkali-oxygen pretreatment capable of improving biomass saccharification for bioethanol production and enabling lignin valorization *via* adsorbents for heavy metal ions under the biorefinery concept. Bioresour Technol 2019; 276: 161-9.
[http://dx.doi.org/10.1016/j.biortech.2018.12.107] [PMID: 30623871]

[34]   Chauhan PS. Lignin nanoparticles: Eco-friendly and versatile tool for new era. Bioresour Technol Rep 2020; 9: 100374.
[http://dx.doi.org/10.1016/j.biteb.2019.100374]

[35]   Fu Y, Liu X, Chen G. Adsorption of heavy metal sewage on nano-materials such as titanate/TiO2 added lignin. Results Phys 2019; 12: 405-11.
[http://dx.doi.org/10.1016/j.rinp.2018.11.084]

[36]   Wang B, Sun D, Wang H-M, Yuan T-Q, Sun R-C. Green and facile preparation of regular lignin nanoparticles with high yield and their natural broad-spectrum sunscreens. ACS Sustain Chem& Eng 2019; 7(2): 2658-66. a
[http://dx.doi.org/10.1021/acssuschemeng.8b05735]

[37]   Shweta K, Jha H. Synthesis and characterization of crystalline carboxymethylated lignin-TEOS nanocomposites for metal adsorption and antibacterial activity. Bioresour Bioprocess 2016; 3(1): 31.
[http://dx.doi.org/10.1186/s40643-016-0107-7]

[38]   Geng J. FeiGu, and Jianmin Chang. Fabrication of magnetic lignosulfonate using ultrasonic-assisted *in situ* synthesis for efficient removal of Cr (VI) and Rhodamine B from wastewater. J Hazard Mater 2019; 375: 174-81.
[http://dx.doi.org/10.1016/j.jhazmat.2019.04.086] [PMID: 31055194]

[39]   Zhang Y, Ni S, Wang X, *et al.* Ultrafast adsorption of heavy metal ions onto functionalized lignin-based hybrid magnetic nanoparticles. Chem Eng J 2019; 372: 82-91.
[http://dx.doi.org/10.1016/j.cej.2019.04.111]

[40]   Wang Z, Chen G, Wang X, Li S, Liu Y, Yang G. Removal of hexavalent chromium by bentonite supported organosolv lignin-stabilized zero-valent iron nanoparticles from wastewater. J Clean Prod 2020; 267: 122009. b
[http://dx.doi.org/10.1016/j.jclepro.2020.122009]

[41]   Klapiszewski Ł, Siwińska-Stefańska K, Kołodyńska D. Preparation and characterization of novel

TiO2/lignin and TiO2-SiO2/lignin hybrids and their use as functional biosorbents for Pb(II). Chem Eng J 2017; 314: 169-81.
[http://dx.doi.org/10.1016/j.cej.2016.12.114]

[42]   Li H, Sun Z, Zhang L, Tian Y, Cui G, Yan S. A cost-effective porous carbon derived from pomelo peel for the removal of methyl orange from aqueous solution. Colloids Surf A Physicochem Eng Asp 2016; 489: 191-9. a
[http://dx.doi.org/10.1016/j.colsurfa.2015.10.041]

[43]   Liu Y, Huang Y, Zhang C, *et al.* Nano-FeS incorporated into stable lignin hydrogel: A novel strategy for cadmium removal from soil. Environ Pollut 2020; 264: 114739.
[http://dx.doi.org/10.1016/j.envpol.2020.114739] [PMID: 32434113]

[44]   Dodson JR, Parker HL, Muñoz García A, *et al.* Bio-derived materials as a green route for precious & critical metal recovery and re-use. Green Chem 2015; 17(4): 1951-65.
[http://dx.doi.org/10.1039/C4GC02483D]

[45]   Peñaranda A JE, Sabino MA, Marcos A. Effect of the presence of lignin or peat in IPN hydrogels on the sorption of heavy metals. Polym Bull 2010; 65(5): 495-508.
[http://dx.doi.org/10.1007/s00289-010-0264-3]

[46]   Yao Q, Xie J, Liu J, Kang H, Liu Y. Adsorption of lead ions using a modified lignin hydrogel. J Polym Res 2014; 21(6): 465.
[http://dx.doi.org/10.1007/s10965-014-0465-9]

[47]   Sun Y, Ma Y, Fang G, Li S, Fu Y. Synthesis of acid hydrolysis lignin-g-poly-(acrylic acid) hydrogel superabsorbent composites and adsorption of lead ions. BioResources 2016; 11(3): 5731-42.
[http://dx.doi.org/10.15376/biores.11.3.5731-5742]

[48]   Chandna, Sanjam, *et al.* Synthesis and applications of lignin-derived hydrogels. Lignin: Biosynthesis and Transformation for Industrial Applications (2020): 231-252.
[http://dx.doi.org/10.1007/978-3-030-40663-9_8]

[49]   Thakur S, Govender PP, Mamo MA, Tamulevicius S, Mishra YK, Thakur VK. Progress in lignin hydrogels and nanocomposites for water purification: Future perspectives. Vacuum 2017; 146: 342-55.
[http://dx.doi.org/10.1016/j.vacuum.2017.08.011]

[50]   Parajuli D, Inoue K, Ohto K, *et al.* Adsorption of heavy metals on crosslinked lignocatechol: a modified lignin gel. React Funct Polym 2005; 62(2): 129-39.
[http://dx.doi.org/10.1016/j.reactfunctpolym.2004.11.003]

[51]   Sun XF, Hao Y, Cao Y, Zeng Q. Superadsorbent hydrogel based on lignin and montmorillonite for Cu(II) ions removal from aqueous solution. Int J Biol Macromol 2019; 127: 511-9.
[http://dx.doi.org/10.1016/j.ijbiomac.2019.01.058] [PMID: 30660568]

[52]   Wang J, Wang S. Preparation, modification and environmental application of biochar: A review. J Clean Prod 2019; 227: 1002-22.
[http://dx.doi.org/10.1016/j.jclepro.2019.04.282]

[53]   Wu P, Cui P, Fang G, Gao J, Zhou D, Wang Y. Sorption mechanism of zinc on reed, lignin, and reed- and lignin-derived biochars: kinetics, equilibrium, and spectroscopic studies. J Soils Sediments 2018; 18(7): 2535-43.
[http://dx.doi.org/10.1007/s11368-018-1928-0]

[54]   Wu F, Chen L, Hu P, Wang Y, Deng J, Mi B. Industrial alkali lignin-derived biochar as highly efficient and low-cost adsorption material for Pb(II) from aquatic environment. Bioresour Technol 2021; 322: 124539.
[http://dx.doi.org/10.1016/j.biortech.2020.124539] [PMID: 33340951]

[55]   Wang B, Ran M, Fang G, Wu T, Ni Y. Biochars from lignin-rich residue of furfural manufacturing process for heavy metal ions remediation. Materials (Basel) 2020; 13(5): 1037. a
[http://dx.doi.org/10.3390/ma13051037] [PMID: 32106506]

[56]   Wang Z, Chen G, Wang X, Li S, Liu Y, Yang G. Removal of hexavalent chromium by bentonite supported organosolv lignin-stabilized zero-valent iron nanoparticles from wastewater. J Clean Prod 2020; 267: 122009. b
[http://dx.doi.org/10.1016/j.jclepro.2020.122009]

[57]   Si R, Wu C, Yu D, Ding Q, Li R. Novel TEMPO-oxidized cellulose nanofiber/polyvinyl alcohol/polyethyleneimine nanoparticles for Cu2+ removal in water. Cellulose 2021; 28(17): 10999-1011.
[http://dx.doi.org/10.1007/s10570-021-04236-4]

[58]   Ciesielczyk F, Bartczak P, Klapiszewski Ł, Jesionowski T. Treatment of model and galvanic waste solutions of copper(II) ions using a lignin/inorganic oxide hybrid as an effective sorbent. J Hazard Mater 2017; 328: 150-9.
[http://dx.doi.org/10.1016/j.jhazmat.2017.01.009] [PMID: 28110149]

[59]   Dhiman V, Kondal N. Colloid Interface Sci. Commun 2021; 41: 100380.

[60]   Ogunsile BO, Bamgboye MO. Biosorption of Lead (II) onto soda lignin gels extracted from Nypa fruiticans. J Environ Chem Eng 2017; 5(3): 2708-17.
[http://dx.doi.org/10.1016/j.jece.2017.05.016]

[61]   Chen H, Qu X, Liu N, Wang S, Chen X, Liu S. Study of the adsorption process of heavy metals cations on Kraft lignin. Chem Eng Res Des 2018; 139: 248-58.
[http://dx.doi.org/10.1016/j.cherd.2018.09.028]

[62]   Shi X, Qiao Y, An X, Tian Y, Zhou H. High-capacity adsorption of Cr(VI) by lignin-based composite: Characterization, performance and mechanism. Int J Biol Macromol 2020; 159: 839-49. a
[http://dx.doi.org/10.1016/j.ijbiomac.2020.05.130] [PMID: 32445824]

[63]   Miretzky, Patricia, and A. Fernandez Cirelli. Cr (VI) and Cr (III) removal from aqueous solution by raw and modified lignocellulosic materials: a review. Journal of hazardous materials 180.1-3 (2010): 1-19.
[http://dx.doi.org/10.1016/j.jhazmat.2010.04.060]

[64]   Gong L, Kong Y, Wu H, Ge Y, Li Z. Sodium alginate microspheres interspersed with modified lignin and bentonite (SA/ML-BT) as a green and highly effective adsorbent for batch and fixed-bed column adsorption of Hg (II). J Inorg Organomet Polym Mater 2021; 31(2): 659-73. a [Ge].
[http://dx.doi.org/10.1007/s10904-020-01757-6]

[65]   Ge Y, Wu S, Qin L, Li Z. Conversion of organosolv lignin into an efficient mercury ion adsorbent by a microwave-assisted method. J Taiwan Inst Chem Eng 2016; 63: 500-5. b
[http://dx.doi.org/10.1016/j.jtice.2016.03.017]

[66]   Jin C, Zhang X, Xin J, *et al.* Clickable synthesis of 1, 2, 4-triazole modified lignin-based adsorbent for the selective removal of Cd (II). ACS Sustain Chem& Eng 2017; 5(5): 4086-93.
[http://dx.doi.org/10.1021/acssuschemeng.7b00072]

CHAPTER 11

Bioremediation using Nanomaterials: A Prospective Approach for Environmental Decontamination

Subham Preetam[1,2,*], Rajeswari Rath[3], Arunima Pandey[4], Arka Ghosh[4] and **Gautam Mohapatra[3,5]**

[1] *Institute of Advanced Materials, IAAM, Gammalkilsvägen 18, Ulrika 59053, Sweden*

[2] *Department of Robotics Engineering, Daegu Gyeongbuk Institute of Science and Technology (DGIST), Daegu 42988, Republic of Korea*

[3] *Centre for Biotechnology, Siksha 'O' Anusandhan (Deemed to be University), Bhubaneswar, Odisha 751030, India*

[4] *KIIT School of Biotechnology, Kalinga Institute of Industrial Technology (KIIT-DU), Bhubaneswar, Odisha 751024, India*

[5] *Department of Biotechnology, Indian Institute of Technology Madras (IITM), Chennai, Tamil Nadu 600036, India*

Abstract: With a focus on environmentally friendly, sustainable technology, new technologies such as the combined process of nanotechnology and bioremediation are urgently needed to accelerate the cost-effective remediation process to alleviate toxic contaminants compared to conventional remediation methods. Numerous studies have shown that nanoparticles possess special qualities, including improved catalysis, adsorption, and increased reactivity. This chapter explores the potential of nanomaterials in bioremediation for environmental decontamination. Bioremediation is a sustainable and eco-friendly approach to remove pollutants from contaminated soil and water. The application of nanotechnology in bioremediation has gained attention due to the unique properties of nanomaterials, such as high surface area, reactivity, and selectivity. The content discusses recent developments in using various nanomaterials, including carbon-based, metal-based, and hybrid nanomaterials, to remove organic and inorganic pollutants from the environment. Also, this highlights the advantages and limitations of using nanomaterials in bioremediation, including their potential toxicity and long-term effects on the environment. Furthermore, we provide insights into the prospects of using nanomaterials in bioremediation and the challenges that need to be addressed for the effective and safe use of nanomaterials in environmental decontamination.

* **Corresponding author Subham Preetam:** Department of Robotics Engineering, Daegu Gyeongbuk Institute of Science and Technology (DGIST), Daegu 42988, Republic of Korea; E-mail: sspritamrath93@gmail.com

Manoranjan Arakha & Arun Kumar Pradhan (Eds.)

Keywords: Bioremediation, Environmental decontamination, Nanomaterials, Nanotechnology, Sustainable technology.

INTRODUCTION

The excessive growth in urbanization and industrial development since the late 19th and early 20th centuries has made it possible for us to achieve new heights in technological and economic progress. This has, in turn, made our quality of life much more advanced. Still, at the same time, the uncontrolled use of industrial techniques, excessive extraction of non-renewable resources, and disposal of toxic waste materials into the environment without proper monitoring have had detrimental effects on our health. It is increasing with every new advancement that humanity makes [1]. The increasing contamination issues are giving rise to numerous complicated infectious diseases, among which respiratory infections and contact dermatitis are the leading ones [2 - 5].

The industrial revolution surrounding the chemical technology industry and its derivatives has given rise to increased levels of trace metals in our surrounding environment [2, 6, 7]. This, in turn, causes the accumulation of metal ions as sediment in the soil and water bodies. Sediment is the primary source of heavy metals in the environment. They can be formed in many ways- 1) Atmospheric deposition of heavy metals like cadmium (Cd), lead (Pb), and mercury (Hg) causes build-up in soil. 2) Wastewater discharge containing zinc (Zn), nickel (Ni), and copper (Cu) can cause hazardous effects on living organisms. 3) Rain leaching refers to a process where heavy metals such as chromium (Cr) and arsenic (As) are transferred to a liquid medium (such as water) from a stable matrix due to the influence of acid rain [8, 9]. Apart from these methods, there are cases where the chemical, as well as the physical characteristics of the water containing the HMs (heavy metals), may change, and this causes the HMs to return from the sediment to the topmost layer of the water in response to the changes occurring in the water. This is known as secondary pollution [10]. The toxicological effects of HMs, the risks of bioaccumulation, and other environmental concerns are why heavy metal contamination has become a severe issue that needs to be solved [11 - 13].

Apart from HMs, various toxic contaminants, like phenol and its derivatives, are found in large quantities in our surrounding environment. The most notable are the halogenated phenols classified as anthropogenic pollutants, meaning that humans have contributed to creating this pollutant [14]. It is also mutagenic and carcinogenic in nature [15]. Various effluent plants and wastewater from different industries, domestic households, chemical spills, and agricultural lands are causing severe contamination of the surrounding water bodies [16]. There are also harmful solid waste materials composed of various dangerous and toxic components. These can be organic or inorganic in nature and are mostly non-biodegradable. These have long-term detrimental effects on soil vitality and fertility and the health of humans and other animals [17, 18].

The pesticides, insecticides, and fertilizer industries produce a considerable amount of solid and liquid waste. The excessive use of pesticides and fertilizers is making the surrounding soil and water poisonous. These toxic and harmful chemicals may enter our food chain through various methods like a) Bioaccumulation- which is the net accumulation of contaminates in an organism [19] and b) Biomagnification- which is the chemical transfer of a toxin from a lower trophic level to higher trophic levels within a food chain resulting in high concentrations of the toxin in question in the apex predators [20]. Given these harmful and adverse conditions, much more research is needed to develop cost-effective decontamination processes. In recent years, new technologies for effective contamination removal have been developed. Among these, bioremediation has gained much interest due to its green and sustainable techniques for treating pollutants [21]. It is also very flexible and can be implemented on a large scale for better waste management [22]. This bio-based approach has profound effects in protecting the environment, and thus, it is considered an 'environmentally appropriate' method. Some examples of bioremediation are- a) Biostimulation- where the growth of the indigenous microorganisms is stimulated, and b) Bioaugmentation- where non-native oil-degrading bacteria are inoculated for the acceleration of the detoxification process in a polluted site with minimal impact on the ecological system [23, 24].

Even though bioremediation is an excellent, practical, and flexible method for dealing with different pollutants, it has a few drawbacks - a) It is often time-consuming as it involves bio-based techniques. b) It performs poorly when there is a high concentration of pollutants, and there is a chance of toxicity in the organism being used for the treatment. c) when dealing with xenobiotics, the treatment efficiencies are drastically low, along with prolonged recovery time [25]. For these reasons, the appropriate selection of the best remediation method is necessary. However, applying only a single technique is not recommended for contamination remediation.

In recent years, the emergence of bio-mimetic nanotechnology has become an innovative strategy to take the bioremediation process forward beyond its current capabilities [26]. This has caused a sudden influx of researchers in this unexplored field. Nanomaterials (NMs) can be defined as materials with the particle's size being 100nn or less in at least one dimension. The new and innovative nano-bioremediation process consists of nanoparticles (NPs) with two or more dimensions greater than 1nm. This novel approach gives rise to various potential strategies that can make waste management processes cost-effective and environment-friendly. The new nano-bioremediation technique makes it possible for biological processes to degrade contaminants in a green and sustainable way. Besides removing contaminants, the NMs also interact with biotic and abiotic elements. These interactions can be either positive or negative in nature. However, through these interactions, scientists are trying to find all possible synergistic effects among the NMs and the various bioremediation particles so that we can fully understand their physio-chemical nature in the water and soil around us [27]. NMs could be used directly or can be used with a combination of microbial enzymes or microbes. These can be used by either immobilizing the enzyme molecules on the NPs or synergistic conjugation of microbes with the NMs [28]. This chapter revolves around the recent developments in nano-bioremediations, the various NMs being used, and the different techniques that are being applied to achieve an environment-friendly decontamination process. Also, the limitations of using this novel approach and the possible remediation, along with the future aspects of this technique, are also discussed in this chapter.

THE SCIENCE OF BIOREMEDIATION WITH NANOMATERIAL

Principles of Nano-bioremediation

Bioremediation is the foremost essential remediation technique for water and soil in the recent century. These operative techniques helped to recover the pollutants from industrial waste, used pesticide residues in soil or water and heavy metal ions present in soil, which causes various health hazards [29, 30]. Recent studies have focused attention on elevating the efficacy of the bioremediation process by using the nano-biotechnology approaches [31]. The mechanism of such a remediation process includes the role of tri-factors, that is, the grade of contaminant, the ability of microbes for remediation, and the type of stimulants (nanomaterial) used to increase the reactivity of microorganisms [32]. Biostimulant nanomaterials are groups of small molecules that can upregulate the bacterial activity and the ability of the microbe to detoxify the polluted water or soil. The size of these nano-particles ranges from at least 1nm to 100nm in the field of bioremediation [33]. The release of toxic waste by the industrial sector is around 10 million tons, and the chemical wastes in soil or water can be more threatened when the reaction byproducts from these waste chemicals form polychlorinated byproducts [34 - 36]. However, nano-bioremediation has advantages over these issues by avoiding the formation of these hazardous intermediate byproducts, which can reduce the toxicity of these wastes, reduce the half-life of the waste, and upregulate the degree of degradation of these waste materials [37]. Bioremediation is more cost-effective than chemical detoxification. Basically, the different processes of bioremediation include bioaccumulation, biosorption stabilization, biosorption, *etc.* These processes include microbes, fungi, bacteria, and plants [38]. The addition of nanoparticles to this plant and microbe-based facilities not only increases the reaction rate of detoxification but also increases the specificity so that it has no or fewer side effects on the environment. The nanoparticle plays a vital role in interacting with toxic soil or water and living organisms. For example, in the case of plants, it is nano-phytoremediation where nanoparticles may help as a nano nutrient to increase the reactivity of and sustainability of the living organism involved with the remediation of toxic soil or water [39 - 41].

Nanomaterial used in Bioremediation

The parameters of nanomaterials play vital roles in increasing the efficacy of biota involved in bioremediation. The smaller the size, the higher the surface area-t--mass ratio and, hence, the greater the reactivity. The quantum effect of such nanoparticles allows them to generate biochemical reactions with less activation energy. These nanoparticles are made up of metals, bimetals (two metals embedded together), and carbon-based materials. These materials can penetrate the site of contaminants compared to other microparticles [42]. The bimetallic nanoparticles act as good oxidants; hence, they work as an effective bioremediation agent compared to granular metals, having a valency of zero, and have a more extraordinary ability as a reactant to the bio-contaminants, which are redox-amenable [43]. The ferrous oxide has an oxide coat that can form complexes with carbon tetrachloride and break into carbon monoxide and CH_4. By using such nanoparticles, current research has developed a strategy to remediate chloride-derived hydrocarbons with an aliphatic chain [43].

Similarly, the oxide of titanium also has the property to form nanotubes and can perform photo electrocatalytic activity and degrade contaminants like pentachlorophenol [44]. However, the field of nano biorobots has been designed with magnetic nanoparticles. Using ferric oxide in bacterial cell immobilization (Pseudomonas species) with ammonium oleate to coat the bacterial surface can help control and concentrate the bacterial cell mass to the target point using the magnetic field applied externally in the bioreactor. These microorganisms help in the desulfurization of sulfur from organic material or fossil fuel [45].

Biological Response During the Combined Application of Nanomaterial

Nanomaterials have a size range of 1-100 nanometers and possess unique physical, chemical, and biological properties compared to their bulk counterparts. Due to their small size and large surface area, nanomaterials can interact with biological systems, such as cells, tissues, and organs, in ways that are different from more extensive materials [46 - 48]. As a result, there has been an increasing interest in using nanomaterials for various biomedical applications, such as drug delivery, imaging, and tissue engineering. However, the use of nanomaterials in biomedical applications is not without challenges. One of the significant challenges is understanding the potential risks associated with their use, such as their possible toxicity and immunogenicity [49]. These risks can be amplified when combining two or more nanomaterials. When two or more types of nanomaterials are combined, they may interact with each other and change their physicochemical properties. For example, the combination of metallic and carbon-based nanomaterials can result in increased cytotoxicity compared to their individual use [50]. This is because metallic nanomaterials can generate reactive oxygen species (ROS), which can react with carbon-based nanomaterials and create more ROS, leading to oxidative stress and cell damage.

The biological response during the combined application of nanomaterials can also depend on the target cells or tissues. Some nanomaterials may cause oxidative stress, inflammation, and cell death, while others may have anti-inflammatory and antioxidant effects [51]. For example, carbon-based nanomaterials, such as graphene oxide, have been shown to have anti-inflammatory effects by inhibiting the production of pro-inflammatory cytokines [52]. However, the same graphene oxide can cause cytotoxicity and ROS production when combined with metallic nanomaterials. The route of exposure can also influence the response. Inhalation of nanomaterials can lead to lung inflammation and fibrosis, while ingestion can cause damage to the gastrointestinal tract [53]. Skin contact can result in skin irritation, inflammation, and oxidative stress. Therefore, the route of exposure should be carefully considered when evaluating the potential risks and benefits of the combined use of nanomaterials.

Moreover, the combined use of nanomaterials with other agents, such as drugs or biomolecules, can also affect their biological response [54 - 56]. The combination of nanomaterials with drugs can enhance drug delivery and improve therapeutic outcomes. For example, the combination of gold nanorods with the chemotherapy drug doxorubicin has been shown to improve drug delivery and increase therapeutic efficacy in cancer treatment. However, combining nanomaterials with biomolecules can also affect their biological behavior. For instance, the combination of carbon-based nanomaterials with proteins can result in protein denaturation and aggregation [57]. Lastly, the biological response during the combined application of nanomaterials is complex. It depends on various factors, such as the type of nanomaterials used, their concentration, size, and surface charge, as well as the cells or tissues they interact with, the route of exposure, and the presence of other agents. It is essential to carefully evaluate the potential risks and benefits of the combined use of nanomaterials and to design safe and effective strategies for their use in biomedical applications. Further research is needed to understand better the mechanisms of interaction between different types of nanomaterials and their biological response.

Bioremediation Current Development of Environmental Nano-applications Based on Molecular Biotechnology

Aside from the nanotechnology applications outlined in environmental biotechnology, further advancements may arise from alternative nanotechnology branches that progress more swiftly, such as medical nanotechnology. The exploration of the functionalization of nanostructures utilizing biomolecules is a promising domain of inquiry within academia. The efficacy of this approach has been evaluated through the utilization of advanced experimental designs that draw inspiration from naturally occurring molecular phenomena. As an alternative to chemically synthesized nanoparticles, functionalized nanoparticles are generated through the conjugation of nanoparticles and naturally derived biomolecules such as nucleic acids, proteins, surfactants, enzymes, polymers, and lipids. This interaction between the nanomaterial and the biomolecule creates a novel hybrid material with surface modifications that are minimally agglomerates and can be used for various biotechnological applications, mainly removing contaminants like aromatic hydrocarbons from aquatic environments. Biofunctionalized nanomaterial has the one-of-a-kind properties of a few combining components in a single compact framework. These would carry multifunctional highlights performing numerous applications [58]. For instance, Bolishetty *et al.* altered the tertiary structure of milk proteins to produce amyloid fibrils capable of trapping various ions *via* the cysteine moieties to avoid the formation of detrimental amyloid protein in the neurons and create membranes containing activated porous carbon and amyloid proteins for the removal and recovery of heavy metals. It is clear from this work that obtaining a cost-efficient source of biomolecules is vital for this sort of advancement.

Moreover, natural proteins exhibit numerous benefits as it is possible to generate them cost-efficiently using a well-known recombinant technology. These entities can accommodate up to twenty diverse amino acids, which offers a considerable range of combinatorial options for interacting with other molecules. Additionally, they can establish new catalytic surfaces and architectural frameworks [15].

Furthermore, the utilization of biotechnology offers the potential for the development of environmentally sustainable approaches for the functionalization of nanoparticles. As it turns out, to create novel cellulose-like polymers functionalized with custom moieties, the bacteria *Komagataeibacter sucrofermentans* has lately been added to the biological toolbox by Gao *et al.* [59]. These bacteria are cultivated using conventional bioreactor techniques, fed glucose monomers that have been chemically modified as needed, and then biologically incorporated into the polymer, which shuns the need for complex stoichiometry, solvent, and the formation of residues that are detrimental to the environment. Surprisingly, the advancement of this biosystem using conventional and cutting-edge biotechnological techniques mutagenesis, protein engineering, or gene editing has the potential to significantly ease the production of a vast number of cellulose-based NMs with a variety of uses [60, 61]. Because they allow a biocompatible and inert microenvironment that influences the enzymes' natural properties the least and aids in maintaining their biological activities, nanoparticles are increasingly being used in enzyme-mediated remediation technology [62]. However, choosing nanomaterials for the remediation process is a crucial step because they could be harmful to the microorganisms that are needed for the repair process [63].

The employment of microbes for bioremediation has emerged as a time-efficient solution due to the intricacy associated with traditional techniques for wastewater management overall, as shown in Fig. (**1**). Moreover, the utilization of bioremediation technology is subject to certain limitations. Specifically, certain types of microorganisms may prove incapable of converting toxic metals into harmless metabolites, thereby inhibiting microbial activity. An advancement in the field of nanotechnology has opened new avenues for the creation of water treatment approaches that are not only cost-effective but also environmentally sustainable. Nanoparticles possess distinct physicochemical characteristics that render them highly desirable for the purpose of wastewater treatment [64]. Numerous nanomaterials have been thoroughly investigated to adsorb heavy metals from wastewater expeditiously. These materials chiefly comprise activated carbon, carbon nanotubes, graphene, ferric oxides, manganese, titanium, magnesium, and zinc oxides. Iron nanoparticles are widely acknowledged as the pioneering nanoparticle in facilitating environmental cleanup efforts, as noted by Tratnyek and Johnson in their 2006 publication [65]. Current advancements in RNA-centered fungicides suggest the technology's viability as a substitute for conventional biochemical fungicides. Double-strand RNAs that target critical mRNAs of fungal pathogens are topically applied onto plant leaves or fruits, resulting in the induction of expression silencing in the pathogen. Nevertheless, the short lifespan of unencapsulated RNAs within the surrounding milieu presents a significant hindrance to be surmounted [66]. Moreover, traditional clay nanosheets were evaluated as double-strand RNA protectors to extend the average life of the biomolecules and lengthen the biocidal effects against the fungus pathogen [67].

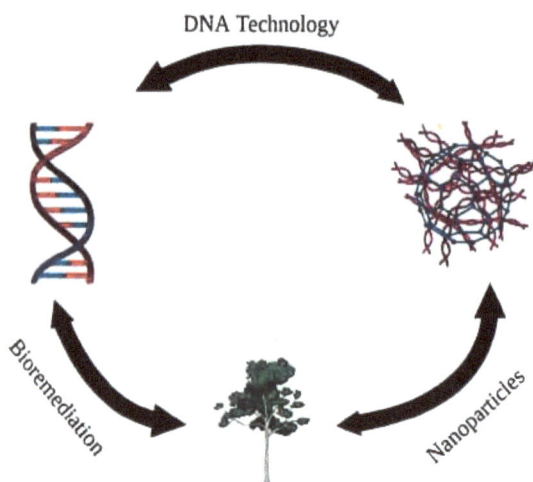

Fig. (1). Application of nanobioremediation in all fields.

The technological advancement in the construction of 3D DNA structures with the aid of DNA hybridization has opened up a frontier research area with vast opportunities for exploration in the field of environmental biotechnology. The coherent but adaptable guidelines for nucleotide recognition across separate strands of DNA may be employed to construct various geometric configurations commonly referred to as "molecular origami", ranging from elementary crossover tiles to polyhedral meshes [68, 69]. They conducted a study delineating the prototypical design of DNA sheets that exhibit intermolecular interactions with the thrombin protein. The ensuing compartmentalization of the protein within the nanosheet is facilitated by other DNA molecules that act as expeditious fasteners. The DNA origami that has transformed a nanotube exhibits a triggered opening mechanism, which exclusively activates upon detection of a specific molecular key in the form of a nucleolin protein of tumorous origin. This event subsequently triggers the release of the enclosed thrombin payload, leading to the induction of tumor-associated coagulation and necrosis. This advancement shows the programmability and intricate mechanics achievable in biomolecular-based biorobots. Whilst this advancement has been targeted towards biomedical applications, it is a remarkable illustration of the biochemical interactions that can be extrapolated to environmental contexts, including developing next-generation pesticides or eradicating antibiotic-resistant bacteria. One issue that arises regarding the utilization of DNA origami structures is their associated cost. Li J *et al.* [68] mentioned in a literature review examined various biotechnological methods for achieving cost-effectiveness in the production of DNA/RNA. Among the methods evaluated were chip synthesis, recombinant bacteria, and naturally occurring bacteria with the capacity to export RNAs.

The utilization of molecular biotechnology has been suggested for extensive implementations, most notably around water desalination. Over a decade ago [55, 70], they conducted a study to evaluate the integration of bacterial protein into a particular system. The researchers incorporated aquaporin Z, a transmembrane protein, into polymeric membranes and demonstrated its effectiveness in the selective exclusion of salt while simultaneously yielding purified water. The present methodology was refined utilizing contemporary protein modeling algorithms and molecular biology methodologies, producing porin proteins that exhibit improved exclusion propensities towards both organic and inorganic water solutes.

The technological utilization of oxygen-sensitive proteins found in both humans and plants to create O_2 biosensors and inducible genetic circuits represents a burgeoning area of study. While initially conceived for *in vivo* implementations, such proteins lend themselves to modification for the purposes of generating functional nanomaterials that can precisely react in proportion to variations in O_2 levels [71]. The concepts examined in relation to proteins have also been exhibited in the case of DNA molecules, which can preclude intricate analytes, such as proteins, thereby creating ample prospects for sensing and purification [72].

In recent literature, the research by Ryu [73] and colleagues and the analysis conducted by Álvarez *et al.* [64] systematically studied the inclusion of transmembrane proteins into membranes and their linkage to diverse transducer mechanisms. Additionally, the research explores the potential applicability of these proteins in areas such as gas monitoring, pesticide detection, microarrays, and energy harvesting.

In subsequent developments, enzymes could potentially integrate into the configurations for the purpose of detecting multifaceted contaminants or their combinations. The notion of biosensors being embedded within an array of materials, including but not limited to patches, temporary tattoos, and wrists, has found significant employment in the field of healthcare surveillance [74, 75]. Notably, the significant economic market pertaining to diabetes management serves as the primary impetus driving the development of these innovative solutions. Hazardous work environments have been a growing concern within the workforce. Safeguarding workers performing precarious duties in radioactive and potentially hazardous workplaces has increasingly emerged as a pertinent issue amongst the labor force. Bionanotechnologies that facilitate real-time monitoring possess the potential to effectively aid in the mitigation of dangerous environments, such as those that are toxic or enclosed.

INTERNATIONAL MARKETS AND REGULATIONS OF NANOTECHNOLOGIES APPLIED IN BIOREMEDIATION

Nanotechnology is a promising science with interdisciplinary applications in electronics, medicine, engineering, remediation, modern medicine, food, microelectronics, pharmaceuticals, and cosmetics [76]. The use of nanotechnologies in bioremediation is anticipated to propel technological advancement for improving environmental conditions in both developed and developing nations [77 - 79]. Extensive studies have been carried out to identify the processes of decontamination and remediation. The application and use of nanomaterials raise significant biosafety concerns due to the limited knowledge and established methods to evaluate their impact on human health, biodiversity depletion [56, 80, 81], bioaccumulation [82], and transport through food chains. Several international institutions, such as the United States Environmental Protection Agency (USEPA), the European Observatory for Nanomaterials (EON), the Organisation for Economic Co-operation and Development (OECD) Working Party on Manufactured Nanomaterials (WPMN), and the International Organisation for Standardisation (ISO) Technical Committee TC 229 "Nanotechnologies" have established collaborative efforts for the purpose of augmenting the implementation of existing regulations.

Moreover, global markets of nanotechnology and bioremediation are anticipated to persistently burgeon, with novel niches being established to ameliorate environmental parameters and the quality of human existence. Surprisingly, it is expected that the international nanotechnology market will surpass a monetary value of US$125 billion by 2024. Its influence on disparate sectors, specifically electronics, energy, biomedical, cosmetics, defense, energy, and agriculture, remains significant [83]. Bioremediation and phytoremediation technologies are expected to experience an annual rate of advancement of US$1.5 billion per year, as per evaluation by the US EPA [80, 84]. Furthermore, new techniques and the use of nanotechnologies will open up new prospects for treating sewage, lakes, rivers, and ponds, among other things, creating new consumers and enhancing global trade.

CONCLUSION

The developments in nano-biotechnology have opened up a vista of avenues for designing various synergistic NP-microorganism systems for the degradation of harmful contaminants. The vast applicability of these techniques in the fields of electronics, medicine, agriculture, and the environment has made it a far-reaching, sustainable process. Using nano-biotechnology, we can customize the size and shape of the desired NPs and make them far more usable. The nano-bioremediation can lower the overall cost of environmental cleanup and provide flexibility to be implemented on a vast scale. In the case of in-situ and ex-situ nano-remediation, the latter is still under research. However, the decontamination levels can even reach near zero values when using the in-situ method. For the decontamination of microbes and solid (organic and inorganic) pollutants, ex-situ nano-remediation techniques are being tested with positive results. Compared to metallic NPs, bio nanoparticles provide various advantages, such as biodegradability and less impact on the surrounding environment. Nano-bioremediation is at a stage where it can be considered a proxy method to combat the adverse effects of environmental pollution in contrast with conventional waste treatment methods. However, there is still a lack of knowledge about the synergistic effects of various NPs among themselves or with other particles like microbes or microbial enzymes. These combined systems have multiple responses to contaminants of diverse nature and thus may give rise to environmental nano-toxicity. Also, safety data on the prolonged usage of NPs in association with various microorganisms is absent, which may cause long-term ecological hazards. The current nanotechnologies that are being used for waste management lack a regulatory framework to follow. That is why numerous scientists are constantly working to improve our understanding of the various interactions of NMs during variable experimentation conditions like changing pH values, temperature, salt concentrations, *etc.* Finally, nano bioremediation has the potential to make a significant contribution to sustainability because it is inexpensive compared to other technologies and offers environmental benefits. Furthermore, the variety of applications for NMs combined with biological treatments has shown high efficacy in the degradation of contaminants, opening up new avenues for addressing numerous environmental issues.

REFERENCES

[1]   Singh R, Manickam N, Mudiam MKR, Murthy RC, Misra V. An integrated (nano-bio) technique for degradation of γ-HCH contaminated soil. J Hazard Mater 2013; 258-259: 35-41.
[http://dx.doi.org/10.1016/j.jhazmat.2013.04.016] [PMID: 23692681]

[2]   Mahapatra M., Pradhan S., Preetam S., Pradhan AK. Role of Biosurfactants in Heavy Metal Removal and Mineral Flotation. Biotechnological Innovations in the Mineral-Metal Industry. Springer 2024; pp. 141-50.
[http://dx.doi.org/10.1007/978-3-031-43625-3_8]

[3]   Rathod S, Preetam S, Pandey C, Bera SP. Exploring synthesis and applications of green nanoparticles and the role of nanotechnology in wastewater treatment. Biotechnol Rep (Amst) 2024; 41: e00830.
[http://dx.doi.org/10.1016/j.btre.2024.e00830] [PMID: 38332899]

[4]   Preetam S, Rath R , Mazumder I, Khan S; Roy C; Ali A, Malik S. Quantitative Methodologies for Determining the Amount and Structure of AOB at the Transcriptional Level in Wastewater Treatment Plants. 2023.
[http://dx.doi.org/10.1039/BK9781837671960-00198]

[5]   Preetam S. Nano revolution: pioneering the future of water reclamation with micro-/nano-robots. Nanoscale Adv 2024; 6(10): 2569-81.

[http://dx.doi.org/10.1039/D3NA01106B] [PMID: 38752135]

[6]   Mukherjee S, Sarkar D, Bhattacharya T, *et al.* Reaction kinetics involved in esterification between the fatty acids in castor oil and furfuryl alcohol. Ind Crops Prod 2024; 213: 118393.
[http://dx.doi.org/10.1016/j.indcrop.2024.118393]

[7]   Mukherjee S, Sengupta A, Preetam S, Das T, Bhattacharya T, Thorat N. Effects of fatty acid esters on mechanical, thermal, microbial, and moisture barrier properties of carboxymethyl cellulose-based edible films. Carbohydrate Polymer Technologies and Applications 2024; p. 100505.
[http://dx.doi.org/10.1016/j.carpta.2024.100505]

[8]   Alonso Castillo ML, Sánchez Trujillo I, Vereda Alonso E, García de Torres A, Cano Pavón JM. Bioavailability of heavy metals in water and sediments from a typical Mediterranean Bay (Málaga Bay, Region of Andalucía, Southern Spain). Mar Pollut Bull 2013; 76(1-2): 427-34.
[http://dx.doi.org/10.1016/j.marpolbul.2013.08.031] [PMID: 24054786]

[9]   Preetam S, Rath R, Khan S, Subudhi PD, Sanyal R. Functionalized exosomes for cancer therapy. Functionalized Nanomaterials for Cancer Research. Elsevier 2024; pp. 167-80.
[http://dx.doi.org/10.1016/B978-0-443-15518-5.00017-3]

[10]   Chen Y, Liu Y, Li Y, Wu Y, Chen Y, Zeng G, Zhang J, Li H. Influence of biochar on heavy metals and microbial community during composting of river sediment with agricultural wastes. Bioresour Technol 2017; 243: 347-55.
[http://dx.doi.org/10.1016/j.biortech.2017.06.100] [PMID: 28683388]

[11]   Gong X, Huang D, Liu Y, Zeng G, Wang R, Wei J. Pyrolysis and reutilization of plant residues after phytoremediation of heavy metals contaminated sediments: For heavy metals stabilization and dye adsorption. Bioresour Technol 2018; 253: 64-71.
[http://dx.doi.org/10.1016/j.biortech.2018.01.018] [PMID: 29328936]

[12]   Bandyopadhyay A, Das T, Nandy S, Sahib S, Preetam S, Gopalakrishnan AV, Dey A. Ligand-based active targeting strategies for cancer theranostics. Naunyn Schmiedebergs Arch Pharmacol 2023; 396(12): 3417-41.
[http://dx.doi.org/10.1007/s00210-023-02612-4] [PMID: 37466702]

[13]   Naser SS, Singh D, Preetam S, Kishore S, Kumar L, Nandi A, *et al.*, Posterity of nanoscience as lipid nanosystems for Alzheimer's disease regression. Mater Today Bio 2023; 21: 100701.
[http://dx.doi.org/10.1016/j.mtbio.2023.100701] [PMID: 37415846]

[14]   Rhind SM. Anthropogenic pollutants: a threat to ecosystem sustainability? Philos Trans R Soc Lond B Biol Sci 2009; 364(1534): 3391-401.
[http://dx.doi.org/10.1098/rstb.2009.0122] [PMID: 19833650]

[15]   Nešvera J, Rucká L, Pátek M. Catabolism of Phenol and Its Derivatives in Bacteria. Adv Appl Microbiol 2015; 93: 107-60.
[http://dx.doi.org/10.1016/bs.aambs.2015.06.002] [PMID: 26505690]

[16]   Lin SH, Juang RS. Adsorption of phenol and its derivatives from water using synthetic resins and low-cost natural adsorbents: A review. J Environ Manage 2009; 90(3): 1336-49.
[http://dx.doi.org/10.1016/j.jenvman.2008.09.003] [PMID: 18995949]

[17]   Dhasmana A, Preetam S, Malik S, Jadon VS, Joshi N, Bhandari G, & Samal SK. Revitalizing elixir with orange peel amplification of alginate fish oil beads for enhanced anti-aging efficacy. Scientific Reports 2024; 14(1): 2040.
[http://dx.doi.org/10.1038/s41598-024-71042-w]

[18]   Abdel-Shafy HI, & Mansour MS. Solid waste issue: Sources, composition, disposal, recycling, and valorization. Egyptian journal of petroleum 2018; 27(4): 1275-90.
[http://dx.doi.org/10.1016/j.ejpe.2018.07.003]

[19]   Bhattacharya T, Preetam S, Mukherjee S, Kar S, Roy DS, Singh H, & Mohapatra, G. Anticancer activity of quantum size carbon dots: opportunities and challenges. Discover Nano. 2024; 19: p.

(1)122.
[http://dx.doi.org/10.1186/s11671-024-04069-7] [PMID: 11300426]

[20]   van Heerwaarden B, Kellermann VM, Hoffmann AA. Environmental Stress and Evolutionary Change. In: Fath B, Ed. Encyclopedia of Ecology. 2nd ed. Oxford: Elsevier 2019; pp. 197-203.
[http://dx.doi.org/10.1016/B978-0-12-409548-9.09136-3]

[21]   Azubuike CC, Chikere CB, Okpokwasili GC. Bioremediation techniques–classification based on site of application: principles, advantages, limitations and prospects. World J Microbiol Biotechnol 2016; 32(11): 180.
[http://dx.doi.org/10.1007/s11274-016-2137-x] [PMID: 27638318]

[22]   Wuana RA, Okieimen FE. Heavy Metals in Contaminated Soils: A Review of Sources, Chemistry, Risks and Best Available Strategies for Remediation. ISRN Ecol 2011; 2011: 1-20.
[http://dx.doi.org/10.5402/2011/402647]

[23]   Tanzadeh J, Ghasemi MF, Anvari M, Issazadeh K. Biological removal of crude oil with the use of native bacterial consortia isolated from the shorelines of the Caspian Sea. Biotechnol Biotechnol Equip 2020; 34(1): 361-74.
[http://dx.doi.org/10.1080/13102818.2020.1756408]

[24]   Preetam S, Jonnalagadda S, Kumar L, Rath R, Chattopadhyay S, Alghamdi BS, *et al.* Therapeutic potential of lipid nanosystems for the treatment of Parkinson's disease. Ageing Res Rev 2023; 89: 101965.
[http://dx.doi.org/10.1016/j.arr.2023.101965] [PMID: 37268112]

[25]   Mapelli F, Scoma A, Michoud G, Borin S, Kalogerakis N, Daffonchio D, *et al.* Biotechnologies for Marine Oil Spill Cleanup: Indissoluble Ties with Microorganisms. Trends Biotechnol 2017; 35(9): 860-70.
[http://dx.doi.org/10.1016/j.tibtech.2017.04.003] [PMID: 28511936]

[26]   Rizwan M, Singh M, Mitra CK, Morve RK. Ecofriendly Application of Nanomaterials: Nanobioremediation. J Nanoparticles 2014; 2014: 1-7.
[http://dx.doi.org/10.1155/2014/431787]

[27]   Cecchin I, Reddy KR, Thomé A, Tessaro EF, Schnaid F. Nanobioremediation: Integration of nanoparticles and bioremediation for sustainable remediation of chlorinated organic contaminants in soils. Int Biodeterior Biodegradation 2017; 119: 419-28.
[http://dx.doi.org/10.1016/j.ibiod.2016.09.027]

[28]   Kumari B, Singh DP. A review on multifaceted application of nanoparticles in the field of bioremediation of petroleum hydrocarbons. Ecol Eng 2016; 97: 98-105.
[http://dx.doi.org/10.1016/j.ecoleng.2016.08.006]

[29]   Sales da Silva IG, Gomes de Almeida FC, da Rocha e Silva NMP, Casazza AA, Converti A, Sarubbo LA. Soil bioremediation: Overview of technologies and trends. Energies 2020; 13(18): p. 4664.
[http://dx.doi.org/10.3390/en13184664]

[30]   Coelho LM, Rezende HC, Coelho LM, Sousa PAR, Melo DFO, Coelho NMM. Bioremediation of polluted waters using microorganisms. 2015. 10: p. 60770.
[http://dx.doi.org/10.5772/60770]

[31]   Kumar, S.R., P.J.H.O.A.I. Gopinath, and H.W. management, Nano-bioremediation applications of nanotechnology for bioremediation. 2017: p. 27-48.

[32]   Vázquez-Núñez E, Molina-Guerrero CE, Peña-Castro JM, Fernández-Luqueño F, de la Rosa-Álvarez MG. Use of nanotechnology for the bioremediation of contaminants. Processes 2020; 8(7): 826.
[http://dx.doi.org/10.3390/pr8070826]

[33]   Singh M, Mitra C, Morve RJJN. Ecofriendly application of nanomaterials. Nanobioremediation 2014; 2014: 740-57.

[34]   Avio, C.G., Gorbi. S and Regoli F. Plastics and microplastics in the oceans: from emerging pollutants

to emerged threat. 2017. 128: p. 2-11.
[http://dx.doi.org/10.1016/j.marenvres.2016.05.012]

[35] Preetam S, Dash L, Sarangi SS, Sahoo MM, Pradhan AK. Application of Nanobiosensor in Health Care Sector. Bio-Nano Interface 2022; pp. 251-70.
[http://dx.doi.org/10.1007/978-981-16-2516-9_14]

[36] Preetam S, *et al.* Preetam S, Nahak BK, Patra S, Toncu DC, Park S, Syväjärvi M *et al.*, Emergence of microfluidics for next generation biomedical devices. Biosens Bioelectron 2022; X: 100106.
[http://dx.doi.org/10.1016/j.biosx.2022.100106]

[37] Kang, J.W. Removing environmental organic pollutants with bioremediation and phytoremediation. 2014. 36: p. 1129-1139.
[http://dx.doi.org/10.1007/s10529-014-1466-9]

[38] Shukla, A., Srivastava S. sustainability, Emerging aspects of bioremediation of arsenic. 2017: p. 395-407.
[http://dx.doi.org/10.1007/978-3-319-49744-7_25]

[39] Franzoni G, Trivellini A, Bulgari R, Cocetta G, Ferrante A. Bioactive molecules as regulatory signals in plant responses to abiotic stresses. Plant signaling molecules. Elsevier 2019; pp. 169-82.
[http://dx.doi.org/10.1016/B978-0-12-816451-8.00010-1]

[40] Malik S, Dhasmana A, Preetam S, Mishra YK, Chaudhary V, Bera SP, *et al.* Exploring Microbial-Based Green Nanobiotechnology for Wastewater Remediation: A Sustainable Strategy. Nanomaterials (Basel) 2022; 12(23): 4187.
[http://dx.doi.org/10.3390/nano12234187] [PMID: 36500810]

[41] Malik S, Kishore S, Nag S, Dhasmana A, Preetam S, Mitra O, *et al.* Ebola Virus Disease Vaccines: Development, Current Perspectives & Challenges. Vaccines (Basel) 2023; 11(2): 268.
[http://dx.doi.org/10.3390/vaccines11020268] [PMID: 36851146]

[42] Liu, W.T. and bioengineering, Nanoparticles and their biological and environmental applications. 2006. 102(1): p. 1-7.
[http://dx.doi.org/10.1263/jbb.102.1]

[43] Nurmi J.T, Tratnyek P.G, Sarathy V, Baer D.R, Amonette J.E, Klaus Pecher K., *et al.*, Characterization and properties of metallic iron nanoparticles: spectroscopy, electrochemistry, and kinetics. 2005. 39(5): p. 1221-1230.
[http://dx.doi.org/10.1021/es049190u]

[44] Quan X, Yang S, Ruan X, Zhao H. Preparation of titania nanotubes and their environmental applications as electrode. 2005. 39(10): p. 3770-3775.
[http://dx.doi.org/10.1021/es048684o]

[45] Shan G, Xing J, Zhang HY, Liu HZ. Biodesulfurization of dibenzothiophene by microbial cells coated with magnetite nanoparticles. 2005. 71(8): p. 4497-4502.
[http://dx.doi.org/10.1128/AEM.71.8.4497-4502.2005]

[46] Karakoti, A.S., L.L. Hench, and S.J.J. Seal, The potential toxicity of nanomaterials—the role of surfaces. 2006. 58: p. 77-82.

[47] Bhattacharjee R, Kumar L, Mukerjee N, Anand U, Dhasmana A, Preetam S *et al.* The emergence of metal oxide nanoparticles (NPs) as a phytomedicine: A two-facet role in plant growth, nano-toxicity and anti-phyto-microbial activity. Biomed Pharmacother 2022; 155: 113658.
[http://dx.doi.org/10.1016/j.biopha.2022.113658] [PMID: 36162370]

[48] Nahak BK, Mishra A, Preetam S, Tiwari A. Advances in organ-on-a-chip materials and devices. ACS Appl Bio Mater 2022; 5(8): 3576-607.
[http://dx.doi.org/10.1021/acsabm.2c00041] [PMID: 35839513]

[49] Chirmule, N., Jawa V, Meibohm BJ. Immunogenicity to therapeutic proteins: impact on PK/PD and efficacy. AAPS J 2012; 14: 296-302.

[50]   Eivazzadeh-Keihan R, Maleki A, de la Guardia M, Bani MS, Chenab KK, Paria Pashazadeh-Panahi P *et al*. Carbon based nanomaterials for tissue engineering of bone: Building new bone on small black scaffolds: A review. J Adv Res 2019; 18: 185-201.
[http://dx.doi.org/10.1016/j.jare.2019.03.011] [PMID: 31032119]

[51]   Pinho RA, Haupenthal DP , Fauser PE, Thirupathi A, Silveira PCL. Gold nanoparticle-based therapy for muscle inflammation and oxidative stress. 2022: p. 3219-3234.
[http://dx.doi.org/10.2147/JIR.S327292]

[52]   Bałaban J, Wierzbicki M, Zielińska-Górska M, Sosnowska M, Daniluk K, Jaworski S. Graphene Oxide Decreases Pro-Inflammatory Proteins Production in Skeletal Muscle Cells Exposed to SARS-CoV-2 Spike Protein. Nanotechnol Sci Appl 2023: 16: 1-18.
[http://dx.doi.org/10.2147/NSA.S391761]

[53]   Sonwani S, Madaan S, Arora J, Suryanarayan S, Deepali Rangra D, Mongia N *et al*. Inhalation exposure to atmospheric nanoparticles and its associated impacts on human health: Front Sustain Cities 2021; 3: 690444.
[http://dx.doi.org/10.3389/frsc.2021.690444]

[54]   Auría-Soro C, Nesma T, Juanes-Velasco P, Landeira-Viñuela A, Fidalgo-Gomez H, Acebes-Fernandez V, *et al*. Interactions of nanoparticles and biosystems: microenvironment of nanoparticles and biomolecules in nanomedicine. Nanomaterials 2019; 9(10): 1365.
[http://dx.doi.org/10.3390/nano9101365]

[55]   Bhattacharjee R, Negi A, Bhattacharya B, Dey T, Mitra P, Preetam S *et al*. Nanotheranostics to Target Antibiotic-resistant Bacteria: Strategies and Applications. OpenNano 2023; p. 100138.
[http://dx.doi.org/10.1016/j.onano.2023.100138]

[56]   Samal SK, Preetam S. *Synthetic Biology: Refining Human Health* 2022.
[http://dx.doi.org/10.1007/978-981-19-3979-2_3]

[57]   Gu Z, Yang Z, Chong Y,i Ge C, Weber JK, Bell DR, Zhou R. Surface curvature relation to protein adsorption for carbon-based nanomaterials. Sci Rep 2015; 5(1): p. 1-9.
[http://dx.doi.org/10.1038/srep10886]

[58]   Basak G, Hazra C, Sen R. Biofunctionalized nanomaterials for in situ clean-up of hydrocarbon contamination: A quantum jump in global bioremediation research. J Environ Manage 2020; 256: 109913.
[http://dx.doi.org/10.1016/j.jenvman.2019.109913] [PMID: 31818738]

[59]   Gao M, Li J, Bao Z, Hu M, Nian R, Feng D, *et al*. A natural in situ fabrication method of functional bacterial cellulose using a microorganism. Nat Commun 2019; 10(1): 437.
[http://dx.doi.org/10.1038/s41467-018-07879-3] [PMID: 30683871]

[60]   Sharma A, Thakur M, Bhattacharya M, Mandal T, Goswami S. Commercial application of cellulose nano-composites – A review. Biotechnol Rep (Amst) 2019; 21: e00316.
[http://dx.doi.org/10.1016/j.btre.2019.e00316] [PMID: 30847286]

[61]   Sharma U, Sharma JG. Nanotechnology for the bioremediation of heavy metals and metalloids. J Appl Biol Biotechnol 2022; 10(5): 34-44.
[http://dx.doi.org/10.7324/JABB.2022.100504]

[62]   Ferreira P, Alves P, Coimbra P, Gil MH. Improving polymeric surfaces for biomedical applications: a review. J Coat Technol Res 2015; 12(3): 463-75.
[http://dx.doi.org/10.1007/s11998-015-9658-3]

[63]   Benjamin SR, Lima FD, Florean EOPT, Guedes MIF. Current trends in nanotechnology for bioremediation. Int J Environ Pollut 2019; 66(1/2/3): 19-40.
[http://dx.doi.org/10.1504/IJEP.2019.104526]

[64]   Hristovski K, Baumgardner A, Westerhoff P. Selecting metal oxide nanomaterials for arsenic removal in fixed bed columns: From nanopowders to aggregated nanoparticle media. J Hazard Mater 2007;

147(1-2): 265-74.
[http://dx.doi.org/10.1016/j.jhazmat.2007.01.017] [PMID: 17254707]

[65]   Tratnyek PG, Johnson RL. Nanotechnologies for environmental cleanup. Nano Today 2006; 1(2): 44-8.
[http://dx.doi.org/10.1016/S1748-0132(06)70048-2]

[66]   Wang M, Weiberg A, Lin FM, Thomma BPHJ, Huang HD, Jin H. Bidirectional cross-kingdom RNAi and fungal uptake of external RNAs confer plant protection. Nat Plants 2016; 2(10): 16151.
[http://dx.doi.org/10.1038/nplants.2016.151] [PMID: 27643635]

[67]   Mitter N, Worrall EA, Robinson KE, Li P, Jain RG, Taochy C, *et al.* Clay nanosheets for topical delivery of RNAi for sustained protection against plant viruses. Nat Plants 2017; 3(2): 16207.
[http://dx.doi.org/10.1038/nplants.2016.207] [PMID: 28067898]

[68]   Li J, Green AA, Yan H, Fan C. Engineering nucleic acid structures for programmable molecular circuitry and intracellular biocomputation. Nat Chem 2017; 9(11): 1056-67.
[http://dx.doi.org/10.1038/nchem.2852] [PMID: 29064489]

[69]   Li S, Jiang Q, Liu S, Zhang Y, Tian Y, Chen S, *et al.* A DNA nanorobot functions as a cancer therapeutic in response to a molecular trigger *in vivo.* Nat Biotechnol 2018; 36(3): 258-64.
[http://dx.doi.org/10.1038/nbt.4071] [PMID: 29431737]

[70]   Kumar M, Grzelakowski M, Zilles J, Clark M, Meier W. Highly permeable polymeric membranes based on the incorporation of the functional water channel protein Aquaporin Z. Proc Natl Acad Sci USA 2007; 104(52): 20719-24.
[http://dx.doi.org/10.1073/pnas.0708762104] [PMID: 18077364]

[71]   Licausi F, Giuntoli B. Synthetic biology of hypoxia. New Phytol 2021; 229(1): 50-6.
[http://dx.doi.org/10.1111/nph.16441] [PMID: 31960974]

[72]   Diederichs T, Pugh G, Dorey A, Xing Y, Burns JR, Nguyen QH, *et al.* Synthetic protein-conductive membrane nanopores built with DNA. Nat Commun 2019; 10(1): 5018.
[http://dx.doi.org/10.1038/s41467-019-12639-y] [PMID: 31685824]

[73]   Ryu H, Fuwad A, Yoon S, ,Jang H, Lee JC, Kim SM, Jeon TJ. Biomimetic membranes with transmembrane proteins: State-of-the-art in transmembrane protein applications. Int J Mol Sci 2019; 20(6): 1437.
[http://dx.doi.org/10.3390/ijms20061437] [PMID: 30901910]

[74]   Kim J, Campbell AS, de Ávila BEF, Wang J. Wearable biosensors for healthcare monitoring. Nat Biotechnol 2019; 37(4): 389-406.
[http://dx.doi.org/10.1038/s41587-019-0045-y] [PMID: 30804534]

[75]   Alvarez PJJ, Chan CK, Elimelech M, Halas NJ, Villagrán D. Emerging opportunities for nanotechnology to enhance water security. Nat Nanotechnol 2018; 13(8): 634-41.
[http://dx.doi.org/10.1038/s41565-018-0203-2] [PMID: 30082804]

[76]   Shastri DV, Arunachalam KD. Nanomaterials for biodeterioration: An introduction. Nano-Bioremediation: Fundamentals and Applications 2022; 23-28.
[http://dx.doi.org/10.1016/B978-0-12-823962-9.00010-6]

[77]   Vázquez Núñez E, de la Rosa-Álvarez G. Environmental behavior of engineered nanomaterials in terrestrial ecosystems: Uptake, transformation and trophic transfer. Curr Opin Environ Sci Health 2018; 6: 42-6.
[http://dx.doi.org/10.1016/j.coesh.2018.07.011]

[78]   Vázquez-Núñez E, Molina-Guerrero CE, Peña-Castro JM, Fernández-Luqueño F, de la Rosa-Álvarez MG. Use of nanotechnology for the bioremediation of contaminants: A review. Processes (Basel) 2020; 8(7): 826.
[http://dx.doi.org/10.3390/pr8070826]

[79]   Bartke S, Hagemann N, Harries N, Hauck J, Bardos P. Market potential of nanoremediation in Europe

– Market drivers and interventions identified in a deliberative scenario approach. Sci Total Environ 2018; 619-620: 1040-8.
[http://dx.doi.org/10.1016/j.scitotenv.2017.11.215] [PMID: 29734582]

[80]   Sun Y, Liang J, Tang L, Li H, Zhu Y, Jiang D, *et al.* Nano-pesticides: A great challenge for biodiversity? Nano Today 2019; 28: 100757.
[http://dx.doi.org/10.1016/j.nantod.2019.06.003]

[81]   Malik S, Dhasmana A, Bora J, Uniyal P, Slama P, Preetam S, *et al.* Ebola virus disease (EVD) outbreak re-emergence regulation in East Africa: preparedness and vaccination perspective. Int J Surg 2023; 109(4): 1029-1031.
[http://dx.doi.org/10.1097/JS9.0000000000000175]

[82]   Lead JR, Batley GE, Alvarez PJJ, Croteau MN, Handy RD, McLaughlin MJ, *et al.* Nanomaterials in the environment: Behavior, fate, bioavailability, and effects—An updated review. Environ Toxicol Chem 2018; 37(8): 2029-63.
[http://dx.doi.org/10.1002/etc.4147] [PMID: 29633323]

[83]   Singh, A., R.C. Kuhad, and O.P. Ward, Biological remediation of soil: an overview of global market and available technologies. Advances in applied bioremediation, 2009: p. 1-19.
[http://dx.doi.org/10.1007/978-3-540-89621-0_1]

[84]   Song Y, Hou D, Zhang J, O'Connor D, Li G,, Gu Q, *et al.* Environmental and socio-economic sustainability appraisal of contaminated land remediation strategies: A case study at a mega-site in China. Sci Total Environ 2018; 610-611: 391-401.
[http://dx.doi.org/10.1016/j.scitotenv.2017.08.016] [PMID: 28806555]

SUBJECT INDEX

A

Acid(s) 32, 42, 70, 92, 94, 110, 111, 112, 126,
    129, 170, 180, 182, 187, 201, 202, 206,
    217, 222
  acrylic 170, 206
  citric 92
  folic 180, 182
  glycolic 170
  lactic 170
  lactic-glycolic 129
  nitric 202
  nucleic 42, 70, 94, 112, 126, 187, 222
  polyacrylic 92, 110
  rain 217
  retinoic 111
  sulfuric 201
Acidic fluids 186
Activation, inflammatory 128
Activity 48, 92, 130, 167, 180, 181, 186, 223
  anti-tubercular 92
  antineoplastic 181
  inhibiting microbial 223
  photocatalytic 48
  therapeutic 167, 180
Acute 73, 146
  myeloid leukaemia (AML) 146
  myocardial infarction (AMI) 73
Adenocarcinoma 162, 163, 177, 178
Adsorption 197, 198, 199, 200, 203, 205, 206,
    207, 208, 209, 210, 216
  kinetic energy 203
  properties 210
  response 206
Agents 46, 134, 149
  biotic 46
  epigenetic 149
  neurotoxic 134
Alzheimer's 52, 73, 101, 113, 129, 134
  disease 52, 73, 101, 113, 134
  disorder 134
  treatment 129

Amalgamated systems 102, 113, 114
Anti-bacterial nanoparticles 132
Anti-inflammatory 127, 129, 131
  activities 131
  applications 127
  cytokines 131
  treatments 129
Anti-metastatic therapeutic techniques 185
Antibodies 42, 61, 64, 68, 69, 70, 72, 73, 75,
    76, 94, 120, 149, 181
  anti-dengue virus 61
  bacterial cell wall 76
  monoclonal 149, 181
Anticancer therapy 187
Antimicrobial 33, 126
  peptides 126
  treatments 33
Antimycobacterial activity 91
Antioxidant effects 221
Antitumor activity 49
Applications 6, 7, 8, 9, 10, 21, 33, 42, 47, 51,
    52, 57, 59, 60, 61, 62, 67, 70, 71, 75,
    144, 166, 222
  bioimaging 9
  biosensing 60
  biosensor 60
  biotechnological 222
  cardiovascular 21
  industrial 7, 10, 71
  medicinal 51, 52
  sensing 60
Aromatic hydrocarbons 222
Arthritis, rheumatoid 67, 129
Artificial neural network (ANN) 168, 175

B

Bacteria 12, 24, 25, 26, 27, 32, 33, 48, 50, 51,
    87, 88, 90, 132, 223, 224
  antibiotic-resistant 50, 224
  pathogenic 48
Bacteriostatic therapy 131

www.ingramcontent.com/pod-product-compliance
Lightning Source LLC
Chambersburg PA
CBHW050824220326
41598CB00006B/309